Photocatalytic Oxidation/Ozonation Processes

Photocatalytic Oxidation/Ozonation Processes

Guest Editors

Fernando J. Beltrán Novillo
Juan F. García-Araya

MDPI

Basel • Beijing • Wuhan • Barcelona • Belgrade • Novi Sad • Cluj • Manchester

Guest Editors

Fernando J. Beltrán Novillo
Ingeniería Química y
Química Física
Universidad de Extremadura
Badajoz
Spain

Juan F. García-Araya
Ingeniería Química y
Química Física
Universidad de Extremadura
Badajoz
Spain

Editorial Office

MDPI AG
Grosspeteranlage 5
4052 Basel, Switzerland

This is a reprint of the Special Issue, published open access by the journal *Catalysts* (ISSN 2073-4344), freely accessible at: https://www.mdpi.com/journal/catalysts/special_issues/photocatalytic_oxidation_ozonation.

For citation purposes, cite each article independently as indicated on the article page online and as indicated below:

Lastname, A.A.; Lastname, B.B. Article Title. *Journal Name* **Year**, *Volume Number*, Page Range.

ISBN 978-3-7258-6672-4 (Hbk)
ISBN 978-3-7258-6673-1 (PDF)
https://doi.org/10.3390/books978-3-7258-6673-1

Contents

About the Editors

Fernando J. Beltrán Novillo

Fernando J. Beltrán Novillo is a full professor at the Departamento de Ingeniería Química y Química Física of the University of Extremadura (Badajoz, Spain). At present, he lectures on chemical engineering subjects such as chemical reactors, chemical engineering, mass and energy balances, unit operations, and water treatment. With over 49 years of university-level teaching experience, he has taught in chemical engineering, chemistry, environmental sciences, and related fields. His research focuses on advanced water treatment processes, including the use of ozone, radiation, photocatalysis, wet air oxidation, and other combined technologies. With an H index of 64, he has authored more than 300 scientific papers and participated in numerous national and international conferences. He has also supervised numerous research projects and doctoral theses.

Juan Fernando García-Araya

Juan Fernando García-Araya is a full professor at the Departamento de Ingeniería Química y Química Física of the University of Extremadura (Badajoz, Spain). At present, he lectures on chemical engineering subjects such as heat transfer, mass transfer petrochemistry, water pollution, and related fields. With over 33 years of university-level teaching experience, he has taught in chemical engineering, chemistry, environmental sciences, and related fields. His research focuses on advanced water treatment processes, including the use of ozone, radiation, photocatalysis, wet air oxidation, and other combined technologies. With an H index of 32, he has authored more than 70 scientific papers and participated in numerous national and international conferences.

catalysts

Editorial

Photocatalytic Oxidation/Ozonation Processes

Juan F. García-Araya * and Fernando J. Beltrán *

Departamento de Ingeniería Química y Química Física, Instituto Universitario de Investigación del Agua, Cambio Climático y Sostenibilidad, Universidad de Extremadura, 06006 Badajoz, Spain
* Correspondence: jfgarcia@unex.es (J.F.G.-A.); fbeltran@unex.es (F.J.B.)

Diffuse pollution and the presence in waters of so-called emerging pollutants, among others, represent a major global environmental problem. Often, the technologies implemented in conventional treatment plants fail to effectively solve this problem [1].

Deepening the study of already known processes (e.g., membranes, adsorption, and chemical oxidation) and some of their variants, or other innovative technologies, could improve water quality indices. Among these technologies are advanced oxidation processes (AOPs) based on the generation of hydroxyl radicals with high oxidizing capacity. These short-lived species can completely eliminate pollutants and form CO_2 and H_2O, unlike classic tertiary processes such as membrane filtration and adsorption, which only transfer contaminants from one phase (water) to another (membrane concentrates, adsorbents, etc.) and do not destroy them.

Ozonation in the presence of homogeneous metallic catalysts (e.g., iron or manganese), hydrogen peroxide, and ultraviolet radiation are widely studied AOPs. Their effectiveness and ease of application make them very attractive for removing pollutants from water.

More innovative are AOPs in which light radiation electronically activates a semiconductor material (catalyst), generating charge carriers (holes and electrons) that eventually form hydroxyl radicals. Works carried out in the laboratory indicate that the process is efficient with UV light as the radiation source and TiO_2 in powder form as the catalyst. However, its implementation in a water treatment plant presents a series of problems, such as the source of UV radiation or the use of powdered TiO_2 (whose separation from treated water is difficult and expensive). Furthermore, the high recombination of charge carriers, which inhibit the oxidation process, is another problem that must be solves. Accordingly, new research focuses on cheap and environmentally sustainable sources of radiation (e.g., sunlight and LEDs) and on the development of efficient catalysts that are easy to use and handle in water treatment plants (e.g., doped TiO_2, supported photocatalysts, photocatalysts with magnetic properties, etc.).

The synergy between ozonation and photocatalytic oxidation (photocatalytic ozonation) constitutes another AOP with application potential due to its various methods of hydroxyl radical formation to enhance the effect of the direct ozonation of pollutants.

This Special Issue focuses on the synthesis and characterization of supported or magnetic photocatalysts and their application in reactions with water pollutants in the presence of visible light (solar or simulated) or radiation from UVA-visible LEDs, with and without the presence of ozone. Additionally, this issue deals with ozone processes in water that involve the simultaneous application of ozone with homogeneous catalysts (e.g., iron and manganese), UV radiation (without catalysts), hydrogen peroxide, etc., which are called ozone advanced oxidation processes. The aim is to highlight conditions that improve photocatalytic oxidation performance and ozone advanced oxidation processes.

The contributions to this issue are summarized below:

Peng et al. [2] synthesized and applied Ag/ZnO nanocomposites in the photocatalytic ozonation of phenol. Their crystal, textural, morphological, optical, and electrochemical properties were investigated in detail using XRD, Raman, SEM, TEM, UV-vis diffuse reflectance spectroscopy (DRS), X-ray photoemission spectroscopy (XPS), and photoluminescence (PL) techniques. The results indicated that silver nanoparticles were well dispersed on the surface of porous ZnO and that intimate contacts were formed at the Ag/ZnO interfaces. This prominently favored the separation and transfer of photoinduced electrons from ZnO to Ag nanoparticles for the activation of ozone to produce hydroxyl and superoxide ion radicals. As a result, significant enhancement in the photocatalytic ozonation of phenol was achieved over Ag/ZnO catalysts. The results also showed a synergistic effect between photocatalysis and ozonation.

Due to the rising threat to aquatic environments, the removal of methyl *tert*-butyl ether (MTBE) is necessary. This pollutant, even at very low concentrations, makes water undrinkable; therefore, an effective treatment must be developed. Šihor et al. [3] carried out photocatalytic oxidation of MTBE in the presence of various TiO_2 photocatalysts with different phase composition prepared using different methods. The authors confirmed that the phase composition of TiO_2 had the most significant influence on the photocatalytic degradation of MTBE. The rutile phase more easily reduced oxygen adsorbed by photogenerated electrons to superoxide radicals, supporting the separation of charge carriers. About 40% of total organic carbon was removed after 1 h of irradiation in the presence of TiO_2–ISOP–C/800 photocatalysts composed of anatase and rutile phase.

Zhang et al. [4] prepared hexagon-shaped potassium ferrite ($K_2Fe_4O_7$) crystals of different sizes using the hydrothermal method. The crystals showed a narrow band gap of 1.44 eV, revealed via UV-visible diffuse reflectance spectroscopy; they were thus used as a heterogeneous Fenton catalyst to degrade methylene blue and crystal violet in the presence of green oxidant H_2O_2 under visible-light irradiation. Among the investigated crystals, the as-prepared one with an average size of 20 μm exhibited better photocatalytic activity due to its high surface area. When this catalyst was used in a photo-Fenton process, 100% methylene blue and 92% crystal violet were degraded within 35 min. Moreover, the catalyst maintained high photocatalytic activity and was stable after four continuous cycles. Trapping experiments showed that the active hydroxyl radical (OH) was dominant in the photo-Fenton reaction. Therefore, this new photo-Fenton catalyst has great potential for the photocatalytic degradation of dye contaminants in water.

Sugiyama et al. [5] investigated which active oxygen species affect the oxidative dehydrogenation of methane by employing photo-catalysts such as TiO_2 and WO_3, which generate active oxygen from UV-LED irradiation conditions under oxygen flow. These photo-catalysts were studied in combination with Sm_2O_3, which is a methane oxidation coupling catalyst. Their constructed reaction system directly irradiated UV-LED to form a solid catalyst via a normal fixed-bed continuous-flow reactor operating under atmospheric pressure. UV-LED irradiation clearly improved the partial oxidation from methane to CO and/or slightly improved the oxidative coupling route from methane to ethylene when binary catalysts consisting of Sm_2O_3 and TiO_2 were used, while negligible UV-LED effects were detected when using individual Sm_2O_3 and WO_3. The authors concluded that Sm_2O_3-TiO_2 coupling catalysts and UV-LED irradiation generated the active oxygen of O_2^-, thus contributing to the oxidative dehydrogenation of methane, while the active oxygen of H_2O_2 from WO_3 under the same conditions afforded only negligible effects on the activation of methane.

Aguilar-Melo et al. [6] evaluated naproxen degradation efficiency via ozonation using nickel oxide films as a catalyst. The nickel oxide films were synthesized via chemical vapor deposition and characterized via X-ray diffraction, scanning electron microscopy, atomic force microscopy, and X-ray photoelectron spectroscopy. Naproxen degradation was conducted for 5 min using 10 films of nickel oxide. The results were compared to those from 100 mg/L nickel oxide powder ozonation and conventional ozonation. Total organic carbon analysis demonstrated mineralization degrees of 12, 35, and 22% with conventional ozonation, nickel oxide powder, and nickel oxide film ozonation, respectively, after 60 min of reaction. The films of nickel oxide were used sequentially four times in ozonation, demonstrating the stability and properties of the synthesized material as a catalyst for ozonation. A proposed modeling strategy using robust parametric identification techniques allows for comparison of the pseudo-monomolecular reaction rates of naproxen decomposition.

Shurbaji et al. [7] conducted a review of visible light photocatalysis for the decomposition of ciprofloxacin, norfloxacin, tetracycline, and sulfonamide antibiotics in wastewater. This review summarized recent studies regarding semiconducting material modifications for antibiotic degradation using visible light irradiation. Antibiotics are chemical compounds that are used to kill bacteria or prevent their growth. They are used in fields such as medicine, agriculture, and veterinary medicine. The removal of antibiotics from wastewater is necessary since they develop antibacterial resistance. Conventional elimination methods are limited due to the high costs and effort involved, as well as their incomplete elimination. Semiconductor-assisted photocatalysis is an effective elimination method for different organic wastes including antibiotics, and various semiconducting materials have been tested to eliminate antibiotics from wastewater; nevertheless, research is still ongoing due to some limitations.

Beltrán et al. [8] analyzed the role of catalytic ozonation processes in the elimination of disinfection byproducts (DBPs) and their precursors in drinking water treatment. Special attention has always been paid to the formation of DBPs in drinking water treatment because of pathogen removal in the preparation of safe water. DBPs are formed by oxidant-disinfectant chemicals (mainly chlorine derivatives such as chlorine, hypochlorous acid, and chloramines), which react with natural organic matter (mainly humic substances). DBPs are usually refractory to oxidation, mainly due to the presence of halogen compounds, so advanced oxidation processes are recommended to deal with their removal. This work reviewed the application of catalytic ozonation processes (with and without the simultaneous presence of radiation), which are moderately recent advanced oxidation processes, for the removal of humic substances (also called DBP precursors) and DBPs themselves. First, a short history of the use of disinfectants in drinking water treatment, DBP formation discovery, and alternative oxidants used is presented. The following sections are dedicated to the application of conventional advanced oxidation processes to remove DBPs and their precursors. Finally, a description is given of the principal research achievements found in the literature regarding the application of catalytic ozonation processes. In this sense, the authors review aspects such as operating conditions, the reactors used, the radiation sources applied, kinetics, and mechanisms.

One of the difficulties in establishing the intrinsic kinetics of photocatalytic oxidation processes is the complexity of the mathematical formula used to determine the rate of photon absorption. To solve this problem, some models have been proposed and checked, such as the six-flux model (SFM) confirmed for central-lamp photoreactors. External-lamp photoreactors are also one of the most-used configurations for studying the photocatalytic oxidation of contaminants in water, and complex mathematical solutions have been reported to solve the rate of photon absorption. Beltrán et al. [9] adapted the SFM equations already reported for the central lamp photoreactor to determine the rate of photon absorption in an external four-lamp photoreactor. The results obtained showed slight differences compared to those of the Monte Carlo method. Additionally, once the rate of photon absorption was validated, the authors determined the intrinsic rate constant and scavenging factor of the photocatalytic oxidation of some contaminant compounds from the already-published results.

Author Contributions: Conceptualization, J.F.G.-A. and F.J.B.; Methodology, J.F.G.-A. and F.J.B.; Writing—original draft preparation, J.F.G.-A.; Writing—review and editing, J.F.G.-A. and F.J.B.; Supervision, J.F.G.-A. and F.J.B.; Funding acquisition, F.J.B. All authors have read and agreed to the published version of the manuscript.

Data Availability Statement: Not applicable.

Acknowledgments: The authors are grateful for the funding provided by the Agencia Estatal de Investigación of Spain through project PID2019-104429RB-I00/AEI/10.13039/501100011033.

Conflicts of Interest: The authors declare no conflict of interest.

References

1. Rivera-Utrilla, J.; Sánchez-Polo, M.; Ferro-García, M.A.; Prados-Joya, G.; Ocampo-Pérez, R. Pharmaceuticals as emerging contaminants and their removal from water. A review. *Chemosphere* **2013**, *93*, 1268–1287. [CrossRef] [PubMed]
2. Peng, J.; Lu, T.; Ming, H.; Ding, Z.; Yu, Z.; Zhang, J.; Hou, Y. Enhanced Photocatalytic Ozonation of Phenol by Ag/ZnO Nanocomposites. *Catalysts* **2019**, *9*, 1006. [CrossRef]
3. Šihor, M.; Reli, M.; Vaštyl, M.; Hrádková, K.; Matějová, L.; Kočí, K. Photocatalytic Oxidation of Methyl *Tert*-Butyl Ether in Presence of Various Phase Compositions of TiO_2. *Catalysts* **2020**, *10*, 35. [CrossRef]
4. Zhang, X.; Geng, Z.; Jian, J.; He, Y.; Lv, Z.; Liu, X.; Yuan, H. Potassium Ferrite as Heterogeneous Photo-Fenton Catalyst for Highly Efficient Dye Degradation. *Catalysts* **2020**, *10*, 293. [CrossRef]
5. Sugiyama, S.; Hayashi, Y.; Okitsu, I.; Shimoda, N.; Katoh, M.; Furube, A.; Kato, Y.; Ninomiya, W. Oxidative Dehydrogenation of Methane When Using TiO_2- or WO_3-Doped Sm_2O_3 in the Presence of Active Oxygen Excited with UV-LED. *Catalysts* **2020**, *10*, 559. [CrossRef]
6. Aguilar-Melo, C.M.; Rodríguez, J.L.; Chairez, I.; Salgado, I.; Andraca Adame, J.A.; Galaviz-Pérez, J.A.; Vazquez-Arenas, J.; Poznyak, T. Enhanced Naproxen Elimination in Water by Catalytic Ozonation Based on NiO Films. *Catalysts* **2020**, *10*, 884. [CrossRef]
7. Shurbaji, S.; Huong, P.T.; Altahtamouni, T.M. Review on the Visible Light Photocatalysis for the Decomposition of Ciprofloxacin, Norfloxacin, Tetracyclines, and Sulfonamides Antibiotics in Wastewater. *Catalysts* **2021**, *11*, 437. [CrossRef]
8. Beltrán, F.J.; Rey, A.; Gimeno, O. The Role of Catalytic Ozonation Processes on the Elimination of DBPs and Their Precursors in Drinking Water Treatment. *Catalysts* **2021**, *11*, 521. [CrossRef]
9. Beltrán, F.J.; Rivas, J.; Garcia-Araya, J.-F. Six Flux Model for the Central Lamp Reactor Applied to an External Four-Lamp Reactor. *Catalysts* **2021**, *11*, 1190. [CrossRef]

catalysts

MDPI

Article

Six Flux Model for the Central Lamp Reactor Applied to an External Four-Lamp Reactor

Fernando J. Beltrán *, Javier Rivas and Juan-Fernando Garcia-Araya

Departamento de Ingeniería Química y Química Física, Instituto Universitario de Investigación del Agua, Cambio Climático y Sostenibilidad (IACYS), Universidad de Extremadura, 06006 Badajoz, Spain; fjrivas@unex.es (F.J.R.); jfgarcia@unex.es (J.F.G.-A.)

* Correspondence: fbeltran@unex.es; Tel.: +34-924-289387

Abstract: One of the difficulties of establishing the intrinsic kinetics of photocatalytic oxidation processes is due to the complex mathematical formula used to determine the rate of photon absorption. To solve this problem, some models have been proposed and checked, such as the Six Flux Model (SFM) confirmed in central lamp photoreactors. External lamp photoreactors are also one of the most used configurations to study the photocatalytic oxidation of contaminants in water, and complex mathematical solutions have been reported to solve the rate of photon absorption. In this work, SFM Equations already reported for the central lamp photoreactor have been adapted to determine the rate of photon absorption in an external four-lamp photoreactor. The results obtained show slight differences from those of the Monte Carlo method. Additionally, once the rate of photon absorption was validated, the intrinsic rate constant and scavenging factor of the photocatalytic oxidation of some contaminant compounds from results already published have been determined.

Keywords: volumetric rate of photon absorption; six flux model; external lamp photoreactor; photocatalytic oxidation kinetics; scavenging factor; water organic contaminants

1. Introduction

It is well known that Advanced Oxidation Processes (AOPs), are the only technologies capable of destroying water contaminants at ambient conditions due to the high oxidizing power of hydroxyl radicals generated in these processes [1,2]. Photocatalytic oxidation (PCO) of water contaminants with ozonation and Fenton oxidation are some of the advanced oxidation processes that have attracted a lot of research interest in the last two decades [3–7]. In this century, literature shows more than 6700 papers on this subject; a figure similar to that of ozonation or Fenton processes, with about 7100 and 5500 papers, respectively, according to the Scope database. PCO consists of the irradiation of semiconductor material to produce positive (holes) and negative (electrons) charges when the energy of radiation at least equals the band gap energy of the semiconductor [8]. Strong oxidant holes and reducing electrons then migrate from the valence and conduction bands, respectively, to the surface of the catalyst to directly, in the case of holes, or indirectly, through species formed from them (hydroxyl, superoxide radicals, etc.), react with the organics present in water [9]. It is evident that absorption of energy is the key step of PCO. The knowledge of the rate of photon absorption is fundamental to establishing the contaminant oxidation rate. The main variables affecting the photon absorption rate concern the geometry of the reacting system (photoreactor), the intensity of radiation and the optical properties of the catalyst, among others [10,11]. When solving the photocatalytic oxidation rate of any contaminant, the radiation transfer equation and the mass balance of the target compound and, in some cases, fluidodynamic aspects are needed, with the former as the most difficult step to be solved [12]. Literature has already reported complicated mathematical models to solve the radiation transfer equation or algorithms based on Monte Carlo methods that cannot be easily applied to any geometry of a given photoreactor [13–15]. This is because these methods and models have to consider the possibilities that one photon can be absorbed, transmitted or dispersed when it reaches a catalyst particle in the reacting medium. More recently, methods based on the probability that one photon can be dispersed after its collision with the catalyst particle by following different directions have also been proposed [16,17]. Among these models, the so-called Six Flux Model (SFM), which is based on the six cartesian directions that the photon can follow after any collision, has been checked and confirmed in a cylindrical photoreactor with a central located UV lamp (central lamp photoreactor). This configuration is likely the preferred option to study the PCO of water contaminants [18,19]. In this photochemical reactor, radiation coming from the central axis, where the lamp is placed, radially travels to reach the reaction media limited between two cylinders. SFM perfectly calculates the local volumetric photon absorption rate (LVRPA) as a function of the axial and radial coordinates. Next, considering the knowledge of the optical catalyst properties, reactor dimensions and the probability function that any photon has to follow the six cartesian directions after colliding with a catalyst particle, the overall absorption rate of photon absorption (OVRPA) of the catalyst can also be calculated. In addition to central lamp photoreactors, external n-lamp photoreactors have also been used in PCO, and their OVRPA was solved by means of Monte Carlo methods [20–22].

Given the similar geometry and radial transport of radiation of the external lamp and central lamp photoreactors, an attempt to apply the SFM equations established for the latter to the former constitutes the first objective of this work. Additionally, a second objective was to determine the rate constant of the reaction of some organics in PCO once the OVRPA has been determined, that is, the intrinsic kinetics of PCO.

2. Results and Discussion

Figures 1 and 2 show the typical schemes of the central lamp and four external lamps photoreactors with their corresponding axial and radial dimension variables, among others.

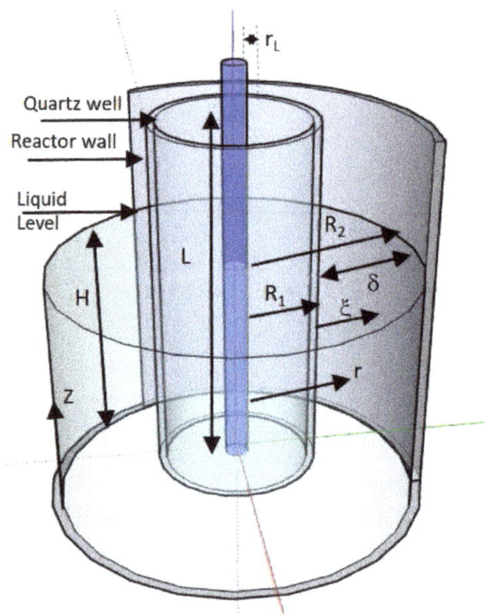

Figure 1. Scheme of the central lamp photoreactor and principal dimensions.

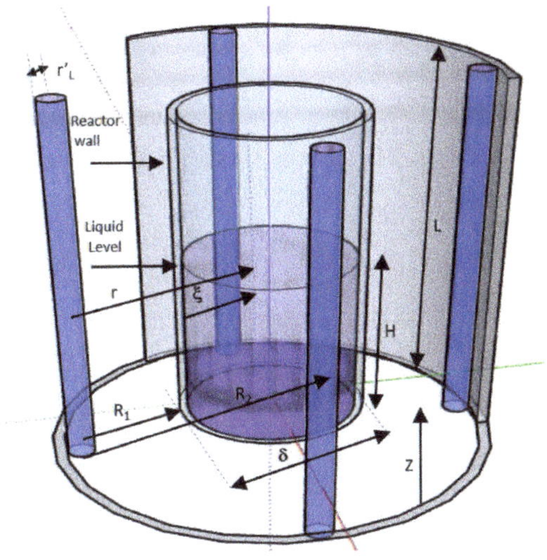

Figure 2. Scheme of the external four lamps photoreactor and principal dimensions.

For the central lamp photoreactor, the SFM yields Equation (1) to determine the local volumetric rate of photon absorption [1]:

$$\text{LVRPA} = \text{K}f(z^*)f(\xi^*) \tag{1}$$

with K, f(z*) and f(ξ*) defined as follows:

$$\text{K} = \frac{r_L I_W \tau(1-\omega)}{2\delta\omega_{cor}} f(\omega,\tau) \tag{2}$$

where r_L is the lamp radius, I_W the incident radiation at the internal wall of the photoreactor or at the radius (R_1) of the well where the lamp is immersed, δ the length path of radiation reaction space or $R_2 - R_1$, τ, the optical density of the catalyst, ω_{cor}, the corrected albedo and f(ω,τ) a function of the scattering albedo (ω) and optical density of the catalyst (see definitions in the Supplementary section),

$$f(z^*) = \arctan\left(\frac{\beta}{2}(2\alpha z^* - \alpha + 1)\right) - \arctan\left(\frac{\beta}{2}(2\alpha z^* - \alpha - 1)\right) \tag{3}$$

where α and β are dimensionless geometrical parameters:

$$\alpha = \frac{H}{L} \tag{4}$$

$$\beta = \frac{L}{R_1} \tag{5}$$

L and H being the length of the lamp and height of the water in the reactor, respectively, and the dimensionless radial, ξ*, and axial, z*, coordinates are defined as follows:

$$\xi^* = \frac{r - R_1}{\delta} \tag{6}$$

$$z^* = \frac{z}{H} \tag{7}$$

where $r - R_1$ and z are the dimension radial and axial coordinates, respectively (see Figure 1),

$$f(\xi^*) = \frac{\exp(-\tau_{ap}\xi^*)}{R_1 + \delta\xi^*} \quad (8)$$

where τ_{ap} is the apparent optical density of the catalyst. Note that ξ^*, the dimensionless radial variable, only just covers the length of the reaction space between R_1 and R_2, that is, δ.

In order to validate SFM Equation (1) for the external four-lamp photoreactor, two assumptions have been considered:

1. Dimension variables of the central lamp photoreactor are substituted by the corresponding variables of the same symbol or name of the external lamp photoreactor so that Figure 2 becomes Figure 1. For instance, the radius of the internal wall of the assumed photoreactor was R_1, that is, the distance between the lamp and the wall of the actual external lamp photoreactor.

2. The radius of the lamp considered in Equation (1) was the diameter of the radius of any lamp of the external lamp photoreactor since it is assumed that only one lamp situated in the central axis of the photoreactor substitutes, as a radiation source, the four lamps of the actual photoreactor.

2.1. LVRPA and OVRPA

To confirm the validity of the SFM model, the external four-lamp photoreactor of reference [21] with corresponding dimensions (see Table S1 and Figure 1) was considered. In reference [21], the photocatalytic oxidation of dichloroacetic acid (DCA) with a catalyst suspension of P25 TiO_2 was studied. In the work, the authors determined the LVRPA and OVRPA by applying the Monte Carlo method, which apparently led to what can be assumed was the exact solution. In the present work, parameter values of SFM related to the optical properties of the catalyst (P25 TiO_2) have also been taken from the literature and are shown in Table S1.

As an example, LVRPA at the central lamp position ($z^* = 0.5$) as a function of the dimensionless radial variable within the reaction space ξ^* and for different catalyst concentrations has been calculated using Equations (1) to (8). Figure 3 shows, as an example, how LVRPA changes with ξ^* for concentrations between 10^{-3} and 5×10^{-2} gL^{-1} (see Figure S1 in the Supplementary information for results at other catalyst concentrations). These results show a high similarity to those previously reported [21]. As can be seen from Figure 3, for a given value of ξ^* close to the reactor wall ($0 < \xi^* < 0.1$), the higher the catalyst concentration, the higher the LVRPA. However, at higher ξ^* values, the opposite situation is observed, though in these cases, LVRPA presents extremely low values or even negligible. This is the result of the increasing scattering radiation as the concentration of catalyst increases. At high catalyst concentrations, the absorption of photons is high as far as they collide with external catalyst particles situated close to the photoreactor's wall, where radiation arrives. Nevertheless, at higher ξ^* values, a low amount of photons reached this reaction media location. As a consequence, absorption rates' negligibility significantly diminished, especially for high catalyst concentration (see Figure S1).

The next step was the determination of OVRPA from the integration of LVRPA into the reaction volume as follows:

$$\text{OVRPA} = \int_0^V \text{LVRPA}\, dV = \int_{R_1}^{R_2}\int_0^H \text{LVRPA}\, 2\pi r dr dz = 2\pi H\delta \int_0^1\int_0^1 (R_1+\delta\xi^*)\text{LVRPA}\ dz^* d\xi^* \quad (9)$$

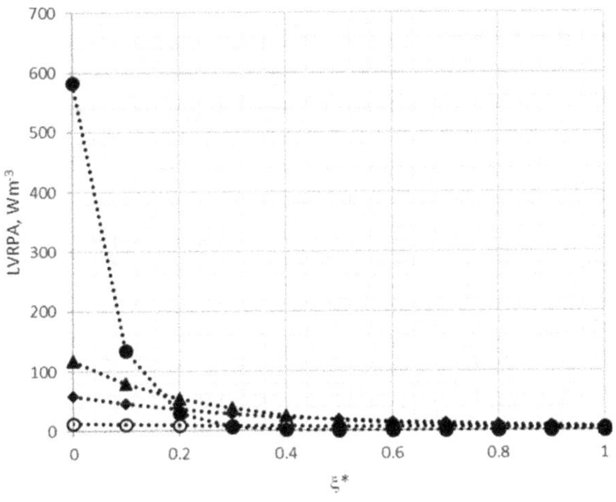

Figure 3. Local Volumetric Rate of Photon Absorption (VRPA) vs. ξ^* from Six Flux Model (SFM) adapted to the external four-lamp photoreactor. Variation with dimensionless radial position within the reaction space and catalyst concentration. Catalyst concentration, gL^{-1}: ○ 10^{-3}, ◆ 5×10^{-3}, ▲ 10^{-2}, ● 5×10^{-2}.

Figure 4 shows the values of OVRPA deduced for different catalyst concentrations.

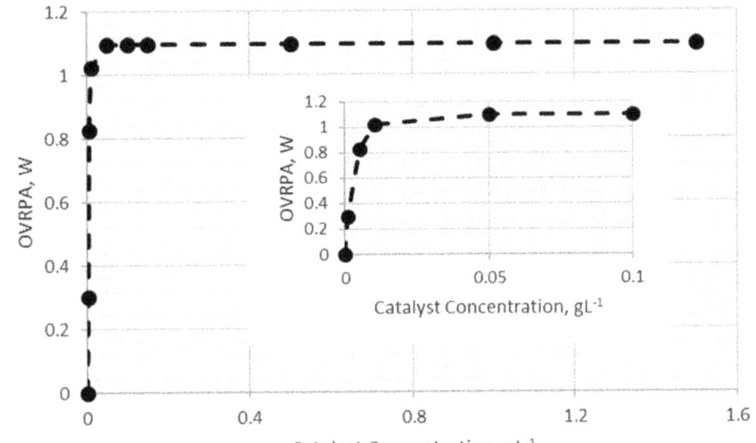

Figure 4. Variation of Overall Volumetric Rate of Photon Absorption (OVRPA) with catalyst concentration.

As can be seen from Figure 4, the increase in the catalyst concentration up to about 0.05 gL^{-1} leads to another corresponding increase in the OVRPA. Catalyst concentrations above this value resulted in a constant value of the OVRPA. This is an expected trend due to the increasing radiation dispersion that the catalyst provokes when its concentration is high. In Figure 5, a comparison of the results obtained with SFM in this work and the Monte Carlo method of Rivas et al. [21] in terms of percentage of absorbed radiation at different catalyst concentrations is shown. As observed from Figure 5, the application of SFM corresponding to the internal lamp photoreactor predicts with deviations lower than 10% the results from the Monte Carlo method for concentrations higher than 0.05 gL^{-1} of catalyst, while it overestimates the results from the Monte Carlo method at low concentrations. In any case, research commonly works on photocatalytic oxidation are carried out above the limit of catalyst concentration, that is, at concentrations leading to the highest photon absorption rate. In this sense, the application of

SFM to the external lamp reactor as if it would behave as a central lamp photoreactor can be a good option to know the overall volumetric rate of photon absorption.

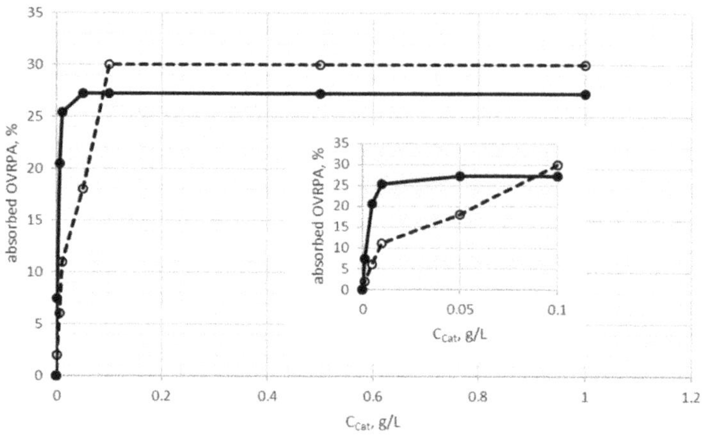

Figure 5. Percentage of absorbed radiation with catalyst concentration from the SFM model (●) and the Monte Carlo method (percentages values from these method deduced from [21] (○).

2.2. Kinetics of Photocatalytic Oxidation of Some Organic Contaminants

The photocatalytic oxidation rate of contaminants in water usually follows Langmuir kinetics [23, 24], which, in many cases, reduces to first-order kinetics when the product of the equilibrium adsorption constant and the concentration of the contaminant is significantly lower than 1. This is the normal scenario in real cases of water treatment where the concentration of the contaminant is low [25]:

$$-r_M = \frac{k_M C_M}{1 + K_M C_M} \simeq k_M C_M \text{ if } K_M C_M << 1 \tag{10}$$

This first-order kinetics, however, can also be confirmed by considering that hydroxyl radical oxidation is the main way to remove contaminants when photocatalytic oxidation is applied. This way of reaction can be produced if hydroxyl radicals are formed from the positive holes in the valence band of the semiconductor:

$$h^+ + H_2O_{ads}(\text{or } -OH_{ads}) \rightarrow HO\bullet \tag{11}$$

and even from successive steps starting with the formation of superoxide ion radicals in the conduction band of the semiconductor:

$$e^- + O_2 \rightarrow O_2^-\bullet \rightarrow \ldots H_2O_2 \rightarrow \ldots HO\bullet \tag{12}$$

Both ways of reaction have been extensively reported in photocatalytic oxidation works [26]. With this assumption and provided that a contaminant M does not directly photodecompose, its photocatalytic oxidation rate would be:

$$-r_M = k_{HOM} C_{HO} C_M \tag{13}$$

where k_{HOM} is the rate constant of the reaction between hydroxyl radicals and M; and the concentration of these free radicals, C_{HO}, is the ratio between their initiation rate, r_{in}, and the sum of the products between the concentration of any hydroxyl radical scavenger substance, C_{Si}, and the corresponding rate constant of its reaction with hydroxyl radicals, k_{HOSi}. This term is usually called the scavenging factor, SF [27]:

$$C_{HO} = \frac{r_{in}}{\sum k_{HOS_i} C_{S_i}} \tag{14}$$

In Equation (14), the initiation rate of radicals is the product between the catalyst quantum yield, ϕ_{cat}, and OVRPA, so that Equation (13) becomes Equation (15), that is, first-order kinetics as in Equation (10):

$$-r_M = k_{HOM}\frac{\phi_{cat}OVRPA}{\sum k_{HOS_i}C_{S_i}}C_M = k_M C_M \tag{15}$$

It should be noted that the SF can change with reaction time because of the changing nature of byproducts formed over time that are usually of increasing hydroxyl radical scavenging character [28].

Determination of the Intrinsic Rate Constant of Photocatalytic Oxidation

According to Equation (15), the ratio between the rate constant k_{HOM} and SF, that is, the intrinsic rate constant of photocatalytic oxidation, can be obtained from k_M once the quantum yield of the catalyst and the OVRPA are known:

$$\frac{k_{HOM}}{\sum k_{HOS_i}C_{S_i}} = \frac{k_M}{\phi_{cat}OVRPA} \tag{16}$$

Notice that this intrinsic rate constant depends on the nature of the aqueous medium since scavenger concentrations affect the oxidation rate of a given contaminant M. In this work, the intrinsic rate constant of a series of contaminants has been determined in ultrapure water. Data on the photocatalytic oxidation of these compounds have been taken from previous works by this research group [20–22]. Table 1 shows the lists of compounds, their k_{HOM} values and the works where the results are published.

Table 1. Contaminants considered and their corresponding hydroxyl radical rate constant values.

Compound	Reference for Results	$k_{HOM} \times 10^{-9}$, $M^{-1}s^{-1}$	Reference for k_{HOM}
Dichloroacetic acid, DCA	[21]	3.9	[29]
2-methyl-4-chlorophenoxyacetic acid MCPA	[20]	6.6	[30]
Metoprolol (MTP)	[22]	7.3	[31]
Diclofenac (DCF)	[22]	7.5	[32]
Ketorolac (KTR)	[22]	7.3	[31]

In these quoted works, an increase of 4 °C in one hour was observed in the reaction medium due to the radiation applied. At first, this can affect the process kinetics. However, in photocatalytic oxidation, the rate constant k_M in Equation (10) depends on the catalyst quantum yield, radiation intensity applied and rate constants of reactions involving free radical and charged species, such as electrons and holes (see Equations (11) and (12)). These parameters, in fact, hardly have a significant influence on photocatalytic kinetics. For instance, Chen et al. [33] carried out photocatalytic runs for the removal of methylene blue, MB, in the 20 to 70 °C temperature interval with minimal differences in MB conversion with time. Accordingly, the temperature effect was neglected in the kinetic study of this work.

Additionally, literature reports negligible adsorption on TiO_2 of Dichloroacetic acid (DCA), Metoprolol (MTP), Ketorolac (KTR) and Diclofenac (DCF) [12,22]. Then, for these compounds, the simplification of Langmuir kinetics to a pseudo-first-order one can be conducted. With respect to 2-methyl-4-chlorophenoxyacetic acid (MCPA), Zertal et al. [34] reported less than 20% adsorption on TiO_2, and taking into account the number of vacant sites of the catalyst [35], a value of the adsorption equilibrium constant K_M = 6303.3 M^{-1} is found, as shown in the Supplementary section. Then, the product $K_M C_M$ at the start of oxidation (see Equation (10)) is 1.57, which is higher than 1. Therefore, Langmuir kinetics cannot be reduced to simple pseudo-first-order kinetics. However, the relationship between k_M and parameters of Equation (15) still hold, that is:

$$k_M = k_{HOM}\frac{\phi_{cat}OVRPA}{\sum k_{HOS_i}C_{S_i}} \tag{17}$$

In Equation (17), the quantum yield of the catalyst, P25 TiO_2, (ϕ_{Cat}) for the concentrations used in the quoted works [20–22], between 0.15 and 0.5 gL^{-1}, and the radiation source emitting between 350 and 410 nm were taken as 3.7×10^{-3} $molEinstein^{-1}$ from Tolosana-Moranchel et al. [36]. Incident radiation at the reactor wall (r = R_1) was 1.19×10^{-5}, 1.15×10^{-6} and 8.33×10^{-7} Einstein$L^{-1}s^{-1}$ for the works from [20–22], respectively.

From the above conclusions, the kinetics of the photocatalytic oxidation of compounds in Table 1 follows the Langmuir model for MCPA and pseudo-first-order for the rest of the compounds. In any case, the external four-lamp photoreactors of quoted references worked as semicontinuous (oxygen was continuously provided) perfectly mixed tanks, so that the mass balance of any compound in these reactors during photocatalytic oxidation is given by Equation (18):

$$\frac{dC_M}{dt} = r_M \tag{18}$$

- Case of first-order kinetics

Solving differential Equation (18) with $r_M = -k_M C_M$ yields the concentration profile of M with time:

$$C_M = C_{M0} \exp(-k_M t) \tag{19}$$

Equation (19) for the photoxidation of the compounds shown in Table 1, which follow first-order kinetics, was confirmed in all cases though it depended on the reaction time (see Figure 6 for the case of DCA at different concentrations as an example).

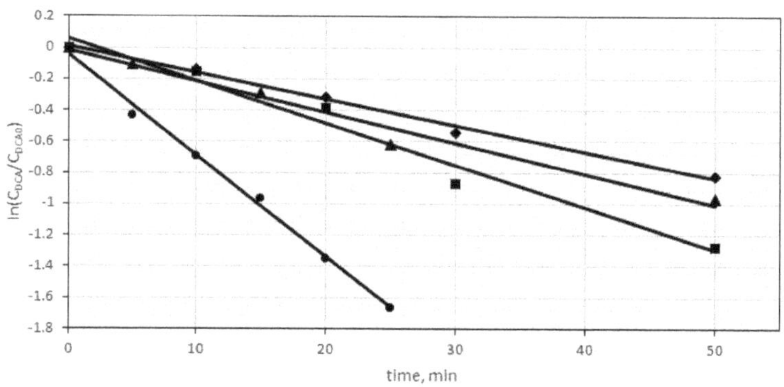

Figure 6. First-order kinetics for the photocatalytic oxidation of Dichloroacetic acid (DCA) (experimental data deduced from [21]). Initial DCA concentration, mgL^{-1}: ● 5; ■ 20; ▲ 50; ♦ 70.

This means that the higher the initial concentration of contaminant, the higher the reaction time where first-order kinetics applies. For instance, for the case of pharmaceuticals MTP, DCF and KTR, the results from [22] for 20 and 0.05 mgL^{-1} initial concentrations show that the first-order kinetics hold during the first 60 and 15 min, respectively. (In the Supplementary section, the confirmation of Equation (18) for MTP, DCF and KTR is also shown in Figure S2.) However, as pointed out before, at higher reaction times, it is likely that first order kinetics still hold but with lower rate constant values due to the increase in SF [28]. In this work, only concentration-time data corresponding to this first period of reaction was considered to determine the intrinsic rate constant.

- Case of Langmuir kinetics

For the case of MCPA, following Langmuir kinetics, the mass balance of Equation (20) applies:

$$\frac{dC_M}{dt} = r_M = -\frac{k_M C_M}{1 + K_M C_M} \tag{20}$$

The separation of variables and integration leads to:

$$\ln\left(\frac{C_M}{C_{M0}}\right) + K_M (C_M - C_{M0}) = Y = -k_M t \tag{21}$$

Figure 7 confirms Equation (21) for photocatalytic oxidation of MCPA, where the slope of the straight line after least squares fitting, as in Figure 6, also corresponds to $-k_M$. It should be highlighted that values of k_M for MCPA photocatalytic oxidation following both the Langmuir model and first-order kinetics are nearly coincident (0.1116 and 0.104 min^{-1}, respectively). In the Supplementary section, the confirmation of Equation (19) of first-order kinetics for the case of MCPA is also presented. Accordingly, for MCPA, in spite of the non-negligible contribution of adsorption, first-order kinetics could be assumed as well.

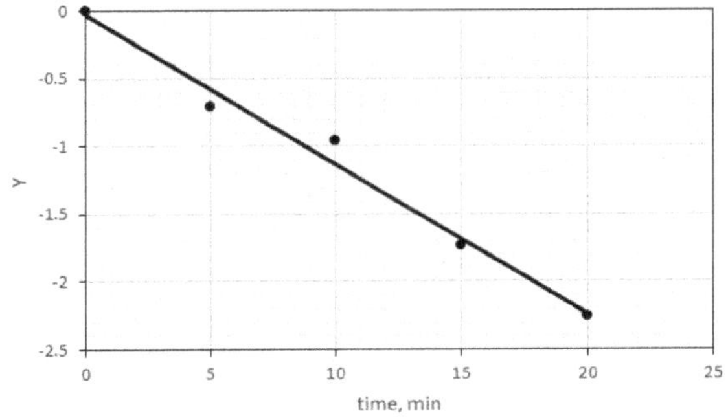

Figure 7. Langmuir kinetics for the photocatalytic oxidation of 2-methyl-4-chlorophenoxyacetic acid (MCPA) (experimental data deduced from [20]).

The values of k_M and the duration of this first initial period (for DCA, MTP, DCF and KTR) where they hold together to the one of MCPA photocatalytic oxidation are given in Table 2. From Table 2, it can be seen that for a given compound, the higher its concentration in water, the lower the intrinsic rate constant and the higher the SF. This is observed in the results of DCA and pharmaceuticals, and it is due to the lesser byproducts formed when the initial concentration was lower.

The complexity of the molecular structure of the compound also has a positive effect on SF as a result of the higher number of intermediates that can be formed from a compound of a more complex structure. This can be seen by comparing the results of DCA and pharmaceuticals at a 20 ppm initial concentration, though the difference in SF is low, 1.04×10^5 against 1.15×10^5 (mean value) s^{-1}, respectively. However, another variable that affects SF is the absorbed radiation, which was 3.24×10^{-6} and 2.27×10^{-7} $EinsteinL^{-1}s^{-1}$ for DCA and pharmaceutical photooxidation, respectively. These values are the reason for the low SF differences in spite of the more complex structure of pharmaceuticals compared to DCA. Additionally, it can be observed that SF values obtained from the simultaneous photooxidation of the three pharmaceutical compounds were similar regardless of their initial concentration. This is an expected result since, in these cases, the formed scavengers belong to the same experiment because the pharmaceuticals were simultaneously treated.

Table 2. Compounds considered, initial concentration, reaction time of first-order kinetics, first-order rate constant, intrinsic rate constant and scavenging factor.

Compound	C_{M0}, mgL^{-1}	Reaction Time, min	k^d, min^{-1}	k_{HOM}/SF [d], M^{-1}	SF^d, s^{-1}
DCA [a]	5	25	6.47×10^{-2}	9.00×10^4	4.33×10^4
	20	50	2.70×10^{-2}	3.76×10^4	1.04×10^5
	50	50	1.98×10^{-2}	2.76×10^4	1.42×10^5
	70	50	1.69×10^{-2}	2.35×10^4	1.66×10^5
MCPA [b]	5	20	1.116×10^{-1}	1.29×10^6	5.13×10^3
KTR [c]	0.05	15	5.47×10^{-2}	1.10×10^6	6.67×10^3
	20	60	3.3×10^{-3}	6.61×10^4	1.10×10^5
DCF [c]	0.05	15	6.11×10^{-2}	1.22×10^6	6.13×10^3
	20	60	3.4×10^{-3}	6.81×10^4	1.10×10^5
MTP [c]	0.05	15	4.62×10^{-2}	0.95×10^6	7.89×10^3
	20	60	2.9×10^{-3}	5.81×10^4	1.26×10^5

[a] From [21], [b] From [20] assuming Langmuir kinetics, [c] Results from photocatalytic oxidation of a contaminant mixture [22] where KTR, DCF and MTP were simultaneously treated, [d] From this work. SF is the Scavenging Factor.

3. Materials and Methods

The materials, methods and experimental set-up are reported in previous works in detail [20–22]. Briefly, the experimental set-up was an external four-lamp reactor that consisted of a 1 L borosilicate glass tank provided with magnetic agitation and inlets for oxygen feed and sampling and outlet for gas exiting. The tank was immersed in a pipe box that also contained four black light lamps that emitted radiation between 350 and 410 nm. The lamps were evenly distributed and attached to the pipe.

4. Conclusions

The main conclusions of this work are:

* The Six Flux Model applied to a central lamp reactor predicts the results of the OVRPA for an external four-lamp photoreactor with deviations lower than 10% with catalyst concentration leading to the highest photon absorption rate. This confirms the results from Monte Carlo methods already published. For low catalyst concentrations, SFM overestimates OVRPA values comparing to those from the Monte Carlo method.

* As expected, the predicted LVRPA values show a radial variation for a given value of the axial dimensionless variable being high at the reactor external wall and practically negligible in the central zone of the photoreactor.

* The OVRPA results confirm the positive effect of catalyst concentration. Thus, for low values of catalyst concentration (<0.1 gL^{-1}), the increase in this variable leads to an increase in OVRPA, while at a higher catalyst concentration, OVRPA goes to a plateau value as a consequence of the higher dispersion or scattering due to the increasing presence of catalyst particles.

* Simplified kinetics allow the determination of the intrinsic reaction rate constant and scavenging factor of hydroxyl radicals. These rate constants increase with the decreasing initial concentration of compounds.

* With the kinetic study, the scavenging factor, SF, of the aqueous solution can also be determined. SF increases with the increase in the initial concentration of the contaminant, and the complex structure of the treated compound formed as a result of the high concentration of byproducts. As expected, for compounds treated with photocatalytic oxidation as a mixture doped to the water, SF yields the same value regardless of the compound results.

Supplementary Materials: The following are available online at https://www.mdpi.com/2073-4344/11/10/1190/s1, Figure S1: LVRPA vs. ξ^* from SFM adapted to the external four lamp photoreactor. Variation with dimensional radial position within the reaction space and catalyst concentration. Catalyst concentration, gL^{-1}: ■ 0.5, ▲ 1, ● 1.5, Figure S2. Confirmation of first order kinetics for KTR, DCF and MTP photocatalytic oxidation. Initial concentration of compounds: 20 mgL^{-1}. Symbols: ■ DCF, ▲ MTP, ● KTR. Experimental results from [4]. Figure S3. Assumed first order kinetics for photocatalytic oxidation of MCPA (experimental results from [5]). Table S1: Values of parameters applied for SFM and dimensions of external lamp photoreactor. Table S2. OVRPA SFM calculated values at different catalyst concentration.

Author Contributions: Conceptualization, F.J.B.; methodology, F.J.B. and J.R.; software, J.R.; validation, F.J.B.; J.R. and J.-F.G.-A.; investigation, F.J.B., J.R. and J.-F.G.-A.; writing—original draft preparation, F.J.B.; writing—review and editing, F.J.B. and J.R.; project administration, F.J.B.; funding acquisition, F.J.B. All authors have read and agreed to the published version of the manuscript.

Funding: This research was funded by Agencia Estatal de Investigación of Spain (PID2019-104429RB-I00/AEI/10.13039/501100011033).

Conflicts of Interest: The authors declare no conflict of interest.

References

1. Sillanpää, M.; Ncibi, M.C.; Matilainen, A. Advanced oxidation processes for the removal of natural organic matter from drinking water sources: A comprehensive review. *J. Environ. Manag.* **2018**, *208*, 56–76. [CrossRef] [PubMed]
2. Deng, Y.; Zhao, R. Advanced Oxidation Processes (AOPs) in Wastewater Treatment. *Curr. Pollut. Rep.* **2015**, *1*, 167–176. [CrossRef]
3. Chatterjee, D.; Dasgupta, S. Visible light induced photocatalytic degradation of organic pollutants. *J. Photochem. Photobiol. C Photochem. Rev.* **2005**, *6*, 186–205. [CrossRef]
4. Beltrán, F.J. *Ozone Reaction Kinetics for Water and Wastewater Systems*; CRC Press: Boca Raton, FL, USA, 2004.

5. Mehrjouei, M.; Muller, S.; Moller, D. A review on photocatalytic ozonation used for the treatment of water and wastewater. *Chem. Eng. J.* **2015**, *263*, 209–219. [CrossRef]
6. Bidga, R.J. Considering Fenton's chemistry for wastewater treatment. *Chem. Eng. Prog.* **1995**, *91*, 62–67.
7. Ribeiro, J.P.; Nunes, M.I. Recent trends and developments in Fenton processes for industrial wastewater treatment—A critical review. *Environ. Res.* **2021**, *197*, 110957. [CrossRef]
8. Mills, A.; Davies, R.H.; Worsley, D. Water purification by semiconductor photocatalysis. *Chem. Soc. Rev.* **1993**, *22*, 417–425. [CrossRef]
9. Schneider, J.; Bahnemann, D.; Ye, J.; Puma, L.G.; Dionysiou, D.D. (Eds.) *Photocatalysis, Vol. 1: Fundamentals and Perspectives*; RSC: London, UK, 2016.
10. Li Puma, G.; Khor, J.N.; Brucato, A. Modelling of an annular photocatalytic reactor for water purification: Oxidation of pesticides. *Environ. Sci. Technol.* **2004**, *38*, 3737–3745. [CrossRef]
11. Cassano, A.E.; Martin, C.A.; Brandi, R.J.; Alfano, O.M. Photoreactor analysis and design: Fundamentals and applications. *Ind. Eng. Chem. Res.* **1995**, *34*, 2155–2201. [CrossRef]
12. Zalazar, C.S.; Romero, R.L.; Martin, C.A.; Cassano, A.E. Photocatalytic intrinsic reaction kinetics I: Mineralization of dichloroacetic acid. *Chem. Eng. Sci.* **2005**, *60*, 5240–5254. [CrossRef]
13. Zazueta, A.L.L.; Destaillats, H.; Li Puma, G. Radiation field modeling and optimization of a compact and modular multi-plate photocatalytic reactor (MPPR) for air/water purification by Monte Carlo method. *Chem. Eng. J.* **2013**, *217*, 475–485. [CrossRef]
14. Alvarado-Rolon, O.; Natividad, R.; Ramirez-Garcia, J.; Orozco-Velazco, J.; Hernandez-Servin, J.A.; Ramirez-Serrano, A. Kinetic modelling of paracetamol degradation by photocatalysis: Incorporating the competition for photons by the organic molecule and the photocatalyst. *J. Photochem. Photobiol. A Chem.* **2021**, *4121*, 113252. [CrossRef]
15. Moreira, J.; Serrano, B.; Ortiz, A.; de Lasa, H. Evaluation of Photon Absorption in an Aqueous TiO_2 Slurry Reactor Using Monte Carlo Simulations and Macroscopic Balance. *Ind. Eng. Chem. Res.* **2010**, *49*, 10524–10534. [CrossRef]
16. Brucato, A.; Cassano, A.E.; Grisafi, F.; Montante, G.; Rizzuti, L.; Vella, G. Estimating radiant fields in flat heterogeneous photoreactors by the six-flux model. *AIChE J.* **2006**, *52*, 3882–3890. [CrossRef]
17. Li Puma, G.; Brucato, A. Dimensionless analysis of slurry photocatalytic reactors using two-flux and six-flux radiation absorption–scattering models. *Cat. Today* **2007**, *122*, 78–90. [CrossRef]
18. Beltrán, F.J.; Aguinaco, A.; García-Araya, J.F.; Oropesa, A. Ozone and photocatalytic processes to remove the antibiotic sulfamethoxazole from water. *Water Res.* **2008**, *42*, 3799–3808. [CrossRef]
19. Mena, E.; Rey, A.; Beltrán, F.J. TiO_2 photocatalytic oxidation of a mixture of emerging contaminants: A kinetic study independent of radiation absorption based on the direct-indirect model. *Chem. Eng. J.* **2018**, *339*, 369–380. [CrossRef]
20. Rivas, J.; Solis, R.R.; Gimeno, O.; Sagasti, J. Photocatalytic elimination of aqueous 2-methyl-4-chlorophenoxyacetic acid in the presence of commercial and nitrogen-doped TiO_2. *Int. J. Environ. Sci. Technol.* **2015**, *12*, 513–526. [CrossRef]
21. Rivas, F.J.; Hidalgo, A.; Solís, R.R.; Tierno, M. Photocatalysis in an external four-lamp reactor: Modelling and validation—dichloroacetic acid photo-oxidation in the presence of TiO_2. *Int. J. Environ. Sci. Technol.* **2019**, *16*, 6705–6716. [CrossRef]
22. Encinas, A.; Rivas, F.J.; Beltrán, F.J.; Oropesa, A. Combination of black-light photocatalysis and ozonation for emerging contaminants degradation in secondary effluents. *Chem. Eng. Technol.* **2013**, *36*, 492–499. [CrossRef]
23. Chong, M.N.; Jin, B.; Chow, C.W.K.; Saint, C. Recent developments in photocatalytic water treatment technology: A review. *Water Res.* **2010**, *44*, 2997–3027. [CrossRef]
24. Liu, B.; Zhao, X.; Terashima, C.; Fujishima, A.; Nakata, K. Thermodynamic and kinetic analysis of heterogeneous photocatalysis for semiconductor systems. *Phys. Chem. Chem. Phys.* **2014**, *16*, 8751. [CrossRef]
25. Malato, S.; Fernandez-Ibañez, P.; Maldonado, M.I.; Blanco, J.; Gernjak, W. Decontamination and disinfection of water by solar photocatalysis: Recent overview and trends. *Catal. Today* **2009**, *147*, 1–59. [CrossRef]
26. Legrini, O.; Oliveros, E.; Braun, A.M. Photochemical processes for water treatment. *Chem. Rev.* **1993**, *93*, 671–698. [CrossRef]
27. Beltrán, F.J.; Rey, A. Free Radical and Direct Ozone Reaction Competition to Remove Priority and Pharmaceutical Water Contaminants with Single and Hydrogen Peroxide Ozonation Systems. *Ozone Sci. Eng.* **2018**, *40*, 251–265. [CrossRef]
28. Beltrán, F.J.; Checa, M.; Rivas, F.J.; García-Araya, J.F. Modeling the Mineralization Kinetics of Visible Led Graphene Oxide/Titania Photocatalytic Ozonation of an Urban Wastewater Containing Pharmaceutical Compounds. *Catalysts* **2020**, *10*, 1256. [CrossRef]
29. Zhai, X.; Chen, Z.; Zhao, S.; Wang, H.; Yang, L. Enhanced ozonation of dichloroacetic acid in aqueous solution using nanometer ZnO powders. *J. Environ. Sci.* **2010**, *22*, 1527–1533. [CrossRef]
30. Benitez, F.J.; Acero, J.L.; Real, F.J.; Roman, S. Oxidation of MCPA and 2,4-D by UV Radiation, Ozone, and the Combinations UV/H_2O_2 and O_3/H_2O_2. *Environ. Sci. Health Part B* **2004**, *39*, 393–409. [CrossRef]
31. Benner, J.; Salhi, E.; Ternes, T.; von Gunten, U. Ozonation of drinking water: Part I. Oxidation kinetics and product formation. *Water Res.* **2008**, *42*, 3003–3012. [CrossRef]
32. Huber, M.M.; Canonica, S.; Park, G.Y.; von Gunten, U. Oxidation of Pharmaceuticals during Ozonation and Advanced Oxidation Processes. *Environ. Sci. Technol.* **2003**, *37*, 1016–1024. [CrossRef]
33. Chen, Y.W.; Hsu, Y.H. Effects of Reaction Temperature on the Photocatalytic Activity of TiO_2 with Pd and Cu Cocatalysts. *Catalysts* **2021**, *11*, 966. [CrossRef]
34. Zertal, A.; Molnár-Gábor, D.; Malouki, M.A.; Sehili, T.; Boule, P. Photocatalytic transformation of 4-chloro-2-methylphenoxyacetic acid (MCPA) on several kinds of TiO2. *Appl. Catal. B Environ.* **2004**, *49*, 83–89. [CrossRef]

35. Rodriguez, R.; Blesa, M.A.; Regazzoni, A.E. Surface Complexation at the TiO2 (anatase)/Aqueous Solution Interface: Chemisorption of Catechol. *J. Colloid Interface Sci.* **1996**, *177*, 122. [CrossRef]
36. Tolosana-Moranchel, A.; Casas, J.A.; Carbajo, J.; Faraldos, M.; Bahamonde, A. Influence of TiO_2 optical parameters in a slurry photocatalytic reactor: Kinetic modelling. *Appl. Catal. B Environ.* **2017**, *200*, 164–173. [CrossRef]

catalysts

MDPI

Article

Enhanced Naproxen Elimination in Water by Catalytic Ozonation Based on NiO Films

Claudia M. Aguilar-Melo [1,2], Julia L. Rodríguez [3,*], Isaac Chairez [2,*], Iván Salgado [4], J. A. Andraca Adame [5], J. A. Galaviz-Pérez [6], Jorge Vazquez-Arenas [7] and Tatyana Poznyak [3]

1 Departamento de Ingeniería de Procesos e Hidráulica, Universidad Autónoma Metropolitana-Itzapalapa, Ciudad de México 09340, Mexico; assiram_12@hotmail.com
2 Departamento de Bioprocesos, UPIBI—Instituto Politécnico Nacional, Ciudad de México 07340, Mexico
3 Laboratorio de Ing. Química Ambiental, ESIQIE—Instituto Politécnico Nacional, Ciudad de México 07340, Mexico; tpoznyak@ipn.mx
4 CIDETEC—Instituto Politécnico Nacional, Ciudad de México 07340, Mexico; isalgador@ipn.mx
5 Departamento Ciencias Básicas, UPIIH—Instituto Politécnico Nacional, Hidalgo 42050, Mexico; andraca1@yahoo.com.mx
6 División Académica Multidisciplinaria de Jalpa de Méndez, Universidad Juárez Autónoma de Tabasco, Villahermosa 86049, Mexico; jorgegalavizperez@gmail.com
7 Catedra Conacyt—Departamento de Química, Universidad Autónoma Metropolitana—Iztapalapa, Av. San Rafael Atlixco No. 186, Ciudad de México 09340, Mexico; jorgegva@hotmail.com
* Correspondence: jlrodriguezs@ipn.mx (J.L.R.); isaac_chairez@yahoo.com (I.C.)

Received: 27 May 2020; Accepted: 28 July 2020; Published: 5 August 2020

Abstract: This study evaluates naproxen (NP) degradation efficiency by ozonation using nickel oxide films ($NiO_{(F)}$) as a catalyst. The NiO films were synthesized by chemical vapor deposition and characterized by X-ray diffraction, scanning electron microscopy, atomic force microscopy and X-ray photoelectron spectroscopy. NP degradation was conducted for 5 min using 10 films of $NiO_{(F)}$ comparing against ozonation using 100 mg/L NiO powder in suspension ($NiO_{(S)}$) and conventional ozonation (O_3-conv). Total organic carbon analysis demonstrated a mineralization degree of 12% with O_3-conv, 35% with NiO as powder and 22% with $NiO_{(F)}$ after 60 min of reaction. The films of $NiO_{(F)}$ were sequentially used 4 times in ozonation demonstrating the stability of the synthesized material, as well as its properties as a catalyst for ozonation. A proposed modeling strategy using robust parametric identification techniques allows the comparison of NP decomposition pseudo-monomolecular reaction rates.

Keywords: naproxen; catalytic ozonation; nickel oxide films; mineralization; reaction kinetics

1. Introduction

The growing interest in catalytic ozonation is a consequence of their remarkable advantages with respect to conventional ozonation whose benefits are (a) enhancing the recalcitrant pollutants removal in water, (b) reducing reaction period, and (c) increasing the mineralization degree [1,2]. The presence of suspended metallic oxides particles in a catalyst breaks down the ozone molecule leading to radical generation ($\cdot OH$ mainly). These radicals are more reactive than molecular ozone due to their higher oxidation potential [3]. Today, different catalysts have been used in the ozonation for decomposing a wide variety of pollutants in water, including suspended metal oxides [4], supported metal oxides [5,6], activated carbon [7], zeolites and clay minerals [8] are the most extended ones due to their fast and efficient radicals generations.

Nickel oxide has shown outstanding results degrading herbicides [9], pharmaceutical compounds [10] and polycyclic aromatic hydrocarbons [11] in combination with ozone.

Most research on heterogeneous catalytic ozonation used suspended particles of metallic oxides as catalysts. This strategy favored the contact frequency between contaminants either with the catalysts surfaces or the oxidants produced [12]. Nevertheless, recuperation, separation and eventual reuse of suspended solid catalysts are more complicated and require additional unitary operations [13].

Catalyst immobilization seems to be an appropriate alternative to avoid some of the known inconveniences of using suspended particles of solid catalysts. There are several options to carry out the catalyst immobilization, to produce different supporting structures (thin films, membranes, covered spheres, etc.). Some conventional methods for thin film synthesis on specific substrates are liquid phase deposition, chemical vapor deposition (CVD), hydrothermal treatment, electrophoretic deposition, among others [14,15]. Each of these methods produces films with diversified characteristics like morphology, surface roughness and reproducibility in the synthesized materials.

The CVD technique produces highly reproducible thin films with coating uniformity and control as well as precise thickness. This method induces chemical reactions (thermal decomposition, reduction, hydrolysis, oxidation, etc.) between precursor and substrate which yields the generation of a solid film over the preheated substrate [16].

Supported catalysts application in some advanced catalytic oxidation processes (photocatalysis, mainly) enhances the decomposition efficiency of complex, toxic and recalcitrant organics contaminants [17,18]. However, the use of catalysts fixed onto a surface (thin film coating) has been scarcely studied, according to our knowledge, in ozonation of toxic and recalcitrant pollutants including diverse pharmaceutical compounds [19].

The presence of pharmaceutical compounds in the environment has been associated with serious problems in lakes and rivers such as mutations, infertility and sexual disorders in the aquatic fauna [20,21]. The increasing concentrations of pharmaceutical compounds in aquatic ecosystems is a consequence of the inappropriate disposal of effluents coming from hospitals, farms, pharmaceutical industries, urine of human beings and animals [22]. Among the regular pharmaceutical pollutants, naproxen (NP) has been detected in different water bodies, since it is a common analgesic, which does not require a medical prescription (Figure 1). NP is highly effective in the treatment of rheumatoid arthritis because of its anti−inflammatory and antipyretic properties [23]. This compound is hardly biodegraded by either aerobic or anaerobic microorganisms [24]. Several treatment methods have been applied to eliminate NP in wastewater, including photocatalysis [25,26], biodegradation, UV photolysis [27,28], sonocatalytic [29,30] and Fenton reaction [31]. These processes have demonstrated low decomposition effectiveness and poor mineralization, as well as other operating drawbacks.

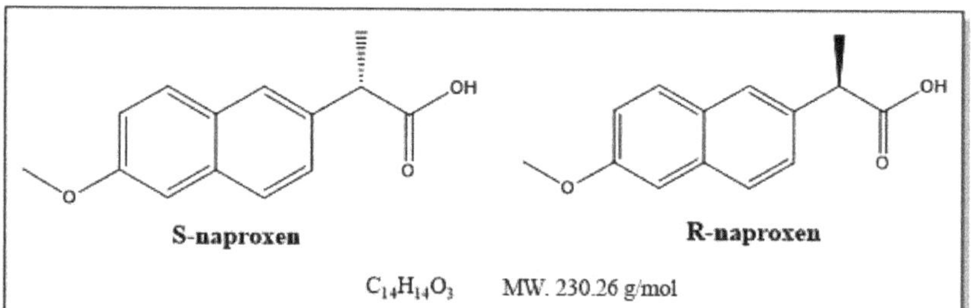

Figure 1. Chemical structures of Naproxen isomers.

One additional problem associated with heterogeneous catalytic ozonation reactions is the lack of accurate models describing the complete process. The understanding of the process conditions and the expected products can contribute to the operation control and quality of the contaminant degradation results [32]. Classical modeling techniques of ozonation include the least square method to obtain an accurate curve that fits the experimental data better. However, high variability presented in such a complex chemical system (including serial and parallel reactions) can lead to inaccurate models, which may limit their applicability for real treatments. To overcome these problems, a robust differentiation technique was implemented.

Catalytic ozonation with NiO in suspension represents an effective alternative to eliminate NP from water [33], but the recovering and eventual reuse of suspended catalysts needs to be resolved. Therefore, it is feasible to propose catalytic ozonation based on NiO thin films aimed to decompose NP in aqueous solutions. Here, NiO films ($NiO_{(F)}$) were supported on glass substrates by the CVD method. A comparison was established between conventional and catalytic ozonation using NiO in powder and films.

2. Results and Discussion

2.1. $NiO_{(F)}$ Characterization

The CVD technique used to deposit NiO allows the obtaining of a film with a uniform gray color. The structural analysis was complemented with the X-ray diffraction (XRD) study. Figure 2 shows the XRD pattern of $NiO_{(F)}$. The analyzed spectrum presents five characteristic peaks at $2\theta = 37.52$, 43.58, 63.14, 75.7 and 79.84°. These signals agree with the commercial $NiO_{(S)}$, corresponding to the (111), (200), (220), (311) and (222) planes, respectively. These signals are indexed to the cubic phase of NiO (Joint Committee on Powder Diffraction Standards No. 78-0423, 73-1519) [34]. The presence of only well-defined five peaks indicates the high purity of the $NiO_{(F)}$. According to the XRD pattern, the $NiO_{(F)}$ obtained by CVD has preferential orientation in 111 direction and the orientation does not change after ozonation. Furthermore, the films present compressive stress due to the slight slippage of the peaks. If we consider (111) planes, $NiO_{(S)}$ presents the peak at position 37.20, while O_3-$NiO_{(F)}$ at 37.36 and $NiO_{(F)}$ at 37.52. The results indicate that $NiO_{(F)}$ has 0.86% compressive stress and O_3-$NiO_{(F)}$ after the ozonation, there is a 0.43% relaxation compared to $NiO_{(S)}$ powders. Additionally, the X-ray diffraction analysis in the software Match3 enables determination of a density of 6.79 g/cm^3 for $NiO_{(F)}$. The crystallinity stability was demonstrated in the $NiO_{(F)}$ XRD pattern after oxidation with ozone (O_3-$NiO_{(F)}$). The well-defined five peaks detected in the sample without ozonation were observed after the treatment. This pattern confirmed the stability of the film after ozonation.

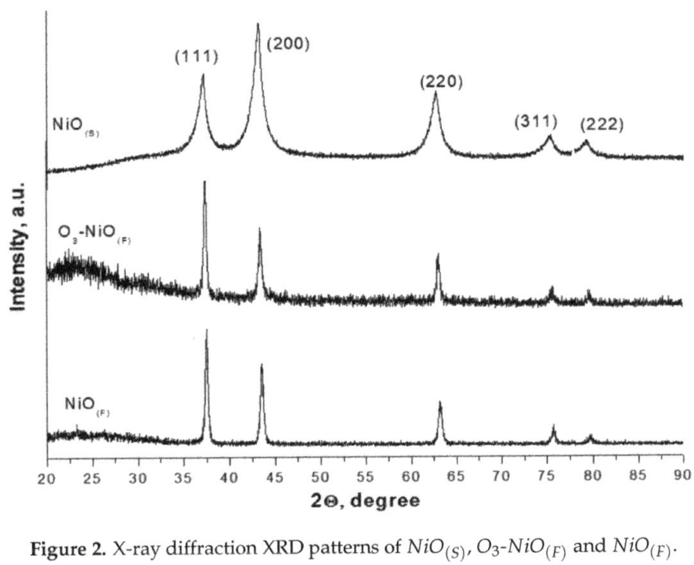

Figure 2. X-ray diffraction XRD patterns of $NiO_{(S)}$, O_3-$NiO_{(F)}$ and $NiO_{(F)}$.

The surface morphology and thickness of $NiO_{(F)}$ were evaluated by Scanning Electron Microscopy (SEM), Figure 3a. The image showed the aggregates of NiO on the glass substrates, revealing a dense and continuous coating.

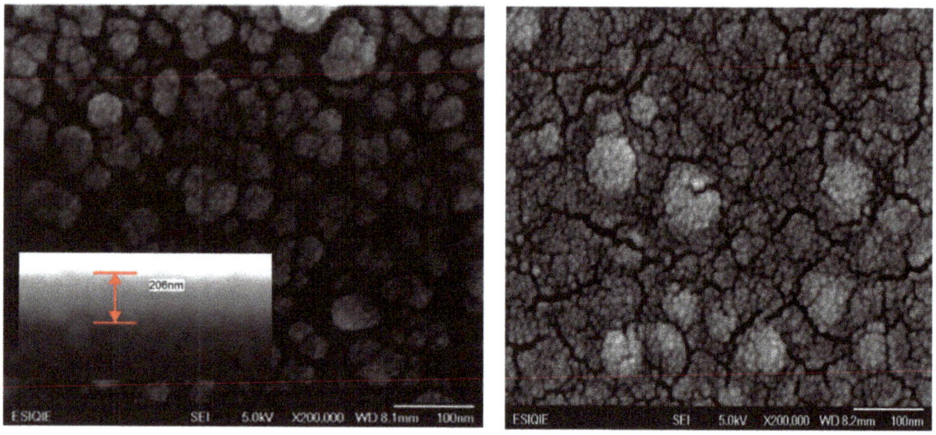

(**a**) (**b**)

Figure 3. SEM images of $NiO_{(F)}$ before and after ozonation (60 min of treatment). Insert Figure 2a presents the cross-sectional image of the $NiO_{(F)}$: (**a**) $NiO_{(F)}$, (**b**) O_3-$NiO_{(F)}$.

The optical image captured at 200,000× demonstrates the presence of NiO aggregates with an equivalent diameter smaller than 100 nm. These aggregates are conformed by little spherical particles (Figure 3a). A similar surface growth was reported in the synthesis of NiO films by electro-deposition on ITO coated [35]. The cross-sectional image of the film (insert Figure 3a) demonstrated the deposition of NiO on the glass slide was around 206 nm of thickness. This value was confirmed by spectroscopic ellipsometry. The average thickness was 240.97 ± 55.56 nm ($n = 3$, an average of five points by sample). After ozonation, $NiO_{(F)}$ (Figure 3b) showed apparent cracks. Lin et al. [35] in the synthesis of $NiO_{(F)}$ by galvanostatic electro-deposition method, observed cracks on the surface. This fact was attributed to the relaxation of the stress in the film. This behavior agrees with the XRD results, moreover, the film stability was demonstrated using the constant diminution of total organic carbon (TOC) when the reused films were used as the catalyst.

Figure 4 shows the Atomic force microscopy (AFM) images of surface morphology of $NiO_{(F)}$ (Figure 4a) and O_3-$NiO_{(F)}$ (Figure 4b). In these figures, no cracks are observed after ozonation. The values of average roughness (Ra) and the root mean square roughness (Rrms) for $NiO_{(F)}$ were 4.66 nm and 5.89 nm, respectively and Ra was 4.82 nm and Rrms was 6.14 nm for O_3-$NiO_{(F)}$. These results unlike that observed in SEM analysis, suggest that the surface is not affected during the ozonation.

X-ray photoelectron spectroscopy (XPS) results for $NiO_{(F)}$ before and after ozonation are shown in Figure 5. The Ni2p region in the XPS spectrum displays characteristic signals attributed to NiO, Figure 5a. Ozone did not change the spectrum in the Ni region which is similar to previous DRX results. Furthermore, the comparison $NiO_{(F)}$ and $NiO_{(S)}$ showed similar spectra. The O1s region (Figure 5b) is decomposed in two main signals for $NiO_{(F)}$ and $NiO_{(S)}$: the first at 529.3 eV corresponding to oxygen atoms lattice of NiO and the second around 531 eV assigned to defective oxygen, for instance oxygen atoms due to nickel vacancies [9]. When the ozonation was carried out, $NiO_{(F)}$ displays two additional peaks (at 531.8 eV (-CO) and 532.3 eV (-COH)) which are attributed to oxygen coordinated to carbon atoms by adsorbed byproducts generating to NP decomposition.

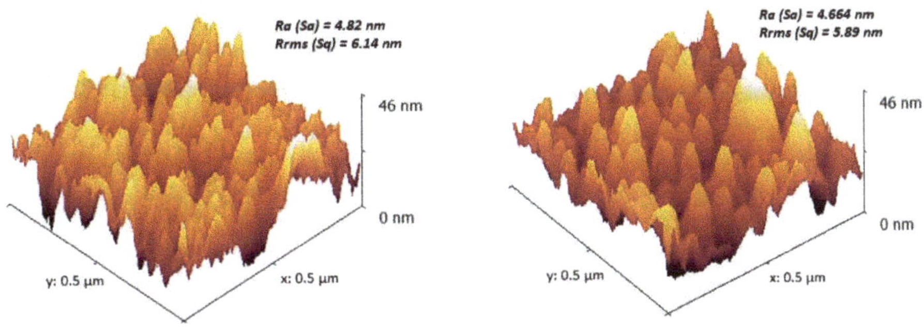

(a) (b)

Figure 4. AFM images of NiO_F before and after ozonation (60 min of ozonation): (**a**) $NiO_{(F)}$. (**b**) O_3-$NiO_{(F)}$.

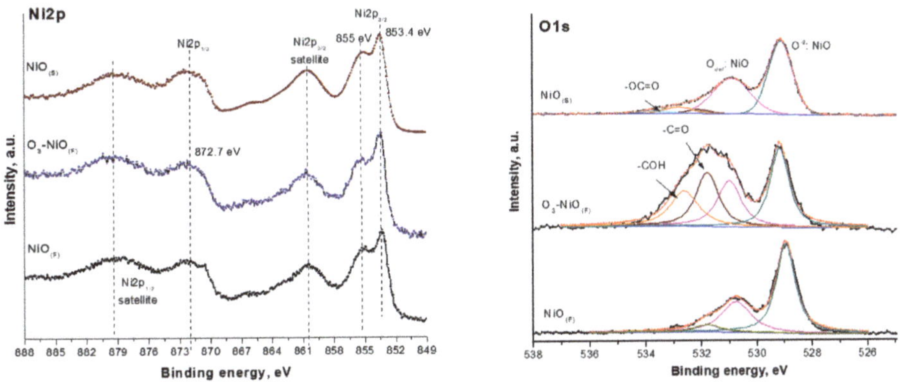

(**a**) Ni2P (**b**) O1S

Figure 5. XPS patterns of $NiO_{(F)}$, $NiO_{(S)}$ and O_3-$NiO_{(F)}$. (**a**) Ni2p and (**b**) O1s regions.

2.2. Evaluation of $NiO_{(F)}$ in the NP Removal Using Ozone

Figure 6a shows the NP degradation profiles (obtained by high efficiency liquid chromatography) for conventional (O_3-conv) and catalytic ozonation with $NiO_{(F)}$ and $NiO_{(S)}$. The NP signal was only detected along the first five minutes of the reaction. In the presence of $NiO_{(F)}$ and $NiO_{(S)}$, the catalyst did not modify the NP degradation profile in comparison with O_3-conv. The determined NP degradation time was faster than the other advanced oxidation process [31,36,37]. The apparent absence of changes in the degradation profile of the parent compound agrees with previous results [38,39]. This fact was attributed to: (1) ozone fast reaction with the initial contaminant (NP) and (2) generation of oxidant species (ozone decomposition with catalyst) and reaction with NP ozonation byproducts.

The pH changed during the ozonation from the initial NP solution which had a pH of 5.0 ± 0.5 to the final pH of 2.5 ± 0.5 after the treatment (60 min). Under these experimental conditions (acid pH), the direct ozonation mechanism was favored and ozone decomposition (formation of oxidant species) was not enforced. On the other hand, the oxalic acid accumulated in ozonation as the main final product which yields the decreasing of pH. This result coincides with the reported results by other researchers and our previous results [38,40].

Few articles have studied NP degradation by catalytic ozonation varying pH. The study proposed in [41] reported NP (50 mg/L) degradation with 0.48 mg/s of ozone at pH of 5, 6, 7, 8 and 9. The treatment at pH = 9 allowed near 100% of NP degradation after 4 min, while only 30% of NP was eliminated with pH = 5 during the same reaction time. This difference was explained considering that under basic pH, direct and indirect mechanisms occur simultaneously. In the case of catalytic ozonation of NP, [33] demonstrated that 1 g/L of TiO_2 yields a higher mineralization degree at pH of 5 and it decreased near 50% at pH 3. Moreover, the same authors proposed two periods of mineralization; the first period, named the fast, corresponding to the fast reaction of easy oxidable compounds; in this period pH plays a main role (while at greater pH = 7 increased the removal of TOC). The second mineralization stage includes the reaction of refractory byproducts and it was independent of pH. Notice that some of these results agree with the outcomes attained in this study.

The NiO effect can be observed in the oxalic acid profile over time (Figure 6b). This organic acid is a recalcitrant compound to O_3-conv [42,43]. Oxalic acid is a main final product formed as a consequence of the aromatic ring outbreak or loss of acid naproxen substituent in the reaction, (mechanism reported by Jallouli et al. [44]). The O_3-$NiO_{(F)}$ duplicates the oxalic acid concentration in comparison with O_3-conv during the first min of reaction. The oxalic acid concentration is similar after 10 min in catalytic and conventional ozonation. The differences in the oxalic acid profile in both ozonation systems justify the proposed inclusion of several oxidant species in the catalytic ozonation. According to [38], the combination of ozone and NiO produced hydroxyl radicals ($\cdot OH$) which favors the elimination of refractory organic compounds due to their oxidation potential (2.8 eV) [45].

Oxalic acid as a NP final product can be generated by two main routes: (1) the breakout of the propionic acid substituent of NP and the release of a 3-carbon fragment, the subsequent 1-carbon addition form the malic acid [44] by the oxidant species in the reaction. This byproduct is susceptible to oxidation with ozone and hydroxyl radicals to produce oxalic acid (a recalcitrant final product) (2). The electrospray ionization mass spectrometry (ESI-Ms) analysis of NP degradation effluents at 5 min showed the formation of aromatic byproducts. The molecules proposed for these ions m/z of 237.11, 217.09, 177.66 and 149.06 only present one aromatic ring.

The fragments obtained in the breakout and degradation of the aromatic compounds can be degraded to oxalic acid. The increase of the oxalic acid concentration in the first two minutes of ozonation is possible due to the first route described above, where the break of aromatic rings is not needed. The subsequent decrease and slight increment of this compound concentration might be related to the aromatic byproduct's degradation after 5 min which was confirmed by the HPLC and ESI-Ms studies. As an example (Figure 7), a signal of a non-identified intermediate byproduct detected by HPLC (12.9 min of retention time) disappeared after 10 min. Its maximal concentration was determined at 2 min.

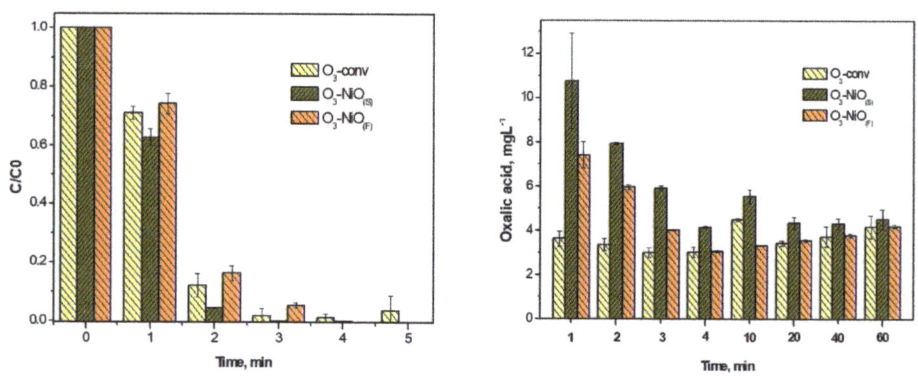

(**a**) NP degradation (**b**) Oxalic acid

Figure 6. NP [20 mg/L] degradation (**a**) and oxalic acid concentration (**b**) generated by O_3-conv, O_3-$NiO_{(S)}$ (100 mg/L) and O_3-$NiO_{(F)}$. $[O_3] = 5.5 \pm 0.5$ mg/L, flow (O_3-O_2) = 0.5 L/min.

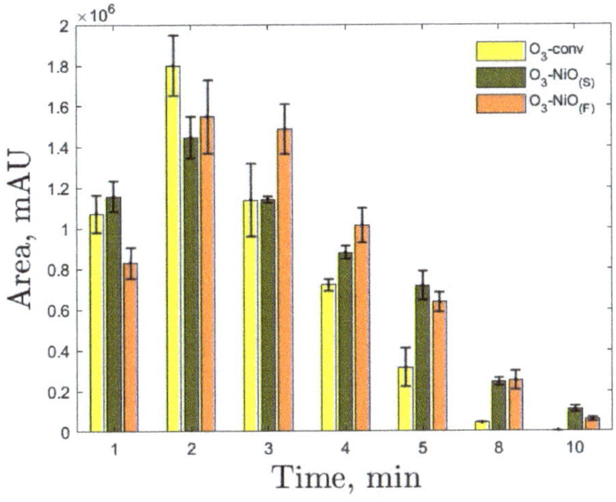

Figure 7. Unidentified compound (12.9 min time retention HPLC) during the NP degradation with conventional and catalytic systems

In a previous article regarding NP degradation with different proportions of water and ethanol as solvents [38], the studies of ESI-Ms and ESI-mass-mass (ESI-Ms-Ms) spectrometry were done in negative mode. Different aromatic byproducts such as

(a) 1-(6-methoxynaphthalene-2-yl)ethylhydroperoxide,
(b) 2-(3-(hydroxymethyl)-4-(2-methoxyethyl)phenyl)propanoic acid
(c) 2-(4-(2-hydroxyvinyl)phenyl)acetic acid and
(d) 4-Methylphenylacetic acid

were detected in the first 5 min of the reaction. However, these byproducts were not detected at 60 min of treatment (conventional and catalytic ozonation). Similar results were obtained with the NiO films for the NP elimination. These results confirm that all these intermediate compounds are close to be eliminated within the time window of the ozonation reaction (This fact is confirmed with the ESI-Ms-Ms analysis in samples taken after 60 min of ozonation). What is more important is that the decomposition of these compounds yields to the formation of organic acids (such as oxalic acid), aldehydes and carbon dioxide. All these compounds are considered to be more biodegradable.

The global efficiency of water treatments can be determined by the TOC variation. The role of $NiO_{(F)}$ was estimated also by TOC analysis (Figure 8). In the O_3-conv, TOC removal after 60 min of the reaction was 12.3%, while introducing 10 films of NiO increases this removal to 22%. The presence of 100 mg/L of $NiO_{(S)}$ approximately achieved 35% of TOC removal, indicating that the solid catalyst was a better system for the NP elimination.

The low and incomplete NP mineralization has been described as a regular condition in catalytic ozonation. Rosal et al. reported near 40% of TOC removal in O_3-conv and around 55% using 1000 mg/L of TiO_2 Degussa P25 after 60 min of reaction. Therefore, the use of TiO_2 as a catalyst increased the TOC removal about 15% [33]. This percentage is similar to the observed in this work with 100 mg/L of NiO in suspension.

The use of $NiO_{(F)}$ showed less TOC removal compared to $NiO_{(S)}$. In this study, to compare correctly the catalytic effectiveness of both catalytic cases, it is worth taking into account what decomposed mass of NP corresponds to the mass of the catalyst in the reaction, it means a class of NP decomposition yield. A rough calculation of the catalyst on the glass slide showed a mass of 1.2 mg per film and about 12 mg over the 10 used films. The calculated amount of catalyst in the treatment is 10 times less than that used in suspension (40 mg). Corresponding, the catalytic activity of $NiO_{(F)}$ is 1.66 mg of the decomposed NP and only 0.5 mg for $NiO_{(S)}$ at 1 mg of catalyst. Therefore, the catalytic activity of $NiO_{(F)}$ is 3.3 times higher than $NiO_{(S)}$. This result shows the possibility of using lower concentrations of the catalyst when depositing it on thin films, in addition to facilitate its recovery after ozonation.

The stability of $NiO_{(F)}$ was studied by TOC removal after 4 consecutive cycles of ozonation, Figure 8. In all cases, $NiO_{(F)}$ removed similar TOC percent after each cycle (22.2% ± 2.9% of TOC removal). Such a result indicates that ozone-NP and ozone-byproducts reactions did not inhibit the activity of $NiO_{(F)}$ after 4 cycles. The use of films as catalysts in the ozonation is scarcely described yet. Guzman et al. (2020) [19], in the 4-chlorophenol and 4-phenolsulfonic acid degradation by ozone (120 min) with six ceria oxide films (26.4 × 76.2 mm) demonstrated a similar TOC removal in five cycles of treatment (about 44% for both compounds). The results confirm the stability of the film in the catalytic ozonation.

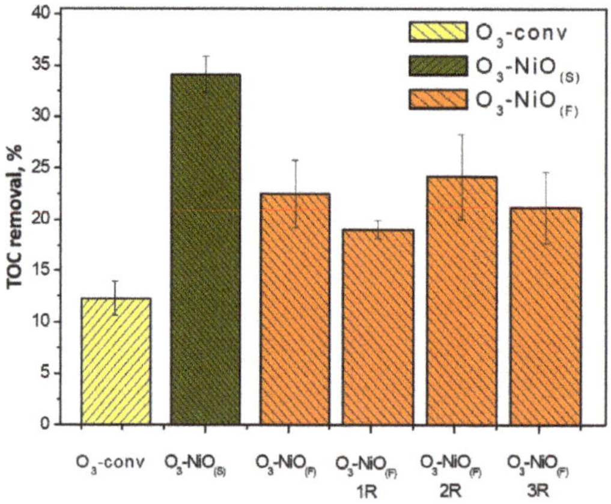

Figure 8. TOC removal (%) in conventional and catalytic ozonation and the effect of reusing films during three cycles with 10 films of *NiO*.

2.3. Mathematical Modeling of the Reaction Dynamics

Table 1 summarizes all parameters used in the Super-Twisting algorithm (STA) to estimate the kinetics of the NP degradation by ozonation. The robustness of STA produces similar values for the needed parameters for tuning the differentiator. Figure 9 shows the time variation of signals produce by the modeling process that yields the kinetics and the approximate model of NP degradation by conventional and catalytic ozonations. In the first step, data are collected from the HPLC. These data are marked with a blue circle in all subfigures. Then, the black continuous line is obtained by an interpolation process (spline functions as an approximation basis). This step is done to have enough data to feed STA applied as a differentiator. The accuracy of the STA is related to the quality of the collected data. The derivative is presented in the second graph. During the first 40 s of simulation, STA presents an overshoot during the adaptation phase, which means it is the time when the value $x_1(t) - C_{NP}(t)$ is reaching the zero value. After this time, the derivative is exact, and one may calculate the maximum rate of NP degradation. This rate is marked with a red circle in each graph and it is reported in Table 1. This process was repeated three times to compare the behavior of the ozonation cases studied in this manuscript.

Figure 9a shows the estimated model for the degradation of NP with conventional O_3, while Figure 9b shows the estimated model for the O_3-$NiO_{(S)}$ and Figure 9c shows the process for O_3-$NiO_{(F)}$. The STA was numerically evaluated with $k_1 = k_2 = 17.7$ for the O_3-conv and $k_1 = k_2 = 15.1$ for the catalytic ozonations.

Notice that all reaction rates constants are not varying significantly. However, their tendencies confirm what one may detect as a critical outcome of this and similar studies on supported catalysts over solid films: films of catalyst may be more efficient than conventional (no catalytic), but less than using the powder catalyst. This condition can be clearly justified by the mass transfer effect which is not limited in the O_3-$NiO_{(S)}$ case, but it could be an issue for the O_3-$NiO_{(F)}$ process. Nevertheless, the process efficiency in terms of the contaminants and byproducts removals concerning the catalyst mass is relevantly higher for the case of thin films. Moreover, the stability of such films regarding reusing experiments is a promising outcome attained here.

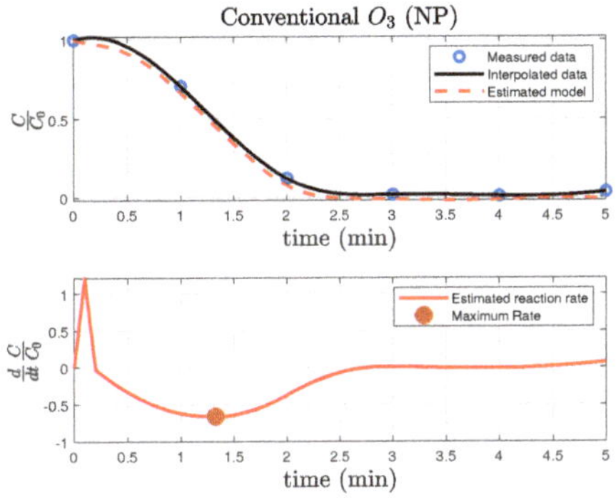

(a)

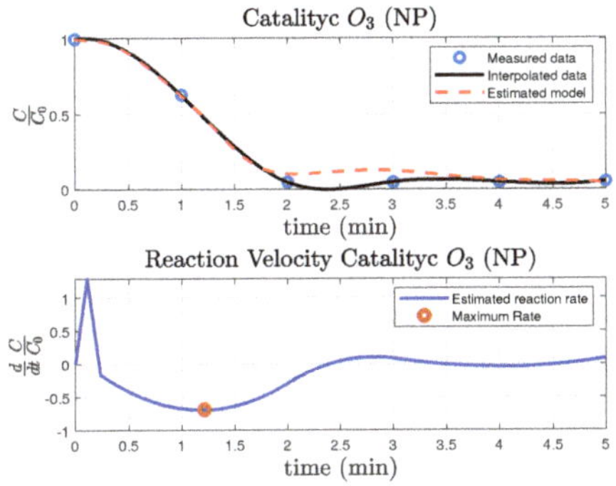

(b)

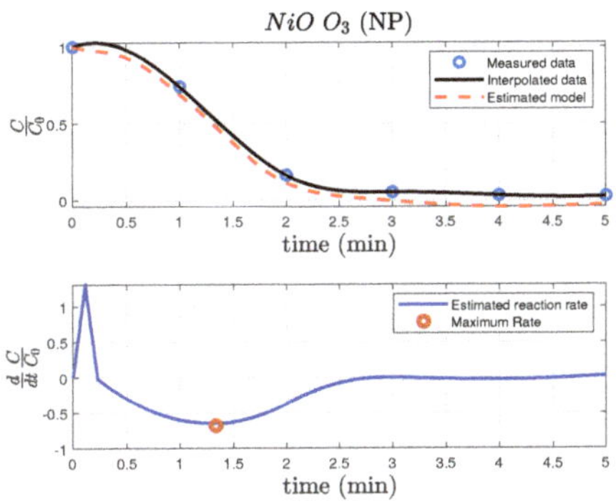

(c)

Figure 9. Time evolution corresponding to the variables of the mathematical model and kinetic constant of NP degradation by ozonation. In all the figures the blue circles represent the data obtained by HPLC, the continuous black line represents the interpolated data and the red dotted line describes the model approximation. The second row of graphs depicts the process of kinetic constant estimation based on the STA method: (**a**) O_3-conv. (**b**) O_3-$NiO_{(S)}$. (**c**) O_3-$NiO_{(P)}$.

Table 1. Reaction rate constants for the NP degradation by conventional and catalytic ozonations.

Parameter	O_3-conv	O_3-$NiO_{(S)}$	O_3-$NiO_{(F)}$
k_{NP}, [min^{-1}]	0.6605	0.6914	0.6784

3. Materials and Methods

This section summarizes all materials and methods used in the catalytic ozonation of NP using supported catalysts forming thin films. Also, the kinetic characterization of catalytic ozonation of NP is proposed using a simplified model based on approximate pseudo-first-order dynamics.

3.1. Synthesis of Nickel Oxide Films

The synthesis of $NiO_{(F)}$ was carried out in a quartz tube within a CVD set-up reported elsewhere. Nickel (II) acetylacetonate $Ni(acac)_2$ (Sigma-Aldrich, Mexico City, Mexico 99.0%) was used as metal-organic precursor of the corresponding metallic oxide. The conditions for NiO deposition were: (a) 200 °C of $Ni(acac)_2$ sublimation temperature, (b) 550 °C of substrate temperature, (c) mixture of Ar-O_2 (60$sccm$-60$sccm$). With these conditions, the allowed pressure in the equipment was 10 mTorr. This parameter was kept constant along the reaction period. The deposition time was 15 min.

The $NiO_{(F)}$ grew up on glass plates of 2.5 × 3.0 cm as the substrate. The substrate was washed previously to the deposition with sonication cycles of 15 min using different solvents: acetone, ethanol, and distillate water. The films were used in ozonation without any additional techniques.

3.2. $NiO_{(F)}$ Characterization

Characterization of $NiO_{(F)}$ was made with the following techniques: XRD, XPS, AFM (Brunker Zeiss Bioscope catalyst), spectroscopic ellipsometry and SEM (JEOL JSM 6701F, Peabody, Massachusetts, USA, 5 kV). These techniques allowed to study the morphology and crystal structure of the obtained $NiO_{(F)}$. The SEM determined the thickness of the NiO film. This parameter was compared with the result obtained by spectroscopic ellipsometry in the HORIBA's UVISEL equipment where the visible-UV spectra were acquired from a range of 1.5–5.5 eV.

XRD analysis was carried out in the Bruker D8 advance equipment with copper radiation operating at 35 kV and 25 mA (Cu$\mathcal{K}\alpha$, $\lambda = 1.54$ Å) in 2θ between 20–90°. The XRD pattern obtained from $NiO_{(F)}$ was compared with $NiO_{(S)}$ with the Bruker D8 AXS equipment using the same operating conditions.

XPS (Thermo Fisher Scientific, Waltham, Massachusetts, USA) was applied to determine changes in chemical binding types (Al $\mathcal{K}\alpha$ X-ray monochromatic source, 1487 eV; charge correction C1s: 284.6 eV).

AFM and NP-10 probes tips (Bruker, www.brukerafmprobes.com) with V-shaped cantilever and 20 nm radius pyramidal-geometry were used. The measurements were done in Tapping mode. The software NanoScope Analysis v1.4 (Bruker, Billerica, Massachusetts, USA, 2018) was used to calculate the average roughness which determines the deviation in height. Root mean square roughness represents the standard deviation of surface heights from AFM images. The thickness of $NiO_{(F)}$ was obtained by spectroscopic ellipsometry.

3.3. Catalytic Ozonation

The stability condition of $NiO_{(F)}$ was studied in catalytic ozonation of NP solutions (Sigma-Aldrich, 99.8%; 20 mg/L dissolved in distillate water). The initial pH of the solution was 5.0 ± 0.5, this parameter was not controlled during the ozonation. In O_3-$NiO_{(F)}$ treatments, ten coated slides were assembled on a Teflon sheet. They were put in a glass bubbled column reactor. This configuration allowed the contact of $NiO_{(F)}$ with NP solution (400 mL) and ozone. The mixture O_3-O_2 (flow: 0.5 L/min and $[O_3] = 5.5 \pm 0.5$ mg/L, HTU500 G ozone generator corona discharge type-AZCO Industries) bubbled through a ceramic porous filter located at the bottom of the self-designed reactor (500 mL). NP degradation by O_3-$NiO_{(F)}$ was compared with O_3-$NiO_{(S)}$ using 100 mg/L of NiO nanopowder <50 nm Sigma-Aldrich, 99.0%.

3.4. Analytical Methods

The monitoring of NP and oxalic acid (the main byproduct formed by the NP decomposition by conventional ozonation) by HPLC was carried out in the Perkin Elmer Flexar equipment. The column for the NP analysis was a Platinum C18 Alltech (250 mm × 4.6 mm) and 0.3 mL/min of water:acetonitrile 50:50 with a pH of 2.5 (phosphoric acid) as mobile phase. The separation followed at 240 nm. The byproducts detection was made using a Prevail TM Alltech–Grace column (150 × 4.6 mm and 5 μm). The mobile phase was a 0.25 mM of KH_2PO_4 Buffer pH of 2.5 (phosphoric acid), the determination followed with 1.0 mL/min at 210 nm. The mineralization degree was calculated using TOC data obtained by the Torch IQOQ Teledyne Tekmar equipment by direct injection. All samples were previously filtered with membranes of 0.2 μm.

3.5. Mathematical Modeling and Parameter Estimation

To compare the NP removing efficiency enforced by the three studied methods, the NP ozonation rate constants were determined. For this study, it is used as a simplified approximation of the kinetic model based on pseudo-first-order dynamics. Then, with the data obtained from the decomposition of NP by ozonation, one can propose a set of finite parameters that correspond to the reaction rate constants. The following kinetic model considers the interaction between the measurable components in the reaction (contaminant, byproducts and ozone).

It is known that the reaction of either molecular ozone or hydroxyl radicals with contaminants satisfies a second-order kinetics, i.e.:

$$\frac{d}{dt}c_{NP}(t) = -\left(k_{NP,O_3}O_{3,l}(t) + k_{NP,OH}OH(t)\right)c_{NP}(t), \tag{1}$$

where k_{NP,O_3}, min^{-1} is the reaction rate constant between ozone in the liquid phase $O_{3,l}$, mole/L and NP while $k_{NP,OH}$ is the reaction rate constant between hydroxide radicals OH, mole/L and NP.

Nevertheless, due to the interaction of ozone with the catalyst leading to the corresponding formation of hydroxyl radicals, the explicit quantification of O_3 and OH independently is a complex task in the presence of NiO films. Therefore, in this case, we propose an alternative method to characterize the decomposition of NP. Consider that $k(t) = k_{O_3}O_3(t) + k_{OH}OH(t)$ is the pseudo-mono molecular time dependent reaction rate parameter. Now, if one considers that $k_{NP} = max_{t \geq 0}k(t)$ it is feasible to propose that the following mathematical structure is given by:

$$\frac{d}{dt}c_{NP}(t) = -k_{NP}c_{NP}(t), \tag{2}$$

where c_{NP} represents the concentration of the contaminant in time during the ozonation and k_{NP} is the kinetic constant associated with the decomposition rate of NP.

Using the differential equations representations of chemical processes and considering that advanced oxidation processes can be sufficiently approximated by pseudo-first-order dynamics. The inherent problem is the available information along with heterogeneous catalytic ozonation reaction. Usually, available measurements along the reaction are concentrations of just a few chemical species including the initial contaminant as well as its byproducts [46]. However, this information is not enough to get an accurate estimation of the first-order reaction rates. Commonly, it is necessary to get estimates of the contaminants' decomposition velocity. With the aim of getting a precise estimation of such decomposition velocity, a robust differentiation technique can be implemented.

Numerical differentiation is a well-studied topic in systems theory. The most common differentiation technique is based on the Euler approximation [47]. However, this method is highly affected by noises in the measurement. This manuscript proposes the application of the STA, which is a robust exact differentiator [48] based on the sliding mode theoretical results. The advantages of this algorithm are its robustness against parametric uncertainties that are common in the chemical process, well-posed discretization equivalents and finite-time convergence. Therefore, by means of the STA, the reaction kinetics are obtained. With the concentration and the information provided by the STA, a least square regression is performed to obtain a more accurate model representation of the heterogeneous catalytic ozonation.

STA [48] can determine the time derivative of a signal even in the presence of certain class of noises and model uncertainties. The algorithm is given by the following set of differential equations

$$\begin{aligned} d_{c_{NP}}(t) &= x_2(t) - k_1|x_1(t) - C_{NP}(t)|^{1/2}\text{sign}(x_1(t) - c_{NP}(t)) \\ \dot{x}_1(t) &= d_{c_{NP}}(t) \\ \dot{x}_2(t) &= -k_2\text{sign}(x_1(t) - c_{NP}(t)) \end{aligned} \tag{3}$$

In Equation (3), $d_{c_{NP}}$ is the actual time derivative of the concentration $c_{NP}(t)$, x_1 and x_2 are internal variables, k_1 and k_2 are positive constants that define the rate of convergence and the quality of estimation of the algorithm. The function sign$(\cdot)$ is defined as

$$\text{sign}(z) = \begin{cases} -1 & if \quad z < 0 \\ \in [-1, 1] & if \quad z = 0 \\ 1 & if \quad z > 0 \end{cases} \tag{4}$$

Notice that $\frac{d}{dt}c_{NP}(t) = d_{c_{NP}}(t)$ when the algorithm has converged to the real concentration $c_{NP}(t)$, the estimation error defined as $x_1(t) - c_{NP}(t)$ is equal zero. Notice the necessity of applying the STA differentiator as many times as the number of compounds identified along with the heterogeneous catalytic ozonation characterization: one for the degradation of NP in the three experimental setups: O_3-conv, O_3-$NiO_{(S)}$ and O_3-$NiO_{(F)}$. The complete procedure to estimate the reaction rates is summarized in the Algorithm 1.

Algorithm 1: Parameter estimation in the NP degradation by ozonation techniques

Data: NP concentration $[c_{NP}]$ during the ozonation process for each methodology
Result: Kinetic constant and model estimation for each methodology

for $i \leftarrow 1$ **to** 3 **do**

1. Data from HPLC study are stored in variable $\bar{c}_{NP}$;
2. An interpolation is made to enhance the data available for analysis $\longrightarrow c_{NP}$;
3. The STA is explicitly discretized using Euler method (See the work in [47]);
4. The value of $d_{NP}(t)$ is the time derivative of c_{NP};
5. The maximum value of $d_{NP}(t)$ corresponds to the maximum reaction rate;
6. A LS method approximates the reaction model.
7. The approximated model is compared with the original data

end

```
/* The for cycle runs three times, one for each procedure (O3-conv, O3-NiO(S)
   and O3-NiO(F).                                                          */
/* The LS method can be applied using the software Matlab®                 */
```

For further references, the theoretical background of the STA working as a differentiator can be found in [47,48].

4. Conclusions

CVD method allowed the synthesis of thin NiO films (240 nm thick) with relevant catalytic characteristics in NP degradation by ozone. NP elimination was carried out in five minutes of reaction with and without *NiO* (powder and film), due to its fast reaction with molecular ozone. The byproduct analysis and TOC removal showed the catalytic activity of the *NiO* in the ozonation. The treatments with 100 mg/L of $NiO_{(S)}$ generated a TOC removal of 35% while in the presence of 10 films of *NiO*, TOC removal increased 10% in comparison with O_3-conv. However, *NiO* mass supported on the films was close to a tenth of the *NiO* used in suspension. The reuse of films (4 cycles) attained similar TOC removal in each cycle ($22 \pm 2.9\%$). This fact showed the stability of the catalyst films. Moreover, the film characterization after ozonation demonstrated some changes in the surface morphology but XRD and XPS patterns confirmed the crystalline stability. The results obtained suggest the possibility of using the deposition of catalyst in the film form in the catalytic ozonation of recalcitrant pollutants with the reduction of several detected disadvantages of the suspended catalyst (elimination and reuse). This study also presented a simplified model characterizing the ozonation kinetics of NP. A novel method yields to estimate a pseudo-monomolecular reaction rate constant which established a comparison between the proposed ozonation of NP with (suspension and films) and without catalysts. This model seems to offer a primary method to explain the reaction kinetics of ozone and NP in the presence of $NiO_{(F)}$ as a catalyst.

Author Contributions: Synthesis and characterization of *NiO* film, J.A.G.-P. and J.A.A.A.; Conventional and Catalytic ozonation, C.M.A.-M. and J.L.R.; Mathematical modeling and parameter estimation, I.S. and I.C.; supervision and writing-review and editing, J.V.-A. and T.P. All authors have read and agreed to the published version of the manuscript.

Funding: This research was funded by Instituto Politécnico Nacional (Project: 20200519) and Consejo Nacional de Ciencia y Tecnología (Posdoctoral studies) and Investigación Científica Básica" 2018 grant no. A1-S-21608.

Acknowledgments: The authors kindly acknowledge to Instituto Politécnico Nacional, Centro de Nanociencias y Micro y Nanotecnologías-IPN, Consejo Nacional de Ciencia y Tecnología (Posdoctoral studies of Claudia M. Aguilar Melo) and PhD. Denisse Fabiola González Ramírez for economical supporting and contribution in this work.

Conflicts of Interest: The authors declare no conflict of interest.

Abbreviations

The following abbreviations are used in this manuscript:

NP	Naproxen
CVD	Chemical vapor deposition
O_3-conv	Conventional ozonation
$NiO_{(F)}$	*NiO* film
O_3-$NiO_{(S)}$	Catalytic ozonation with suspension *NiO*
O_3-$NiO_{(F)}$	Catalytic ozonation with film *NiO*

References

1. Wang, J.; Chen, H. Catalytic ozonation for water and wastewater treatment: Recent advances and perspective. *Sci. Total Environ.* **2020**, *704*, 135249. [CrossRef] [PubMed]
2. Assadi, M.H.N.; Hanaor, D.A. The effects of copper doping on photocatalytic activity at (101) planes of anatase TiO_2: A theoretical study. *Appl. Surf. Sci.* **2016**, *387*, 682–689. [CrossRef]
3. Zhang, S.; Quan, X.; Wang, D. Catalytic Ozonation in Arrayed Zinc Oxide Nanotubes as Highly Efficient Mini-Column Catalyst Reactors (MCRs): Augmentation of Hydroxyl Radical Exposure. *Environ. Sci. Technol.* **2018**, *52*, 8701–8711. [CrossRef] [PubMed]
4. Rosal, R.; Gonzalo, S.; Santiago, J.; Rodríguez, A.; Perdigón-Melón, J.A.; García-Calvo, E. Kinetics and Mechanism of Catalytic Ozonation of Aqueous Pollutants on Metal Oxide Catalysts. *Ozone Sci. Eng.* **2011**, *33*, 434–440. [CrossRef]
5. Chen, G.; Wang, Z.; Lin, F.; Zhang, Z.; Yu, H.; Yan, B.; Wang, Z. Comparative investigation on catalytic ozonation of VOCs in different types over supported MnOx catalysts. *J. Hazard. Mater.* **2020**, *391*, 122218. [CrossRef]

6. Beltrán, F.J.; Álvarez, P.M.; Gimeno, O. Graphene-Based Catalysts for Ozone Processes to Decontaminate Water. *Molecules* **2019**, *24*. [CrossRef]
7. Wang, B.; Zhang, H.; Wang, F.; Xiong, X.; Tian, K.; Sun, Y.; Yu, T. Application of Heterogeneous Catalytic Ozonation for Refractory Organics in Wastewater. *Catalysts* **2019**, *9*, 241. [CrossRef]
8. Hu, J.; Li, Y.; Nan, S.; Yoza, B.A.; Li, Y.; Zhan, Y.; Wang, Q.; Li, Q.X.; Guo, S.; Chen, C. Catalytic Ozonation of Nitrobenzene by Manganese-Based Y Zeolites. *Front. Chem.* **2020**, *8*, 80. [CrossRef] [PubMed]
9. Rodríguez, J.L.; Valenzuela, M.A.; Poznyak, T.; Lartundo, L.; Chairez, I. Reactivity of NiO for 2,4-D degradation with ozone: XPS studies. *J. Hazard. Mater.* **2013**, *262*, 472–481. [CrossRef] [PubMed]
10. Aguilar, C.M.; Vazquez-Arenas, J.; Castillo-Araiza, O.O.; Rodríguez, J.L.; Chairez, I.; Salinas, E.; Poznyak, T. Improving ozonation to remove carbamazepine through ozone-assisted catalysis using different NiO concentrations. *Environ. Sci. Pollut. Res. Int.* **2020**. [CrossRef] [PubMed]
11. Aguilar, C.M.; Rodríguez, J.L.; Chairez, I.; Tiznado, H.; Poznyak, T. Naphthalene degradation by catalytic ozonation based on nickel oxide: study of the ethanol as cosolvent. *Environ. Sci. Pollut. Res. Int.* **2017**, *24*, 25550–25560. [CrossRef] [PubMed]
12. Biard, P.F.; Werghi, B.; Soutrel, I.; Orhand, R.; Couvert, A.; Denicourt-Nowicki, A.; Roucoux, A. Efficient catalytic ozonation by ruthenium nanoparticles supported on SiO2 or TiO_2: Towards the use of a non-woven fiber paper as original support. *Chem. Eng. J.* **2016**, *289*, 374–381. [CrossRef]
13. Shahamat, Y.D.; Farzadkia, M.; Nasseri, S.; Mahvi, A.H.; Gholami, M.; Esrafili, A. Magnetic heterogeneous catalytic ozonation: a new removal method for phenol in industrial wastewater. *J. Environ. Health Sci. Eng.* **2014**, *12*, 50. [CrossRef] [PubMed]
14. Battiato, S.; Giangregorio, M.M.; Catalano, M.R.; Nigro, R.L.; Losurdo, M.; Malandrino, G. Morphology-controlled synthesis of NiO films: The role of the precursor and the effect of the substrate nature on the films' structural/optical properties. *RSC Adv.* **2016**, *6*, 30813–30823. [CrossRef]
15. Xi, Y.Y.; Li, D.; Djurišić, A.B.; Xie, M.H.; Man, K.Y.K.; Chan, W.K. Hydrothermal Synthesis vs. Electrodeposition for High Specific Capacitance Nanostructured NiO Films. *Electrochem. Solid State Lett.* **2008**, *11*, D56. [CrossRef]
16. Vargas Garcia, J.R.; Lazcano Ugalde, E.M.; Hernandez Santiago, F.; Hallen Lopez, J.M. Nanostructured nickel oxide films prepared by chemical vapor deposition and their electrochromic properties. *J. Nanosci. Nanotechnol.* **2008**, *8*, 2703–2706. [CrossRef]
17. Pellegrino, F.; De Bellis, N.; Ferraris, F.; Prozzi, M.; Zangirolami, M.; Petriglieri, J.R.; Schiavi, I.; Bianco-Prevot, A.; Maurino, V. Evaluation of the Photocatalytic Activity of a Cordierite-Honeycomb-Supported TiO_2 Film with a Liquid-Solid Photoreactor. *Molecules* **2019**, *24*, 4499. [CrossRef]
18. Tekin, D.; Tekin, T.; Kiziltas, H. Photocatalytic degradation kinetics of Orange G dye over ZnO and Ag/ZnO thin film catalysts. *Sci. Rep.* **2019**, *9*, 1–7. [CrossRef]
19. Guzmán, I.C.; Rodríguez, J.L.; Poznyak, T.; Chairez, I.; Hernández, I.; Hernández, R.T. Catalytic ozonation of 4-chlorophenol and 4-phenolsulfonic acid by CeO2 films. *Catal. Commun.* **2020**, *133*, 105827. [CrossRef]
20. Muir, D.; Simmons, D.; Wang, X.; Peart, T.; Villella, M.; Miller, J.; Sherry, J. Bioaccumulation of pharmaceuticals and personal care product chemicals in fish exposed to wastewater effluent in an urban wetland. *Sci. Rep.* **2017**, *7*, 16999. [CrossRef]
21. Arnold, K.E.; Brown, A.R.; Ankley, G.T.; Sumpter, J.P. Medicating the environment: assessing risks of pharmaceuticals to wildlife and ecosystems. *Philos. Trans. R. Soc. Biol. Sci.* **2014**, *369*, 20130569. [CrossRef]
22. Emmanuel, E.; Perrodin, Y.; Keck, G.; Blanchard, J.M.; Vermande, P. Ecotoxicological risk assessment of hospital wastewater: a proposed framework for raw effluents discharging into urban sewer network. *J. Hazard. Mater.* **2005**, *117*, 1–11. [CrossRef]
23. Wojcieszyńska, D.; Guzik, U. Naproxen in the environment: its occurrence, toxicity to nontarget organisms and biodegradation. *Appl. Microbiol. Biotechnol.* **2020**, *104*, 1849–1857. [CrossRef]
24. Domaradzka, D.; Guzik, U.; Wojcieszyńska, D. Biodegradation and biotransformation of polycyclic non-steroidal anti-inflammatory drugs. *Rev. Environ. Sci. Biotechnol.* **2015**, *14*, 229–239. [CrossRef]
25. Li, Z.; Liu, G.; Su, Q.; Lv, C.; Jin, X.; Wen, X. UV-Induced Photodegradation of Naproxen Using a Nano γ-FeOOH Composite: Degradation Kinetics and Photocatalytic Mechanism. *Front. Chem.* **2019**, *7*. [CrossRef]
26. Ray, S.K.; Dhakal, D.; Lee, S.W. Rapid degradation of naproxen by AgBr-α-$NiMoO_4$ composite photocatalyst in visible light: Mechanism and pathways. *Chem. Eng. J.* **2018**, *347*, 836–848, [CrossRef]

27. Arany, E.; Szabó, R.K.; Apáti, L.; Alapi, T.; Ilisz, I.; Mazellier, P.; Dombi, A.; Gajda-Schrantz, K. Degradation of naproxen by UV, VUV photolysis and their combination. *J. Hazard. Mater.* **2013**, *262*, 151–157. [CrossRef]
28. Liu, Y.; Tang, Y.; Wu, Y.; Feng, L.; Zhang, L. Degradation of naproxen in chlorination and UV/chlorine processes: kinetics and degradation products. *Environ. Sci. Pollut. Res. Int.* **2019**, *26*, 34301–34310. [CrossRef]
29. Karaca, M.; Kıranşan, M.; Karaca, S.; Khataee, A.; Karimi, A. Sonocatalytic removal of naproxen by synthesized zinc oxide nanoparticles on montmorillonite. *Ultrason. Sonochemistry* **2016**, *31*, 250–256, [CrossRef]
30. Tizhoosh, N.Y.; Khataee, A.; Hassandoost, R.; Soltani, R.D.C.; Doustkhah, E. Ultrasound-engineered synthesis of WS2@ CeO2 heterostructure for sonocatalytic degradation of tylosin. *Ultrason. Sonochemistry* **2020**, *67*, 105114. [CrossRef]
31. Dulova, N.; Kattel, E.; Trapido, M. Degradation of naproxen by ferrous ion-activated hydrogen peroxide, persulfate and combined hydrogen peroxide/persulfate processes: The effect of citric acid addition. *Chem. Eng. J.* **2017**, *318*, 254–263. [CrossRef]
32. van Kollenburg, G.H.; van Es, J.; Gerreten, J.; Lanters, H.; Bouman, R.; Koelewijn, W.; Davies, A.N.; Buydens, L.; van Manen, H.J.; Jansen, J.J. Understanding Chemical Production Processes by using PLS Path Model Parameters as Soft Sensors. *Comput. Chem. Eng.* **2020**, *139*, 106841. [CrossRef]
33. Rosal, R.; Rodríguez, A.; Gonzalo, M.; García-Calvo, E. Catalytic ozonation of naproxen and carbamazepine on titanium dioxide. *Appl. Catal. Environ.* **2008**, *84*, 48–57, [CrossRef]
34. Patil, V.P.; Pawar, S.; Chougule, M.; Godse, P.; Sakhare, R.; Sen, S.; Joshi, P. Effect of Annealing on Structural, Morphological, Electrical and Optical Studies of Nickel Oxide Thin Films. *J. Surf. Eng. Mater. Adv. Technol.* **2011**, *1*, 720–726. [CrossRef]
35. Lin, S.H.; Chen, F.R.; Kai, J.J. Electrochromic properties of nano-composite nickel oxide film. *Appl. Surf. Sci.* **2008**, *254*, 3357–3363. [CrossRef]
36. Kanakaraju, D.; Motti, C.A.; Glass, B.D.; Oelgemöller, M. TiO_2 photocatalysis of naproxen: Effect of the water matrix, anions and diclofenac on degradation rates. *Chemosphere* **2015**, *139*, 579–588. [CrossRef]
37. Mohamed, A.; Salama, A.; Nasser, W.S.; Uheida, A. Photodegradation of Ibuprofen, Cetirizine, and Naproxen by PAN-MWCNT/TiO_2-NH2 nanofiber membrane under UV light irradiation. *Environ. Sci. Eur.* **2018**, *30*, 47. [CrossRef]
38. Aguilar, C.M.; Chairez, I.; Rodríguez, J.L.; Tiznado, H.; Santillán, R.; Arrieta, D.; Poznyak, T. Inhibition effect of ethanol in naproxen degradation by catalytic ozonation with NiO. *RSC Adv.* **2019**, *9*, 14822–14833. [CrossRef]
39. Magallanes, D.; Rodríguez, J.L.; Poznyak, T.; Valenzuela, M.A.; Lartundo, L.; Chairez, I. Efficient mineralization of benzoic and phthalic acids in water by catalytic ozonation using a nickel oxide catalyst. *New J. Chem.* **2015**, *39*, 7839–7848. [CrossRef]
40. Fuentes, I.; Rodriguez, J.L.; Tiznado, H.; Romo-Herrera, J.M.; Chairez, I.; Poznyak, T. Terephthalic acid decomposition by photocatalytic ozonation with VxOy/ZnO under different UV-A LEDs distributions. *Chem. Eng. Commun.* **2020**, *207*, 263–277. [CrossRef]
41. Patel, S.; Majumder, S.K.; Das, P.; Ghosh, P. Ozone microbubble-aided intensification of degradation of naproxen in a plant prototype. *J. Environ. Chem. Eng.* **2019**, *7*, 103102. [CrossRef]
42. Liu, J.; Ke, L.; Liu, J.; Sun, L.; Yuan, X.; Li, Y.; Xia, D. Enhanced catalytic ozonation towards oxalic acid degradation over novel copper doped manganese oxide octahedral molecular sieves nanorods. *J. Hazard. Mater.* **2019**, *371*, 42–52. [CrossRef] [PubMed]
43. Huang, Y.; Sun, Y.; Xu, Z.; Luo, M.; Zhu, C.; Li, L. Removal of aqueous oxalic acid by heterogeneous catalytic ozonation with MnOx/sewage sludge-derived activated carbon as catalysts. *Sci. Total Environ.* **2017**, *575*, 50–57. [CrossRef] [PubMed]
44. Jallouli, N.; Elghniji, K.; Hentati, O.; Ribeiro, A.R.; Silva, A.M.; Ksibi, M. UV and solar photo-degradation of naproxen: TiO_2 catalyst effect, reaction kinetics, products identification and toxicity assessment. *J. Hazard. Mater.* **2016**, *304*, 329–336. [CrossRef]
45. Wu, K.; Zhang, F.; Wu, H.; Wei, C. The mineralization of oxalic acid and bio-treated coking wastewater by catalytic ozonation using nickel oxide. *Environ. Sci. Pollut. Res.* **2018**, *25*, 2389–2400. [CrossRef]
46. Zhang, X.; Zou, Y.; Li, S.; Xu, S. A weighted auto regressive LSTM based approach for chemical processes modeling. *Neurocomputing* **2019**, *367*, 64–74. [CrossRef]

47. Salgado, I.; Chairez, I.; Bandyopadhyay, B.; Fridman, L.; Camacho, O. Discrete-time non-linear state observer based on a super twisting-like algorithm. *IET Control Theory Appl.* **2014**, *8*, 803–812. [CrossRef]
48. Levant, A. Robust exact differentiation via sliding mode technique. *Automatica* **1998**, *34*, 379–384. [CrossRef]

catalysts

Article

Oxidative Dehydrogenation of Methane When Using TiO_2- or WO_3-Doped Sm_2O_3 in the Presence of Active Oxygen Excited with UV-LED

Shigeru Sugiyama [1,*], Yasunori Hayashi [2], Ikumi Okitsu [2], Naohiro Shimoda [1], Masahiro Katoh [1], Akihiro Furube [3], Yuki Kato [4] and Wataru Ninomiya [4]

1 Department of Applied Chemistry, Tokushima University, Minamijosanjima, Tokushima-shi, Tokushima 770-8506, Japan; shimoda@tokushima-u.ac.jp (N.S.); katoh@tokushima-u.ac.jp (M.K.)
2 Department of Chemical Science and Technology, Tokushima University, Minamijosanjima, Tokushima-shi, Tokushima 770-8506, Japan; hysys1110@gmail.com (Y.H.); c.1935.okkeyy@gmail.com (I.O.)
3 Department of Optical Science, Tokushima University, Minamijosanjima, Tokushima-shi, Tokushima 770-8506, Japan; furube.akihiro@tokushima-u.ac.jp
4 Hiroshima R&D Center, Mitsubishi Chemical Corporation, 20-1, Miyuki-cho, Otake-shi, Hiroshima 739-0693, Japan; kato.yuki.ma@m-chemical.co.jp (Y.K.); ninomiya.wataru.me@m-chemical.co.jp (W.N.)
* Correspondence: sugiyama@tokushima-u.ac.jp; Tel.: +81-88-656-7432

Received: 14 April 2020; Accepted: 15 May 2020; Published: 18 May 2020

Abstract: There are active oxygen species that contribute to oxidative coupling or the partial oxidation during the oxidative dehydrogenation of methane when using solid oxide catalysts, and those species have not been definitively identified. In the present study, we clarify which of the active oxygen species affect the oxidative dehydrogenation of methane by employing photo-catalysts such as TiO_2 or WO_3, which generate active oxygen from UV-LED irradiation conditions under an oxygen flow. These photo-catalysts were studied in combination with Sm_2O_3, which is a methane oxidation coupling catalyst. For this purpose, we constructed a reaction system that could directly irradiate UV-LED to a solid catalyst via a normal fixed-bed continuous-flow reactor operated at atmospheric pressure. Binary catalysts prepared from TiO_2 or WO_3 were either supported on or kneaded with Sm_2O_3 in the present study. UV-LED irradiation clearly improved the partial oxidation from methane to CO and/or slightly improved the oxidative coupling route from methane to ethylene when binary catalysts consisting of Sm_2O_3 and TiO_2 are used, while negligible UV-LED effects were detected when using Sm_2O_3 and WO_3. These results indicate that with UV-LED irradiation the active oxygen of O_2^- from TiO_2 certainly contributes to the activation of methane during the oxidative dehydrogenation of methane when using Sm_2O_3, while the active oxygen of H_2O_2 from WO_3 under the same conditions afforded only negligible effects on the activation of methane.

Keywords: methane; oxidative dehydrogenation; active oxygen; UV-LED; TiO_2; WO_3; Sm_2O_3

1. Introduction

The conversion of methane to high value-added chemicals is an important issue in the field of catalyst research. In recent years, research on the catalytic reaction of methane has been actively conducted due to progress in the production technology of natural gas, which consists mainly of methane gas [1–3]. Although methane has the potential for conversion to a variety of important chemicals, its application as a raw material in catalytic reactions has been limited due to chemical stability. Therefore, methane is still used mainly as fuel.

To overcome the stability problem, many researchers are studying the direct conversion of methane to value-added chemicals such as methanol [4], carbon monoxide [5], ethylene [6], and aromatic compounds [7]. The direct conversion of methane is considered the most efficient way to use methane gas because the desired product requires only a one-step catalytic reaction. In particular, the oxidative coupling of methane (OCM) to ethylene and ethane has been the subject of much research over the past three decades since these C_2 hydrocarbons are the most widely used petrochemicals in the world. In the OCM reaction, methane reacts with oxygen exothermically on a solid oxide catalyst to produce these C_2 hydrocarbons together with water [8]. It is generally accepted that gaseous oxygen and active oxygen derived from a solid oxide catalyst could contribute to the oxidative conversion of methane [9–11]. Furthermore, the OCM is believed to consist of both heterogeneous and homogeneous reactions. First, in a heterogeneous reaction, active oxygen in the catalyst extracts hydrogen from methane to generate methyl radicals. The methyl radicals are then dimerized to C_2 hydrocarbons by a homogeneous gas-phase reaction (Scheme 1) [12–15]. Contributions have been proposed from active oxygen species such as O_2^-, OH, H_2O_2, or $^1O_2$ (singlet oxygen) together with gas-phase oxygen (O_2) or catalytic lattice oxygen (O^{2-}), but exactly what kind of active oxygen species contribute to the oxidative dehydrogenation of methane is yet to be clarified [16–21].

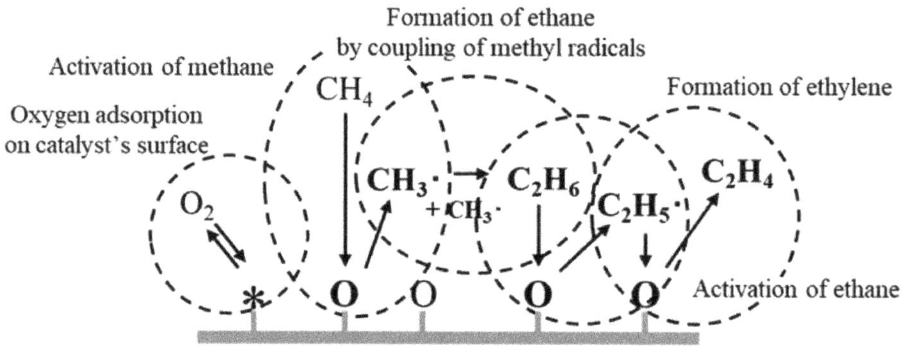

Scheme 1. Mechanism for the oxidative coupling of methane.

In the present study, we focused on the characteristics of photo-catalysts. Photo-catalysts such as titanium oxide (TiO_2) and tungsten oxide (WO_3) activate oxygen when electrons (e^-) are excited by irradiation from an excitation light (UV-LED in the present study) and holes (h^+) are sequentially generated, which results in the formation of active oxygen (Scheme 2).

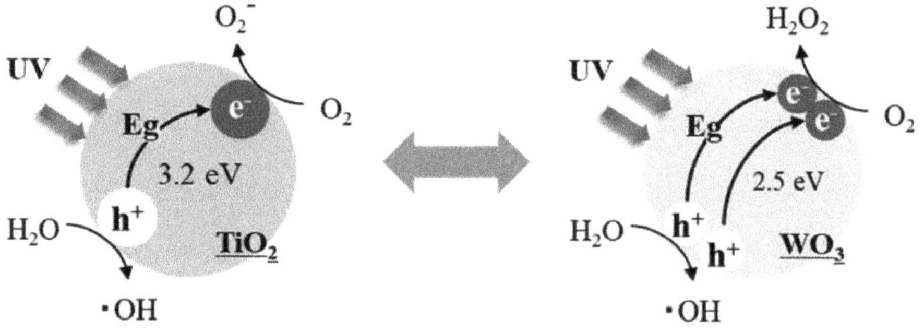

Scheme 2. Production of active oxygen species via UV irradiation of TiO_2 and WO_3.

Based on Scheme 2, the oxidative dehydrogenation of methane was studied via contact with samarium oxide (Sm_2O_3; OCM-catalyst) and by examining the active oxygen species generated via irradiating UV-LED irradiation of either TiO_2 or WO_3 (photo-catalyst) under a gaseous O_2 atmosphere. It is generally accepted that O_2^- is generated from a one-electron reduction of TiO_2, and H_2O_2 is generated from a two-electron reduction of WO_3 [22]. When the active oxygen species derived from either TiO_2 or WO_3 contacted Sm_2O_3 during the oxidative dehydrogenation of methane, the product distribution was expected to depend on the presence or absence of UV-LED irradiation. The purpose of this study was to confirm and clarify the contributions of each of the active oxygen species. It is noteworthy that titanium and tungsten have been used as the active species in various catalysts for the oxidative coupling of methane [23,24].

2. Results and Discussion

This study involved both mixed- and supported-catalysts that consisted of Sm_2O_3 together with TiO_2 or WO_3. Based on our preliminary experiments, the loading of photo-catalysts such as TiO_2 and WO_3 was fixed at 5 wt.%. First, the mixed-catalyst activity using 5 wt.% TiO_2 + Sm_2O_3 was tested together with that of either Sm_2O_3 or TiO_2. The specific surface areas of Sm_2O_3, TiO_2, and 5 wt.% TiO_2 + Sm_2O_3 were 7, 47, and 22 m^2/g, respectively. Figure 1 shows the effect that UV-LED irradiation exerted on the oxidative dehydrogenation of methane at T = 898 K; $P(CH_4)$ = 28.7 kPa; and $P(O_2)$ = 2.03 kPa ($P(CH_4)/P(O_2)$ = 14.2). Since stable catalytic activity was detected on all catalysts used to the point of 4.5 h on-stream, the activity at 0.75 h on-stream was discussed in the present study. As shown in Figure 1, UV-LED irradiation of Sm_2O_3 showed no advantageous effects on either C_2 yield or on the conversions of O_2 and CH_4, while CO selectivity was slightly changed from 9.4% to 10.5% by the irradiation. A similar effect of UV-LED on TiO_2 yielded CO selectivity of 83.8% to 85.2%. It should be noted that the conversions of CH_4 and O_2 were not influenced by the irradiation of UV-LED due to the oxygen-limiting conditions. When adding 5 wt.% TiO_2 into Sm_2O_3 (5 wt.% TiO_2 + Sm_2O_3), the unique nature of Sm_2O_3 that allows coupling with methane was mostly masked by the nature of TiO_2 that allows the partial oxidation of methane, and this resulted in a slight formation of C_2H_6 on 5 wt.% TiO_2 + Sm_2O_3. Furthermore, an evident improvement in CO selectivity of from 19.9% to 28.9% was detected followed by a suppression of CO_2 selectivity of from 78.6% to 67.2% after UV-LED irradiation of the mixed-catalyst. It should be noted that C_2H_6 selectivity was also slightly improved from 1.5% to 3.9% via UV-LED irradiation using 5 wt.% TiO_2 + Sm_2O_3. Therefore, O_2^- generated via the UV-LED irradiation of TiO_2 under a gaseous O_2 atmosphere seemed to contribute to the acceleration of the partial oxidation of CH_4 to CO together with the oxidative dehydrogenation of CH_4 to C_2H_6. No enhancement was detected from either the partial oxidation or the oxidative dehydrogenation of methane using 5 wt.% TiO_2 + Sm_2O_3 via UV-LED at a $P(O_2)$ as high as 4.05 kPa, which indicated that the presence of large amounts of reactant oxygen may obliterate the effects of O_2^- due to the small amount of active oxygen.

Figure 2 shows the effect of UV-LED irradiation on the oxidative dehydrogenation of methane over Sm_2O_3, WO_3, and 5 wt.% WO_3 + Sm_2O_3 as a mixed-catalyst under the same reaction conditions as those used for obtaining the results shown in Figure 1. The specific surface areas of WO_3 and 5 wt.% WO_3 + Sm_2O_3 were 5 and 6 m^2/g, respectively. As shown in Figure 2, WO_3 produced CO alone via partial oxidation of methane regardless of the use of UV-LED irradiation while O_2 conversion was increased from 6% to 13%. In the present case, the addition of 5 wt.% WO_3 into Sm_2O_3 did not completely mask the unique nature of Sm_2O_3 in the oxidative coupling of methane. The effects of UV-LED irradiation on the catalytic activity of 5 wt.% WO_3 + Sm_2O_3 were rather small. Slight decreases were detected for CH_4 conversion, C_2 yield, C_2H_6 selectivity, CO selectivity, and CO_2 selectivity together with slight increases in O_2 conversion and C_2H_4 selectivity that ranged from 12.2% to 12.9%. Therefore, the effect of H_2O_2 generated by UV-LED irradiation on WO_3 under a gaseous O_2 atmosphere could have been negligible while those of H_2O_2 seemed to slightly contribute to an acceleration of the oxidative dehydrogenation of C_2H_6 to C_2H_4.

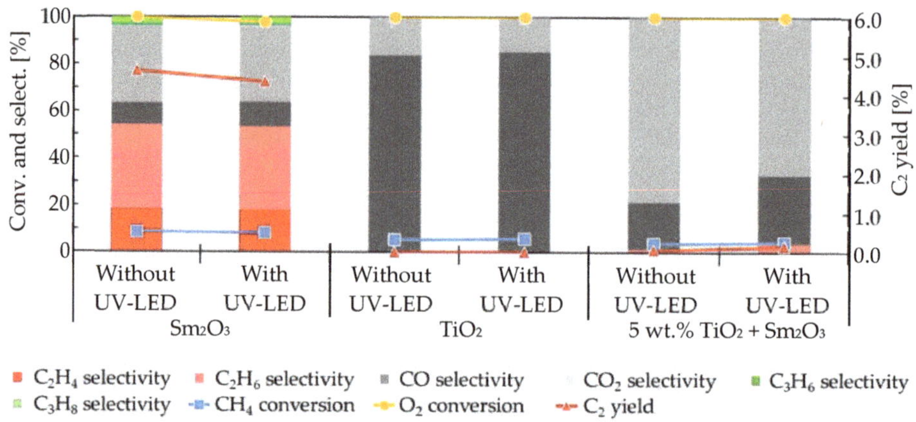

Figure 1. Effect of UV-LED irradiation on the oxidative dehydrogenation of methane when using Sm_2O_3, TiO_2, and 5 wt.% TiO_2 + Sm_2O_3.

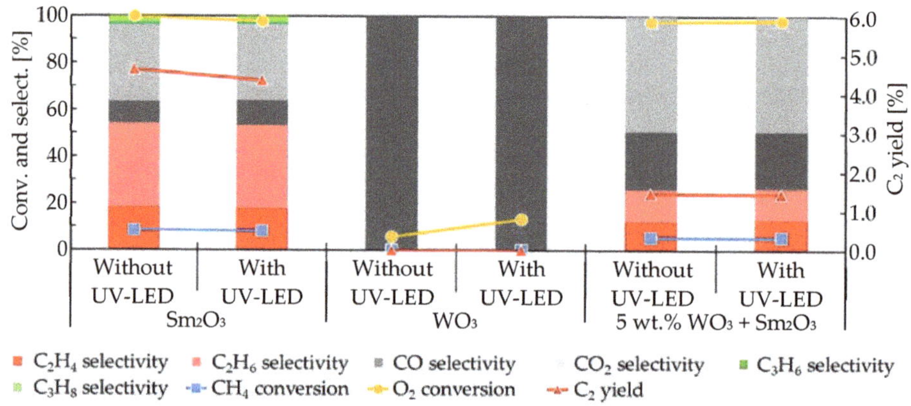

Figure 2. Effects of UV-LED irradiation on the oxidative dehydrogenation of methane when using Sm_2O_3, WO_3, and 5 wt.% WO_3 + Sm_2O_3.

An effect from UV-LED irradiation was not evident when using mixed-catalysts. Therefore, supported-catalysts were used in the present study. In Figure 3, the use of UV-LED irradiation on the oxidative dehydrogenation of methane when using 5 wt.% TiO_2 + Sm_2O_3 and 5 wt.% WO_3 + Sm_2O_3 mixed-catalysts is compared with the results over 5 wt.% TiO_2/Sm_2O_3 and 5 wt.% WO_3/Sm_2O_3 supported-catalysts under the same reaction conditions as those used to obtain the results shown in Figures 1 and 2. The specific surface areas of supported-catalysts 5 wt.% TiO_2/Sm_2O_3 and 5 wt.% WO_3/Sm_2O_3 were 9 and 6 m^2/g, respectively. Figure 3 compares the effect of UV-LED irradiation using 5 wt.% TiO_2 + Sm_2O_3 with that using 5 wt.% TiO_2/Sm_2O_3, and the effect was more evident when using the supported catalyst. For example, CH_4 conversion, C_2 yield, C_2H_4 selectivity, C_2H_6 selectivity, and CO selectivity when using the supported catalyst all were enhanced by UV-LED irradiation from 3.6%, 0.3%, 0.0%, 9.0%, and 26.5% when using 5 wt.% TiO_2 + Sm_2O_3 to 5.6%, 0.7%, 2.4%, 10.0%, and 47.7% when using 5 wt.% TiO_2/Sm_2O_3. By contrast, during deep oxidation, CO_2 selectivity was suppressed by UV-LED irradiation from 64.4% to 39.8%. It is noteworthy that the catalytic activity on the 5 wt.% TiO_2 + Sm_2O_3 catalyst (Figure 3) was higher than that of TiO_2 itself, because activities such as the methane conversion and C_2 selectivity on Sm_2O_3 were higher than that on TiO_2, as shown in Figure 1. As shown in Figure 3, the conversions of both CH_4 and O_2 were insensitive to the irradiation of UV-LED due to the oxygen-limiting conditions. It was evident that UV-LED irradiation enhanced the formation of C_2 compounds and CO and suppressed the deep oxidation to CO_2. A comparison of the activity when using 5 wt.% WO_3 + Sm_2O_3 with the use of 5 wt.% WO_3/Sm_2O_3 revealed a negligible effect from UV-LED. Additionally, an increase in C_2H_4 selectivity from 0.0% to 1.2% by UV-LED was detected when using 5 wt.% WO_3/Sm_2O_3, which was similar to the use of 5 wt.% WO_3 + Sm_2O_3, as shown in Figure 2.

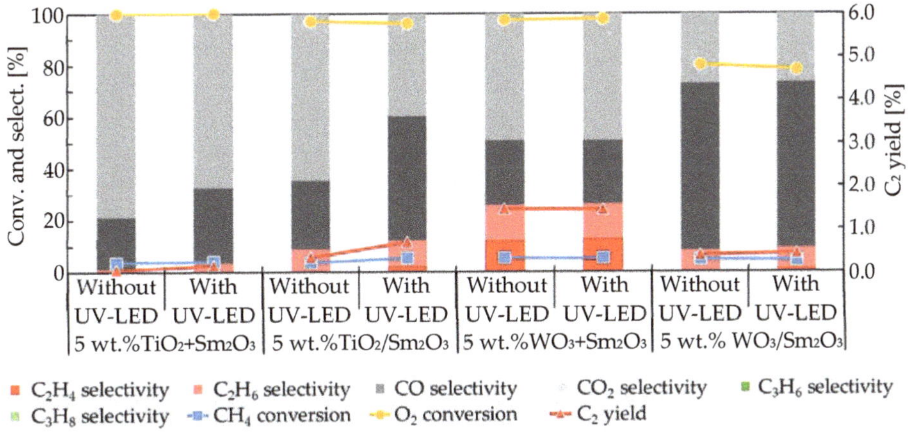

Figure 3. Comparison of the effects of UV-LED irradiation of the oxidative dehydrogenation of methane when using 5 wt.% TiO_2 + Sm_2O_3, 5 wt.% TiO_2/Sm_2O_3, 5 wt.% WO_3 + Sm_2O_3, and 5 wt.% WO_3/Sm_2O_3.

Based on Figure 3, the effect of UV-LED irradiation was more evident when using the supported-catalysts than when the mixed-catalysts were used. Table 1 summarizes the effect of UV-LED irradiation on the selectivities for CO, CO_2, C_2H_4, and C_2H_6 obtained from the oxidative dehydrogenation of methane over the mixed- and supported-catalysts using the data shown in Figure 3. The positive values in Table 1 indicate that the selectivity for each product was enhanced by UV-LED irradiation, while the negative values indicate that the selectivity was suppressed. Values less than 1.0 in Table 1 indicate that UV-LED irradiation had little effect on the corresponding selectivity.

Table 1. Effect of UV-LED irradiation on the selectivity for each of the products when using the binary catalysts in the present study.

Catalyst	ΔCO Selectivity [%]	$ΔCO_2$ Selectivity [%]	$ΔC_2H_6$ Selectivity [%]	$ΔC_2H_4$ Selectivity [%]
5 wt.% TiO_2 + Sm_2O_3	9.0	−11.4	2.4	Not detected
5 wt.% TiO_2/Sm_2O_3	21.2	−24.6	1.0	2.4
5 wt.% WO_3 + Sm_2O_3	−0.3	−0.1	−0.2	0.7
5 wt.% WO_3/Sm_2O_3	−0.7	−0.5	0.0	1.2

Although the effect of UV-LED irradiation was not evident for either 5 wt.% WO_3 + Sm_2O_3 or 5 wt.% WO_3/Sm_2O_3, Table 1 is used here to discuss the effects of UV-LED irradiation. Active oxygen such as O_2^- is generated when using both 5 wt.% TiO_2 + Sm_2O_3 and 5 wt.% TiO_2/Sm_2O_3 due to the presence of TiO_2 in the binary catalysts [22]. When using these catalysts, the selectivities for CO, C_2H_6, and/or C_2H_4 were improved by UV-LED irradiation, while the selectivity for CO_2 was suppressed. Therefore, the formation of O_2^- by UV-LED when using the binary catalysts seems to have contributed to an enhancement of the formation of partial oxidation products, while the deep oxidation production of CO_2 was suppressed. When using 5 wt.% WO_3 + Sm_2O_3 and 5 wt.% WO_3/Sm_2O_3, active oxygen such as H_2O_2 is generated due to the presence of WO_3 in the binary catalysts [22]. Although the effect of UV-LED irradiation was rather small or negligible when using these catalysts compared with that when using TiO_2-loading catalysts, a small but rather negligible enhancement of the selectivity to C_2H_4 was detected with the use of 5 wt.% WO_3 + Sm_2O_3 and 5 wt.% WO_3/Sm_2O_3. Therefore, the formation of H_2O_2 from UV-LED when using these binary catalysts may slightly contribute to the oxidative dehydrogenation of C_2H_6 to C_2H_4. Based on these results, it is possible to summarize the influence that active oxygen species exert on the present catalyst system, as shown in Scheme 3.

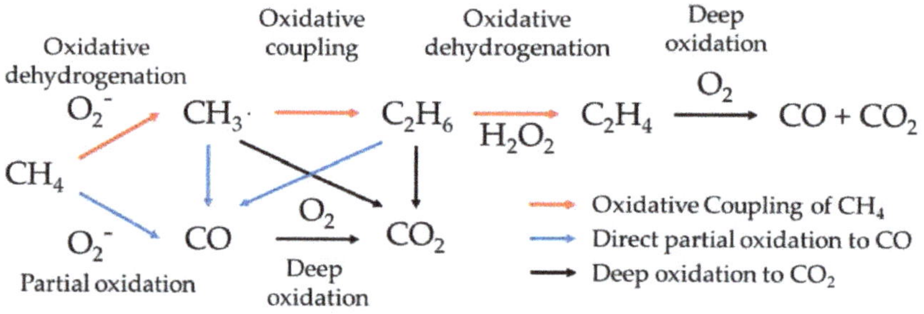

Scheme 3. Proposed contribution of active oxygen in the present binary catalysts.

The active oxygen of O_2^- that formed when using TiO_2 + Sm_2O_3 and TiO_2/Sm_2O_3 contributed to the positive effect for the formations of CO, C_2H_6, and C_2H_4 together with a suppression of the deep oxidation of C_2H_4 to CO and CO_2. Furthermore, as shown in the results for TiO_2, the O_2^- formed on TiO_2 alone directly contributed to the partial oxidation of CH_4 to CO. The active oxygen of H_2O_2 that formed when using both 5 wt.% WO_3 + Sm_2O_3 and 5 wt.% WO_3/Sm_2O_3 showed a negligible contribution to the conversion of C_2H_6 to C_2H_4 via oxidative dehydrogenation. It should be noted that H_2O_2 is an active species for other partial oxidations such as the epoxidation of alkenes. Therefore, the WO_3 system may be one of the most plausible candidates for the epoxidation of alkenes under UV-LED irradiation. Gaseous O_2 is the main contributor to the deep oxidation to CO_2.

Finally, the catalysts used in the present study were analyzed using XRD. XRD patterns of the single oxides of Sm_2O_3 and WO_3 were matched to the reference patterns for the corresponding oxide (PDF 01-078-4055 and 01-083-0950, respectively; not shown). For 5 wt.% WO_3 + Sm_2O_3 and 5 wt.% WO_3/Sm_2O_3, the XRD peaks due to Sm_2O_3 were detected alone (not shown). As shown in Figure 4A, before the reaction, TiO_2 was a mixture of anatase- and rutile-type TiO_2 (PDF 00-064-0863 and 01-086-0148, respectively). The anatase-type remained after the reaction, regardless of the UV-LED irradiation. Furthermore, Figure 4B,C shows that 5 wt.% TiO_2 + Sm_2O_3 and 5 wt.% TiO_2/Sm_2O_3 contained a trace amount of anatase-type TiO_2 together with Sm_2O_3 before the reaction. However, after the reaction with and without UV-LED irradiation, peaks due to Sm_2O_3 were detected together with a trace amount of anatase-type TiO_2. Based on these XRD results, we concluded that anatase-type TiO_2 remained during the reaction and the effect of UV-LED on the reaction came from the contribution of the anatase-type TiO_2 [25].

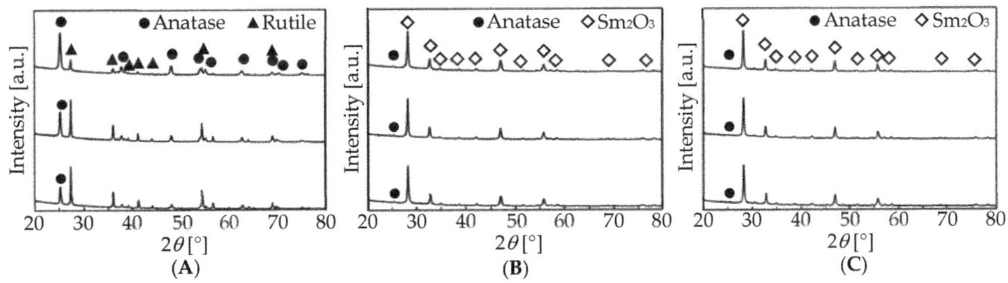

Figure 4. XRD of (**A**) TiO_2, (**B**) 5 wt.% TiO_2 + Sm_2O_3, and (**C**) 5 wt.% WO_3/Sm_2O_3. Upper—before the reaction. Middle and lower—after the reaction without and with UV-LED.

3. Materials and Methods

Mixed-catalysts (TiO_2 + Sm_2O_3 and WO_3 + Sm_2O_3) were prepared via the kneading of Sm_2O_3 (Wako Pure Chemical Industries, Ltd., Osaka, Japan) with either TiO_2 (JRC-TIO-15, a reference catalyst supplied from The Catalysis Society of Japan, Tokyo, Japan) or WO_3 (Wako Pure Chemical Industries, Ltd.) for 30 min. For the preparation of 5 wt.% TiO_2 + Sm_2O_3, 0.018 g of TiO_2 was kneaded with 0.350 g of Sm_2O_3 for 30 min. Supported-catalysts (TiO_2/Sm_2O_3 and WO_3/Sm_2O_3) were prepared via impregnation. The preparation of 5 wt.% TiO_2/Sm_2O_3 began with 20 mL of 2-propanol (Wako Pure Chemical Industries, Ltd.)) into which we dissolved 0.592 g of titanium tetraisopropoxide (Wako Pure Chemical Industries, Ltd.) and 3.00 g of Sm_2O_3, followed by the further addition of 35 mL of distilled water. The resultant suspension was then evaporated and dried at 333 K for 24 h. Finally, the resultant solid was calcined at 973 K for 3 h. The preparation of 5 wt.% WO_3/Sm_2O_3 began with 20 mL of aqueous solution into which we dissolved 0.174 g of ammonium (para)tungstate hydrate (Sigma-Aldrich Japan Co. LLC, Tokyo, Japan) and 3.00 g of Sm_2O_3. The resultant suspension was treated in a manner similar to the preparation of TiO_2/Sm_2O_3. In order to analyze those catalysts, X-ray diffraction (XRD) patterns were obtained using a SmartLab/R/INP/DX (Rigaku Co., Osaka Japan) with a Cu Kα radiation monochromator at 45 kV and 150 mA. In order to estimate the specific surface areas of those catalysts via BET, nitrogen adsorption isotherms of the catalysts pretreated at 473 K for 5 h were measured using a BELSORPmax12 (MicrotracBEL, Osaka, Japan) at 77 K.

The catalytic experiments were performed in a fixed-bed continuous-flow quartz reactor, which was placed in an electric furnace with an optical window, and operated at atmospheric pressure and 898 K (Scheme 4). As a light source for UV-LED irradiation, a Lightningcure LC-L1V3 (Hamamatsu Photonics K.K., Shizuoka, Japan) was used. This light source emits UV light at a wavelength of 365 nm for an average maximum irradiation intensity of 14,000 mW/cm^2 and a maximum output of 450 mW, which is sufficient for the activation of O_2 when using TiO_2 and WO_3 under the present reaction conditions.

The temperature of the catalyst (0.350 g and 0.368 g for single and binary oxide catalysts, respectively) was increased to 898 K under a flow of He. After the reaction temperature was stabilized, the catalyst was treated with a flow of O_2 (15 mL/min) for 1 h. Activity tests were then carried out under 15 mL/min of a reactant gas flow that consisted of CH_4 and O_2 diluted with He. In the present study, partial-pressure ratios of 7.1 and 14.2 were employed for CH_4/O_2, and the partial pressures were then adjusted to $P(CH_4)/P(O_2)$ = 28.7 kPa/4.05 kPa and 28.7 kPa/2.03 kPa. Under these conditions, homogeneous reactions were not detected. The reaction was monitored using an on-line gas chromatograph (GC-8APT, Shimadzu Corp., Kyoto, Japan) that involved the use of a thermal conductivity detector (TCD). The columns in the TCD-GC consisted of a Molecular Sieve 5A (0.3 m × Φ 3 mm) for the detection of O_2, CO, and CH_4 at 318 K and a Porapak Q (6 m × Φ 3 mm) for the detection of CO_2, C_2, and C_3 species at the column temperatures between 318 and 493 K with a heating rate of 10 K/min. The conversion and the selectivity were estimated on a carbon basis.

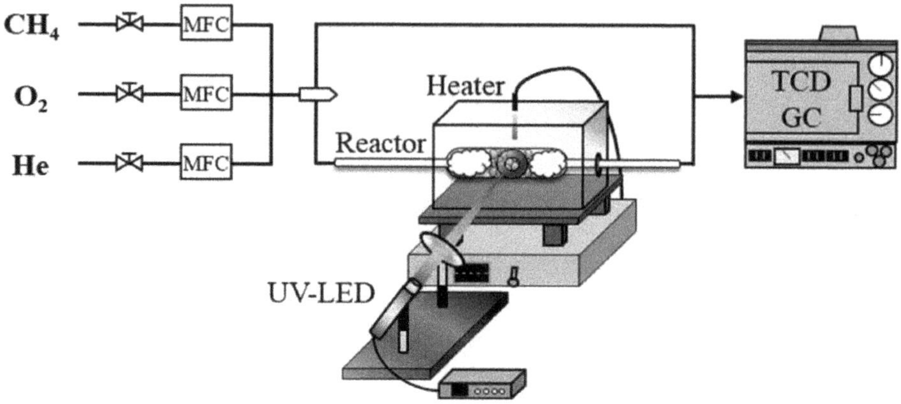

Scheme 4. Fixed-bed continuous-flow quartz reactor with UV-LED.

4. Conclusions

In order to investigate the active oxygen effect that O_2^- and H_2O_2 exert on the catalytic oxidative dehydrogenation of methane, binary oxide consisting of Sm_2O_3, which is an oxidative coupling catalyst for methane, and TiO_2 or WO_3, which generate O_2^- or H_2O_2, respectively, when irradiated with UV-LED, were prepared using kneading and impregnation methods. Regardless of the preparation methods, O_2^- generated from TiO_2 under UV-LED irradiation promoted the partial oxidation of methane to CO and oxidative conversion to C_2 compounds, while it suppressed complete oxidation to CO_2. By contrast, regardless of the preparation methods, H_2O_2 generated from WO_3 under UV-LED irradiation had no evident effect on the oxidation of methane. It is noteworthy that the use of MgO instead of Sm_2O_3 had no effect on the results of UV-LED irradiation. Therefore, it is suggested that the use of any oxide catalyst with great redox properties equal to those of Sm_2O_3 would produce the above-mentioned advantageous effects via UV-LED irradiation.

Author Contributions: Conceptualization and methodology, S.S., A.F., Y.K., and W.N.; validation, Y.H., I.O., N.S., and M.K.; formal analysis and investigation, S.S., Y.Y., and I.O.; writing—original draft preparation, S.S.; writing—review and editing, S.S., N.S., M.K., A.F., Y.K., and W.N.; and, supervision, S.S. All authors have read and agreed to the published version of the manuscript.

Funding: This research was funded by JSPS KAKENHI Grant Number JP17K19014 and by the Research Clusters Program of Tokushima University (1702001).

Acknowledgments: The authors gratefully acknowledge Toshihiro Okamoto of the Institute of Post-LED Photonics, Tokushima University for his valuable suggestions concerning photo-catalysts.

Conflicts of Interest: The authors declare no conflicts of interest.

References

1. Tang, P.; Zhu, Q.; Wu, Z.; Ma, D. Methane activation: The past and future. *Energy Environ. Sci.* **2014**, *7*, 2580–2591. [CrossRef]
2. Lunsford, J.H. Catalytic conversion of methane to more useful chemicals and fuels: A challenge for the 21st century. *Catal. Today* **2000**, *63*, 165–174. [CrossRef]
3. Galadima, A.; Muraza, O. Revisiting the oxidative coupling of methane to ethylene in the golden period of shale gas: A review. *J. Ind. Eng. Chem.* **2016**, *37*, 1–13. [CrossRef]
4. Zakaria, Z.; Kamarudin, S.K. Direct conversion technologies of methane to methanol: An overview. *Renew. Sustain. Energy Rev.* **2016**, *65*, 250–261. [CrossRef]
5. Sugiyama, S.; Minami, T.; Higaki, T.; Hayashi, H.; Moffat, J.B. High selective conversion of methane to carbon monoxide and the effects of chlorine additives in the gas- and solid-phases on the oxidation of methane on strontium hydroxyapatites. *Ind. Eng. Chem. Res.* **1997**, *36*, 328–334. [CrossRef]

6. Liu, Y.; Li, D.; Wang, T.; Liu, Y.; Xu, T.; Zhang, Y. Efficient conversion of methane to aromatics by coupling methylation reaction. *ACS Catal.* **2016**, *6*, 5366–5370. [CrossRef]
7. Gesser, H.D.; Hunter, N.R.; Prakash, C.B. The direct conversion of methane to methanol by controlled oxidation. *Chem. Rev.* **1985**, *85*, 235–244. [CrossRef]
8. Lee, J.Y.; Jeon, W.; Choi, J.W.; Suh, Y.W.; Ha, J.M.; Suh, D.J.; Park, Y.K. Scaled-up production of C-2 hydrocarbons by the oxidative coupling of methane over pelletized $Na_2WO_4/Mn/SiO_2$ catalysts: Observing hot spots for the selective process. *Fuel* **2013**, *106*, 851–857. [CrossRef]
9. Zhang, H.B.; Lin, G.D.; Wan, H.L.; Liu, Y.D.; Weng, W.Z.; Cai, J.X.; Shen, Y.F.; Tsai, K.R. Active-oxygen species on non-reducible rare-earth-oxide-based catalysts in oxidative coupling of methane. *Catal. Lett.* **2001**, *73*, 141–147. [CrossRef]
10. Lunsford, J.H. The catalytic oxidative coupling of methane. *Angew. Chem. Int. Ed. Engl.* **1995**, *34*, 970–980. [CrossRef]
11. Wang, H.; Cong, Y.; Yang, W. Oxidative coupling of methane in $Ba_{0.5}Sr_{0.5}Co_{0.8}Fe_{0.2}O_{3-\delta}$ tubular membrane reactors. *Catal. Today* **2005**, *104*, 160–167. [CrossRef]
12. Lee, M.R.; Park, M.J.; Jeon, W.; Choi, J.W.; Suh, Y.W.; Suh, D.J. A kinetic model for the oxidative coupling of methane over $Na_2WO_4/Mn/SiO_2$. *Fuel Process. Technol.* **2012**, *96*, 175–182. [CrossRef]
13. Sun, J.; Thybaut, J.W.; Marin, G.B. Microkinetics of methane oxidative coupling. *Catal. Today* **2008**, *137*, 90–102. [CrossRef]
14. Beck, B.; Fleischer, V.; Arndt, S.; Hevia, M.G.; Urakawa, A.; Hugo, P.; Schomäcker, R. Oxidative coupling of methane—A complex surface/gas phase mechanism with strong impact on the reaction engineering. *Catal. Today* **2014**, *228*, 212–218. [CrossRef]
15. Gao, Z.; Ma, Y. Direct oxidation of methyl radicals in OCM process deduced from correlation of product selectivities. *J. Nat. Gas Chem.* **2010**, *19*, 534–538. [CrossRef]
16. Osada, Y.; Koike, S.; Fukushima, T.; Ogasawara, S.; Shikada, T.; Ikariya, T. Oxidative coupling of methane over Y,O,-CaO catalysts. *Appl. Catal.* **1990**, *59*, 59–74. [CrossRef]
17. Yang, T.L.; Feng, L.B.; Shen, S.K. Oxygen species on the surface of La_2O_3/CaO and its role in the oxidative coupling of methane. *J. Catal.* **1994**, *145*, 384–389. [CrossRef]
18. Spinicci, R.; Marini, P.; De Rossi, S.; Faticanti, M.; Porta, P. Oxidative coupling of methane on $LaAlO_3$ perovskites partially substituted with alkali or alkali-earth ions. *J. Mol. Catal. A* **2001**, *176*, 253–265. [CrossRef]
19. Jeon, W.; Lee, J.Y.; Lee, M.; Choi, J.; Ha, J.; Suh, D.J.; Kim, I.W. Oxidative coupling of methane to C_2 hydrocarbons on the Mg–Ti mixed oxide-supported catalysts at the lower reaction temperature: Role of surface oxygen atoms. *Appl. Catal. A* **2013**, *464–465*, 68–77. [CrossRef]
20. Schwach, P.; Pan, X.; Bao, X. Direct Conversion of Methane to Value-added Chemicals over Heterogeneous Catalysts: Challenges and Prospects. *Chem. Rev.* **2017**, *117*, 8497–8520. [CrossRef]
21. Gambo, Y.; Jalil, A.A.; Triwahyono, S.; Abdulrasheed, A.A. Recent Advances and Future Prospect in Catalysts for oxidative coupling of Methane to Ethylene: A Review. *J. Ind. Eng. Chem.* **2018**, *59*, 218–229. [CrossRef]
22. Hoffmann, M.R.; Martin, S.T.; Choi, W.; Bahnemann, D.W. Environmental applications of semiconductor photocatalysis. *Chem. Rev.* **1995**, *95*, 69–96. [CrossRef]
23. Yunarti, R.T.; Gu, S.; Choi, J.-W.; Jae, J.; Sun, D.J.; Ha, J.-M. Oxidative Coupling of Methane Using Mg/Ti-Doped SiO_2-Spported Na_2WO_4/Mn Catalysts. *ACS Sustain. Chem. Eng.* **2017**, *5*, 3667–3674. [CrossRef]
24. Gu, S.; Oh, H.-S.; Choi, J.-W.; Suh, D.J.; Jae, J.; Choi, J.; Ha, J.-M. Effects of Metal or Metal Oxide Additives on Oxidative Coupling of Methane Using Na_2WO_4/SiO_2 Catalysts: Reducibility of Metal Additives to Manipulate the Catalytic Activity. *Appl. Catal. A* **2018**, *562*, 114–119. [CrossRef]
25. Chen, X.; Mao, S.S. Titanium dioxide nanomaterials: synthesis, properties, modifications, and applications. *Chem. Rev.* **2007**, *107*, 2891–2959. [CrossRef]

catalysts

MDPI

Article

Potassium Ferrite as Heterogeneous Photo-Fenton Catalyst for Highly Efficient Dye Degradation

Xinghui Zhang [1], Zhibin Geng [1], Juan Jian [2], Yiqiang He [1], Zipeng Lv [1], Xinxin Liu [3] and Hongming Yuan [1,*]

1 State Key Laboratory of Inorganic Synthesis and Preparative Chemistry, College of Chemistry, Jilin University, Changchun 130012, China; zhangxinghuih@163.com (X.Z.); gengzb14@mails.jlu.edu.cn (Z.G.); yiqianghe199510@163.com (Y.H.); lvzp17@mails.jlu.edu.cn (Z.L.)

2 Key Laboratory of Preparation and Applications of Environmental Friendly Material of the Ministry of Education, College of Chemistry, Jilin Normal University, Changchun 130103, China; jianjuan1528241015@163.com

3 Institute of Catalysis for Energy and Environment, College of Chemistry and Chemical Engineering, Shenyang Normal University, Shenyang 110034, China; liuxinxin1114@163.com

* Correspondence: hmyuan@jlu.edu.cn; Tel.: +86-0431-8516-8318

Received: 18 January 2020; Accepted: 3 March 2020; Published: 4 March 2020

Abstract: In this work, hexagon-shaped potassium ferrite ($K_2Fe_4O_7$) crystals with different sizes were prepared using the hydrothermal method. The crystals showed a narrow band gap of 1.44 eV, revealed by UV-visible diffuse reflectance spectroscopy, and was thus used as a heterogeneous Fenton catalyst to degrade methylene blue (MB) and crystal violet (CV) in the presence of green oxidant H_2O_2 under visible-light irradiation. Among the investigated crystals, the as-prepared one with an average size of 20 μm (KFO-20) exhibited better photocatalytic activity due to its high surface area. When it was used as a photo-Fenton catalyst, 100% MB and 92% CV were degraded within 35 min. Moreover, the catalyst maintained high photocatalytic activity and was stable after four continuous cycles. The trapping experiments showed that the active hydroxyl radical (·OH) was dominant in the photo-Fenton reaction. Therefore, this new photo-Fenton catalyst has great potential for the photocatalytic degradation of dye contaminants in water.

Keywords: $K_2Fe_4O_7$; visible light; Fenton catalyst; degradation; methylene blue and crystal violet

1. Introduction

Dyes are widely used in many industries, including textiles, printing, pulp, and paper. However, they also cause great environmental pollution that threatens public health due to their complicated constitution and high chemical stability. With the rapid development of the global economy, it is estimated that each year almost 2×10^5 tons of dyes are discharged directly into lakes, rivers, and groundwater [1–4]. Therefore, dye treatment in water systems is increasingly important.

Over the past few decades, many wastewater-treatment technologies have been developed for the removal or degradation of dye contaminants, such as traditional biological treatments, adsorption, coagulation/flocculation, membrane separation, precipitation, ion exchange, and advanced oxidation processes (AOPs). Among these methods, AOPs that include photochemical, catalytic, sonochemical, ozone, electrochemical, and Fenton oxidation can convert or degrade such contaminants into small molecules [5–7]. In particular, photo-Fenton oxidation has been proven to be highly effective for dye treatments [8,9]. However, homogeneous photo-Fenton technology exhibits limitations such as loss of catalysts and large iron sludge. To overcome these drawbacks, one potential strategy is using heterogeneous photo-Fenton catalysts to degrade dye contaminants. So far, several iron oxides and iron hydroxides utilised as photo-Fenton catalysts, such as Fe_3O_4, α-Fe_2O_3, α-FeOOH, and β-Fe_2O_3, have been used to catalyse or degrade various kinds of dye molecules due to their good photocatalytic activity, stable structure, and narrow band gap [10–14]. The catalytic activity of metal oxides is closely related to the crystal orientation, crystallinity, nanostructure, morphology, particle size, and surface properties of the material [15–20]. To further enhance their catalytic activity, some with nanometre sizes were prepared using nanotechnology [11,21,22]. Other transition metals were introduced into their structure to improve their catalytic activity [23–27]. Recently, iron oxide and carbon composites have also been used to effectively degrade organic dyes [28–33].

At present, there is an urgent need for low-cost, resource-friendly, stable iron-based heterogeneous Fenton catalysts for the degradation of dye contaminants in wastewater. The Fenton catalytic properties of iron oxides are strongly dependent on the crystal structural characters, band gap, and oxidation states of iron ions at octahedral and tetrahedral sites. All of the iron oxides mentioned above are constructed from iron octahedral and tetrahedral unites, and adopt either corundum or spinel structures. The search for an iron-based heterogeneous photo-Fenton catalyst with a new crystal structure is a key step towards more effectively degrading dye contaminants. Recently, we successfully synthesised potassium ferrite ($K_2Fe_4O_7$) using the hydrothermal process [34]. This environmentally friendly iron oxide exhibited superhigh ion conductivity and excellent thermal stability. Its 3D structure consisted of octahedral FeO_6 and tetrahedral FeO_4 with shared vertices and edges. Therefore, the structure may be developed for use as a powerful candidate for a heterogeneous photo-Fenton catalyst for the degradation of dye molecules.

In this work, we prepared the hexagonal $K_2Fe_4O_7$ crystals with different sizes using the hydrothermal method. Its band gap was determined by UV-visible diffuse reflectance spectroscopy. Methylene blue (MB) and crystal violet (CV) were selected as models to evaluate the catalytic activity of this iron oxide under photo-Fenton conditions. We then studied the effect of pH value, the concentration of hydrogen peroxide, and particle size on MB degradation. We also evaluated the reusability and stability of the catalyst.

2. Results and Discussion

2.1. Characterisation

SEM images of KFO-20 (Figure 1 a, b), KFO-80 (Figure 1 c, d), and KFO-180 (Figure 1 e, f) show that the obtained $K_2Fe_4O_7$ crystals each had the same hexagonal shape and smooth surface except for crystal size. The powder XRD patterns of KFO-20, KFO-80, and KFO-180 are shown in Figure 2, where it is clear that all diffraction peaks were consistent with that of $K_2Fe_4O_7$ reported previously [34]. This indicates that each sample's structure remained intact and was in a pure phase. Studies showed that, as a result of higher crystallinity, there were fewer lattice defects, which are more conductive to the conduction of the charge carrier. This result is consistent with photoluminescence (PL).

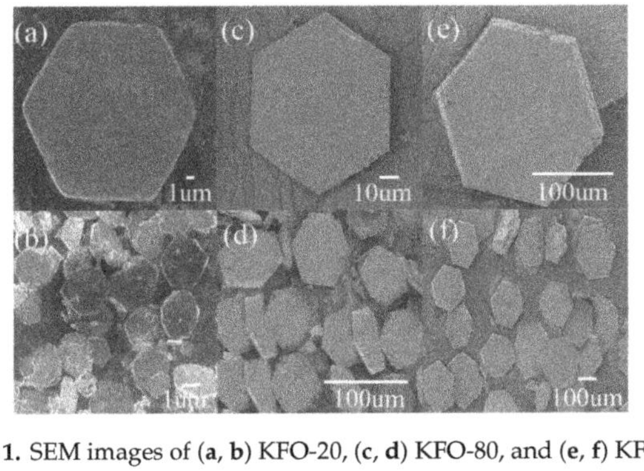

Figure 1. SEM images of (**a**, **b**) KFO-20, (**c**, **d**) KFO-80, and (**e**, **f**) KFO-180.

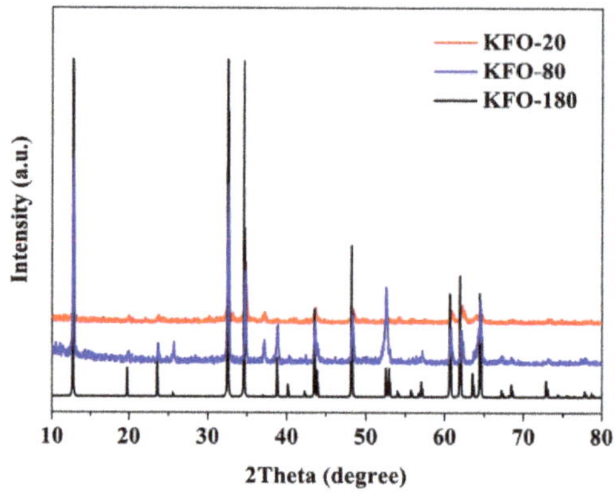

Figure 2. Powder XRD patterns of KFO-20, KFO-80, and KFO-180.

The band gaps of three samples were determined by UV-visible diffuse reflectance spectroscopy measurement. They exhibited the same broad absorption with an absorption edge at about 858 nm (Figure 3a). Band-gap energy was calculated as 1.44 eV according to the equation Eg = hc/λ, where Eg is the energy band gap (eV) and hc = 1240 eV [35]. This narrow band gap was less than those of Fe_3O_4 [14], α-Fe_2O_3 [12], α-FeOOH [12], and $NiFe_2O_4$ [11], and was thus a favourable candidate to effectively degrade dye molecules in photo-Fenton catalytic processes under low-energy visible-light irradiation. The comparison of various catalytic systems for MB degradation under different photo-Fenton processes is shown in Table 1.

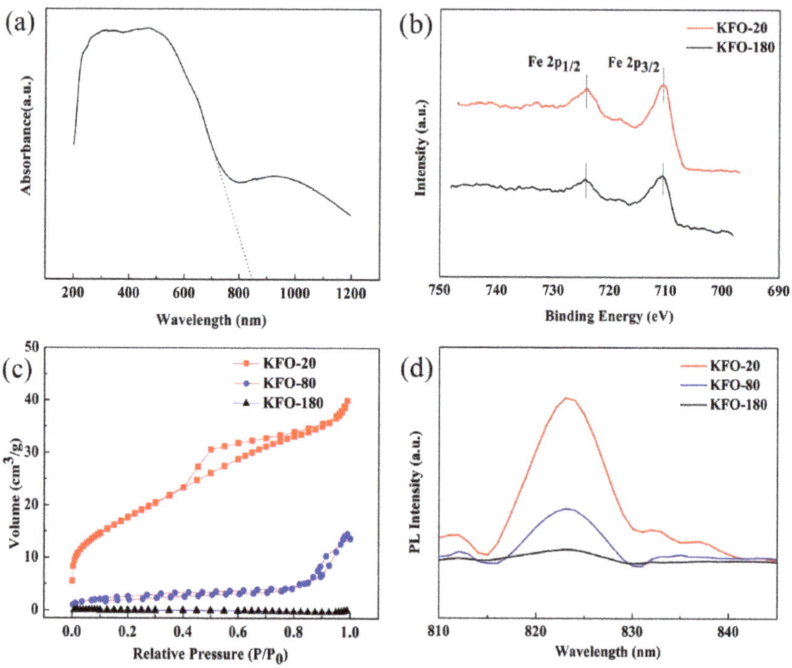

Figure 3. (**a**) UV-visible diffuse reflectance spectroscopy; (**b**) XPS spectra of Fe $2p_{1/2}$ and Fe $2p_{3/2}$ for KFO-20 and KFO-180; (**c**) N_2 adsorption–desorption isotherms measured at 77K of KFO; and (**d**) photoluminescence (PL) spectra of KFO.

Table 1. Comparison of various catalytic systems for methylene blue (MB) degradation under different photo-Fenton processes.

Catalyst	Reaction Condition	Degradation Rate	Time	Reference
$NiFe_2O_4$	[Catalyst] = 0.2 g/L, $[H_2O_2]$ = 5 mM, [MB] = 30 mg/L and light irradiation	98.5%	50 min	[11]
Fe_3O_4	[Catalyst] = 4 g/L, $[H_2O_2]$ = 50 mM, [MB] = 100 mg/L, and UV irradiation	20%	60 min	[14]
α-Fe_2O_3	[Catalyst] = 0.025 g/L, $[H_2O_2]$ = 1.10 mM, [MB] = 40 mg/L and UV irradiation	94.7%	80 min	[12]
$CuFe_2O_4$	[Catalyst] = 0.1 g/L, $[H_2O_2]$ = 20 mM, [MB] = 30 mg/L and light irradiation	80%	80 min	[36]
ZnO	[Catalyst] = 0.02 g/L, $[H_2O_2]$ = 5 mM, [MB] = 20 mg/L and light irradiation	4.1%	60 min	[37]
TiO_2	[Catalyst] = 1.0 g/L, [MB] = 50 mg/L and light irradiation	30%	60 min	[38]
$K_2Fe_4O_7$	[Catalyst] = 0.03 g/L, $[H_2O_2]$ = 5 mM, [MB] = 20 mg/L and light irradiation	100%	35 min	This article

The XPS spectra of KFO-20 and KFO-180 are shown in Figure 3b. There were two obvious peaks with binding energies of 724.2 and 711.3 eV, which corresponded, respectively, to Fe $2p_{1/2}$ and Fe $2p_{3/2}$. XPS results were in good agreement with those of the reported crystal [34], suggesting that the oxidation state of iron ion in the as-prepared KFO was +3.

The Brunauer-Emmett-Teller (BET) surface area was measured via nitrogen adsorption–desorption experiments. As shown in Figure 3c, the adsorption–desorption isotherms of KFO corresponded to Type IV on the basis of the Brunauer-Deming-Teller classification [39,40]. The specific surface area of KFO-20, KFO-80, and KFO-180 was 63.79, 9.63, and 0.42 m^2/g, respectively, which increased with the decrease of particle size. In this work, we found that crystallinity contributed little to MB degradation comparing to the specific surface area.

The photoluminescence (PL) spectrum was used to investigate the recombination behaviour of photogenerated electrons and holes in KFO-20, KFO-80, and KFO-180. The PL emission spectra of the three crystals at an excitation wavelength of 475 nm are shown in Figure 3d. A strong peak near 820 nm caused by the band–band PL phenomenon was observed, which was approximately equal to the band-gap energy of KFO (1.44 eV). However, peak intensity decreased with the decrease of crystal size due to shorter diffusion distance, suggesting that KFO-20 with a lower recombination rate of photogenerated electron-hole pair was expected to have higher photocatalytic activity.

2.2. Photo-Fenton MB Degradation

We first selected KFO-20 as a Fenton catalyst to perform the degradation of MB at pH = 2, 20 mg/L of MB at room temperature in the absence and presence of visible light, and with or without hydrogen peroxide. Degradation results are shown in Figure 4a. With visible-light radiation in the presence of KFO-20 and H_2O_2 (curve A), 100% of the MB was degraded within 35 min. However, under darker conditions, 46% of the MB was degraded within the same timeframe (curve B), indicating that its Fenton catalytic activity was significantly enhanced by introducing visible light. Only 12% of MB was degraded when the catalyst was not employed due to the self-sensitisation of MB (curve C). MB concentration had no apparent change when without hydrogen peroxide (curve D), confirming that MB degradation was attributed to the photo-Fenton process. The catalytic activity of the catalyst was superior to conventional photo-Fenton catalysts mentioned above [11–14]. Therefore, it can be concluded that the excellent catalytic activity of this iron oxide is assigned to its structural feature and the nature of the iron ion.

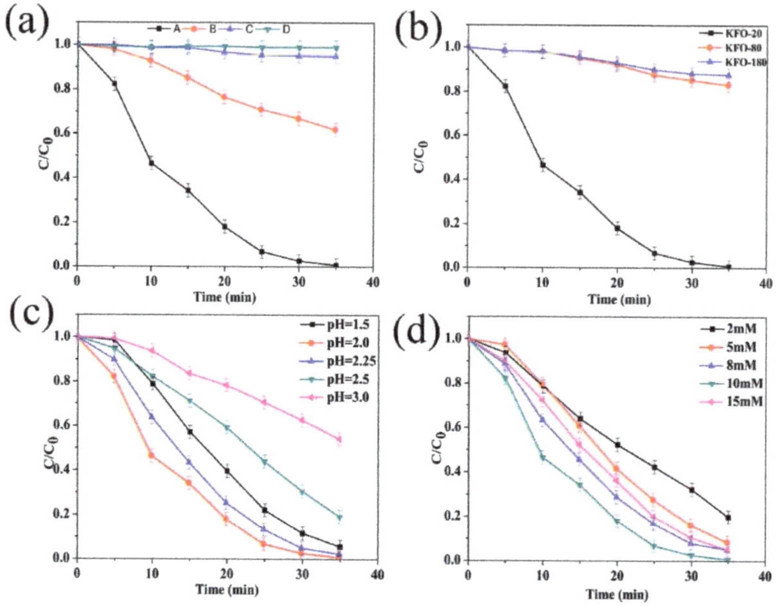

Figure 4. (**a**) Degradation efficiency of methylene blue (MB) under different conditions. A: MB + KFO-20 + H_2O_2 + light; B: MB + KFO-20 + H_2O_2; C: MB + H_2O_2 + light; D: MB + KFO-20 + light; (**b**) degradation efficiency of MB using KFO catalysts with different sizes; (**c**) degradation efficiency of MB at different pH values; (**d**) degradation efficiency of MB at various H_2O_2 concentrations. Reaction conditions: MB concentration, 20 mg/L; KFO mass, 30 mg; H_2O_2, 10 mM; reaction time, 35 min.

2.2.1. Effect of Catalyst Size

In general, the catalytic activity of heterogeneous Fenton catalysts was proportional to their specific surface area. A larger specific surface area can provide more active sites for a photo-Fenton catalytic process. Therefore, we investigated the effect of different KFO sizes on the degradation of MB under the same conditions. As shown in Figure 4b, about 34% and 24% of degradation were achieved within 35 min for KFO-80 and KFO-180, respectively, due to their different specific surface areas, which was consistent with the results of BET surface-area measurements.

2.2.2. Effect of pH Value

The Fenton oxidation reaction was sensitive to the pH value of the reacting solution. Therefore, the effect of the initial pH value on the degradation of MB (20 mg/L) was investigated by using the KFO-20 photo-Fenton catalyst. Figure 4c gives the removal efficiency of MB at different pH values within 35 min. At pH = 1.5, about 98% of MB degradation was obtained, and 100% of MB was degraded when we increased the pH value to 2.0 because excess H^+ acted as a scavenger for the hydroxyl radical (·OH), as shown in Equation (1) [40]. As pH value increased, the removal percentage decreased to 90% at pH = 2.5. Therefore, the photo-Fenton catalyst can be used to effectively degrade MB in the pH range of 1.5–2.5.

$$\cdot OH + H^+ + e^- \rightarrow H_2O \quad (1)$$

2.2.3. Effect of H_2O_2 Dose

In the heterogeneous Fenton reaction, redox cycling between Fe^{3+} and Fe^{2+} was achieved by H_2O_2. Figure 4d shows the effect of the H_2O_2 concentration on MB degradation in the presence of the KFO-20 catalyst and visible light. MB degradation increased to 100% when the molar concentration of

H_2O_2 increased from 2 to 10 mM. This could be due to the formation of much ·OH in the photo-Fenton reaction (Equation (2)) [41]. However, 98% of MB was degraded at the concentration of 15 mM, because hydroperoxyl radicals (HO·) that were generated through excessive H_2O_2 reacting with the ·OH did not contribute to MB degradation (Equations (3)–(5)) [40,42–46]. This scavenging effect has been observed in many photo-Fenton processes.

$$H_2O_2 + h\upsilon \rightarrow 2{\cdot}OH \quad (2)$$

$$H_2O_2 + {\cdot}OH \rightarrow H_2O + H_2O \quad (3)$$

$$H_2O_2 + HO_2{\cdot} \rightarrow OH + H_2O + O_2 \quad (4)$$

$$H_2O + HO{\cdot} \rightarrow H_2O + O_2 \quad (5)$$

2.2.4. Dynamic Degradation

In the photo-Fenton reaction system, a pseudo-first-order kinetic constant model was used to quantitatively evaluate the photocatalytic process. The rate constant was obtained by plotting −ln (C/C_0) versus time, as shown in Figure 5a, where C_0 and C are the initial concentration and the concentration at time t, respectively. The specific surface area S (m^2/g), rate constant, and k (min^{-1}) of KFO are shown in Table 2. The rate constants of KFO-20, KFO-80, and KFO-180 were 0.155, 0.007 and 0.006 min^{-1}, respectively. The rate constant of KFO-20 was 26 times that of KFO-180, and the rate constant of the catalyst increased with the increase of specific surface area.

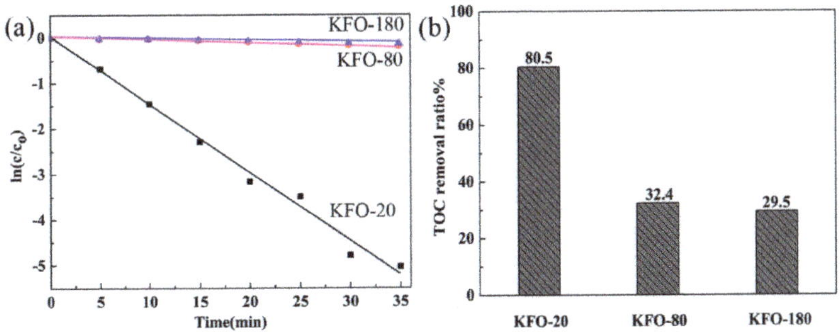

Figure 5. (**a**) Pseudo-first-order kinetics degradation of MB; (**b**) total organic carbon (TOC) removal of MB in presence of differently sized KFOs. Reaction conditions: MB concentration, 20 mg/L; KFO mass, 30 mg; H_2O_2, 10 mM; reaction time, 35 min.

Table 2. KFO specific surface area S (m^2/g) and rate constant k (min^{-1}).

	S (m^2/g)	k (min^{-1})
KFO-20	63.79	0.155
KFO-80	9.63	0.007
KFO-180	0.42	0.006

2.2.5. Total Organic Carbon

In order to determine the degree of mineralisation of MB, the total organic carbon (TOC) removal rates of different KFO sizes were measured using MB as a substrate under the same conditions (Figure 5b). The removal rate of TOC increased with the decrease of KFO sizes. The removal rate of KFO-20 reached 80.5%, which indicated that the KFO-20 effectively mineralised MB.

2.2.6. Universal Applicability of Catalyst

In order to evaluate the photo-Fenton KFO catalysts for other types of dye degradation under the same conditions, experiments of crystal violet (CV) degradation were carried out using catalysts with different sizes. As shown in Figure 6a, KFO-180 and KFO-80 did not produce high degradation efficiency within 35 min. However, 92% of CV was degraded by KFO-20. On the basis of the aforementioned results, the KFO-20 catalyst displayed the best photo-Fenton catalytic activity.

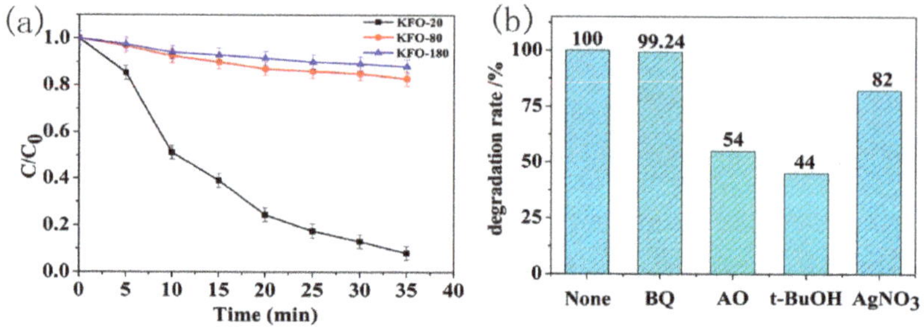

Figure 6. (**a**) Degradation efficiency of crystal violet (CV) with KFO catalysts at pH 2; (**b**) effects of different scavengers on MB degradation at pH 2 in presence of KFO-20, H_2O_2, and light.

2.2.7. Trapping Experiments

To determine active species that were dominant for the degradation of MB, several scavengers, including 10 mmol/L of p-benzoquinone (BQ), ammonium oxalate (AO), tertiary butyl alcohol (t-BuOH), and silver nitrate ($AgNO_3$), were used to trap these active species; results are shown in Figure 6b. BQ could trap the superoxide anion ($\cdot O_2^-$). When BQ was added into the solution, 100% of the MB was degraded within 35 min, indicating that the $\cdot O_2^-$ radical has little contribution to the degradation process. After t-BuOH was added as an ·OH scavenger, the degradation rate of the MB was 45%, and the degradation process was significantly inhibited. This indicated that the ·OH radical played a key role in the degradation process. When AO and $AgNO_3$ were used to capture h^+ and e^-, respectively, 55% and 82% of the MB were degraded. The difference between them came from the contribution of additional ·OH radicals generated by photogenerated holes reacting with H_2O_2 (Equation (2)).

2.3. *Possible Photo-Fenton Catalytic Mechanism*

On the basis of the above experiment results, we proposed the photo-Fenton catalytic mechanism of KFO. The decomposed H_2O_2 and generated hydroxyl radicals on the surface of KFO particles can be described in the following reactions (Equations (2)–(7)). Fe^{3+} was reduced by adsorbed H_2O_2 to generate Fe^{2+} and peroxide hydroxyl radicals (HOO·) on the surface of a catalyst (Equation (6)). Meanwhile, HOO· reacted with hydrogen peroxide in an aqueous solution to produce a highly active hydroxyl radical (·OH) (Equation (4)). The generation of ·OH, which can efficiently degrade organic compounds in photo-Fenton-like reactions, can be described through the reaction between the formed Fe^{2+} and H_2O_2 (Equation (7)). Under visible-light irradiation, electron-hole pairs were generated on the surface of KFO (Equation (8)). Subsequently, the photogenerated electrons reduced hydrogen peroxide and reacted with dissolved oxygen to produce HO·, HO^-, and $\cdot O_2^-$ (Equations (9) and (10)) [44,45,47,48]. Additional ·OH were generated by photogenerated holes that reduced water and hydrogen peroxide (Equations (11) and (12)) [44,45,47,48]. Scheme 1 gives a graphical representation of photo-Fenton MB degradation.

$$Fe^{3+} + H_2O_2 \rightarrow Fe^{2+} + HO_2\cdot + H^+ \quad (6)$$

$$Fe^{2+} + H_2O_2 \rightarrow Fe^{3+} + HO\cdot + OH^- \quad (7)$$

$$KFO + hv \rightarrow KFO\ (h^+ + e^-) \quad (8)$$

$$e^- + H_2O_2 \rightarrow HO^- + HO \quad (9)$$

$$e^- + O_2 \rightarrow O_2^- \quad (10)$$

$$h^+ + H_2O \rightarrow H^+ + HO \tag{11}$$

$$h^+ + HO^- \rightarrow HO \tag{12}$$

$$HO\cdot + MB \rightarrow \text{degradation products} \tag{13}$$

$$h^+ + MB \rightarrow \text{degradation products} \tag{14}$$

Therefore, $HO\cdot$ and the h^+ hole dominantly oxidised the MB molecule (Equations (13) and (14)).

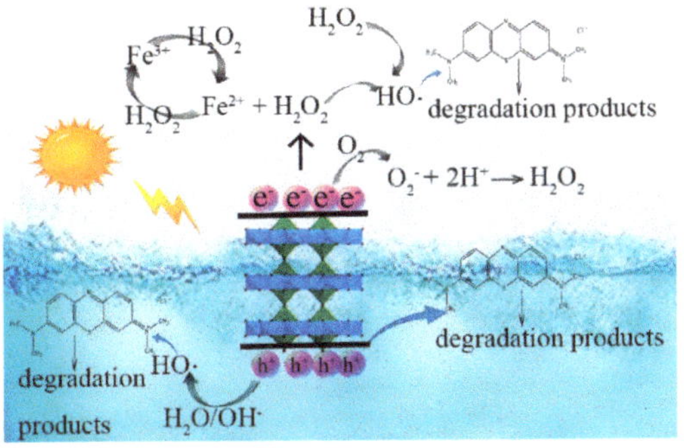

Scheme 1. MB degradation by KFO-20.

2.4. Reusability and Chemical Stability

From the view of industrial applications, a catalyst's chemical stability and reusability are both important. To evaluate the possibility of catalyst reuse, photo-Fenton reaction for MB degradation was successively performed using the KFO-20 catalyst (measurements are described in Supplementary Materials). As exhibited in Figure 7a, it is clearly seen that the catalytic activity of KFO-20 still remained efficient after four cycles under the same conditions. Moreover, after the reaction, the surface morphology of KFO-20 was almost indistinguishable from that of a fresh KFO-20 (Figure 7b). In addition, the powder XRD pattern of the used sample agreed with that of the as-prepared sample (Figure S1), suggesting that it had good chemical stability.

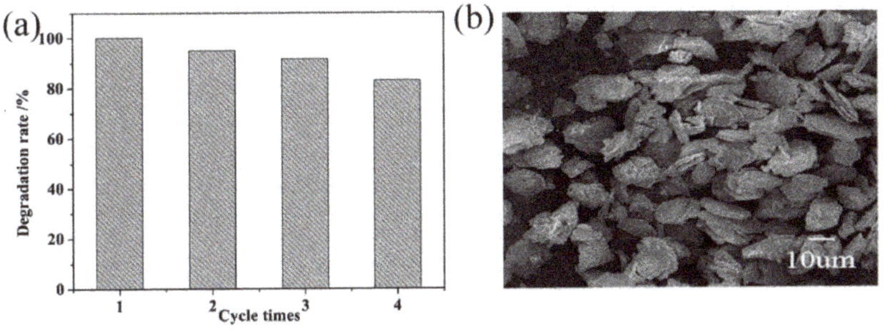

Figure 7. (**a**) Degradation efficiency of MB for KFO-20 at different cycles; (**b**) SEM image for KFO-20 after four successive cycles.

3. Materials and Methods

3.1. Materials

Analytical-grade ferric nitrate ($FeNO_3{\cdot}9H_2O$), potassium hydroxide (KOH), sodium hydroxide (NaOH), barium sulfate ($BaSO_4$), ammonium oxalate (AO), and hydrogen peroxide (H_2O_2) were purchased from Sinopharm Chemical Reagent Co., Ltd. (Changchun, China). P-benzoquinone (BQ) was supplied by Aladdin Ltd. (Changchun, China). Methylene blue (MB) and crystal violet were obtained from Xinzhong Chemical Reagent Co., Ltd. (Changchun, China). Tertiary butyl alcohol (t-BuOH) and hydrochloric acid (HCl) were acquired from Shanghai Runjie Chemical Reagent Co., Ltd. (Shanghai, China). Silver nitrate ($AgNO_3$) was purchased from Shanghai Chemical Reagent Co., Ltd. (Shanghai, China). Potassium bromide was obtained from Shanghai Macklin Biochemical Co., Ltd. (Changchun, China). All chemicals could be directly used without further treatment.

3.2. Preparation

KFO-20 was prepared using the hydrothermal process on the basis of our ever-reported method [34]. First, we added 4 g of ferric nitrate into 32 mL of deionised water to form a clarifying solution. We slowly added 60 g of potassium hydroxide into a ferric nitrate solution while stirring continuously until the colour became light brown. The mixture was directly added into a Teflon-lined stainless-steel autoclave and heated at 180 °C for 2 h. After cooling down to room temperature, the crystal was washed 10 times with deionised water until its pH was neutralised. The crystal was then dried overnight in a vacuum drying oven at 60 °C. The preparation method of $K_2Fe_4O_7$ with average sizes of 80 (KFO-80) and 180 μm (KFO-180) can be found in our supporting information.

3.3. Material Characterisation

Powder XRD experiments were performed on a Rigaku D-Max 2550 diffractometer (Tokyo, Japan) with Cu-Kα radiation (λ = 1.5418 Å). SEM images were obtained on a JEOL-6700 scanning electron microscope (Tokyo, Japan). UV-visible diffuse reflectance spectroscopy was performed with a Shimadzu UV-2450 spectrometer (Kyoto, Japan) XPS analysis was performed on a VG Scienta R3000 spectrometer (New York, N. Y., USA) with Al Kα (1486.6 eV) as the X-ray source.

3.4. MB Degradation

The degradation of MB was carried out in the presence of visible light in a cylindrical Pyrex vessel with a PL-300 xenon lamp (Beijing China) (λ > 400 nm) as the analogue sunlight source. During the experiment, the distance between light source and container was kept at 10 cm to ensure the same light intensity. The reaction vessel was equipped with a cooling device to ensure a constant reaction temperature. The initial experiment conditions were as follows: a mixture of 0.03 g of the micron scale catalyst and 100 mL of MB were stirred in dark conditions for 60 min to reach adsorption equilibrium. Then, 10 mM of H_2O_2 (30% w/v) was added into the solution, and the pH was adjusted to 2. The pH value was adjusted by adding 0.1 M HCl and 0.1 M NaOH. Then, 2 mL of the reaction solution was taken out with a 5 mL disposable syringe at regular intervals, and the catalyst in the solution was removed with a 0.2 μm filter. The concentration of MB remaining in the solution was determined by UV-visible spectroscopy. The degradation rate was calculated using Equation (15), where C_0 and C were the initial concentration and the concentration at time t, respectively.

$$Dr = C/C_0 \tag{15}$$

4. Conclusions

Hexagonal $K_2Fe_4O_7$ crystals with different sizes were successfully prepared under hydrothermal conditions; their narrow band gap was 1.44 eV. Three KFO crystals had specific surface areas of 63.79, 9.63, and 0.42 m^2/g, respectively. A small-crystal KFO-20 with a high surface area was used as a photo-Fenton catalyst for the degradation of MB and CV in the presence of green oxidant H_2O_2 under visible-light irradiation. The degradation rates of MB and CV reached 100% and 92% within 35 min, respectively. Moreover, KFO-20 was shown to have good reusability and stability in the photo-Fenton degradation of MB. The ·OH radical was dominant in the photo-Fenton catalytic reaction. Combining the KFO catalyst and carbon materials with a catalyst of nanometre size could significantly enhance heterogeneous Fenton catalytic activity and be used as an efficient and stable catalyst for the removal of dye contaminants in wastewater in the future.

Supplementary Materials: The following are available online at http://www.mdpi.com/2073-4344/10/3/293/s1, Figure S1: Powder XRD patterns for KFO-20 after four cycles.

Author Contributions: Conceptualisation, Z.G.; software, J.J.; visualisation, Y.H.; investigation, Z.L.; data curation, X.L.; writing—original-draft preparation, X.Z.; writing—review and editing, H.Y. All authors have read and agreed to the published version of the manuscript.

Funding: This research was supported by funds granted by the Natural Science Foundation of Liaoning Province of China (2019-ZD-0481) and the Project of Education Office of Liaoning Province (LQN201909).

Acknowledgments: This work was financially supported by the State Key Laboratory of Inorganic Synthesis and Preparative Chemistry, the Natural Science Foundation of Liaoning Province of China (2019-ZD-0481), and the Project of Education Office of Liaoning Province (LQN201909).

Conflicts of Interest: The authors declare no conflict of interest.

References

1. Jegatheesan, V.; Pramanik, B.K.; Chen, J.Y.; Navaratna, D.; Chang, C.Y.; Shu, L. Treatment of textile wastewater with membrane bioreactor: A critical review. *Bioresour. Technol.* **2016**, *204*, 202–212. [CrossRef] [PubMed]
2. Megala, S.; Prabhu, S.; Harish, S.; Navaneethan, M.; Sohila, S.; Ramesh, R. Enhanced photocatalytic dye degradation activity of carbonate intercalated layered Zn, ZnNi and ZnCu hydroxides. *Appl. Surf. Sci.* **2019**, *481*, 385–393. [CrossRef]
3. Phaltane, S.A.; Vanalakar, S.A.; Bhat, T.S.; Patil, P.S.; Sartale, S.D.; Kadam, L.D. Photocatalytic degradation of methylene blue by hydrothermally synthesized CZTS nanoparticles. *J. Mater. Sci. Mater. Electron.* **2017**, *28*, 8186–8191. [CrossRef]
4. Hachem, C.; Bocquillon, F.; Zahraa, O.; Bouchy, M. Decolourization of textile industry wastewater by the photocatalytic degradation process. *Dyes Pigments* **2001**, *49*, 117–125. [CrossRef]
5. Zhou, T.; Lu, X.H.; Wang, J.; Wong, F.S.; Li, Y.Z. Rapid decolorization and mineralization of simulated textile wastewater in a heterogeneous Fenton like system with/without external energy. *J. Hazard. Mater.* **2009**, *165*, 193–199. [CrossRef] [PubMed]
6. Mahamallik, P.; Pal, A. Degradation of textile wastewater by modified photo-Fenton process: Application of Co(II) adsorbed surfactant-modified alumina as heterogeneous catalyst. *J. Environ. Chem. Eng.* **2017**, *5*, 2886–2893. [CrossRef]
7. Moreira, F.C.; Boaventura, R.A.R.; Brillas, E.; Vilar, V.J.P. Electrochemical advanced oxidation processes: A review on their application to synthetic and real wastewaters. *Appl. Catal. B Environ.* **2017**, *202*, 217–261. [CrossRef]
8. Babuponnusami, A.; Muthukumar, K. A review on Fenton and improvements to the Fenton process for wastewater treatment. *J. Environ. Chem. Eng.* **2014**, *2*, 557–572. [CrossRef]
9. Gogate, P.R.; Pandit, A.B. A review of imperative technologies for wastewater treatment II: Hybrid methods. *Adv. Environ. Res.* **2004**, *8*, 553–597. [CrossRef]
10. El-Khouly, S.M.; Fathy, N. Multi-walled carbon nanotubes supported amorphous Fe_2O_3 and Ag_2O–Fe_2O_3 as Fenton catalysts for degradation of maxilon red dye. *Asia-Pac. J. Chem. Eng.* **2018**, *13*, e2184. [CrossRef]
11. Teodoro, A.; Boncz, M.Á.; Júnior, A.M.; Paulo, P.L. Disinfection of greywater pre-treated by constructed wetlands using photo-Fenton: Influence of pH on the decay of Pseudomonas aeruginosa. *J. Environ. Chem. Eng.* **2014**, *2*, 958–962. [CrossRef]
12. Guo, X.; Wang, D. Photo-Fenton degradation of methylene blue by synergistic action of oxalic acid and hydrogen peroxide with $NiFe_2O_4$ hollow nanospheres catalyst. *J. Environ. Chem. Eng.* **2019**, *7*, 102814. [CrossRef]
13. Wang, Y.B.; Zhao, H.Y.; Li, M.F.; Fan, J.Q.; Zhao, G.H. Magnetic ordered mesoporous copper ferrite as a heterogeneous Fenton catalyst for the degradation of imidacloprid. *Appl. Catal. B Environ.* **2014**, *147*, 534–545. [CrossRef]
14. Zhang, Y.M.; Zhang, N.S.; Wang, T.T.; Huang, H.T.; Chen, Y.; Li, Z.S.; Zou, Z.G. Heterogeneous degradation of organic contaminants in the photo-Fenton reaction employing pure cubic β-Fe_2O_3. *Appl. Catal. B Environ.* **2019**, *245*, 410–419. [CrossRef]

15. Hu, Z.; Oh, W.-D.; Liu, Y.; Yang, E.-H.; Lim, T.-T. Controllable mullite bismuth ferrite micro/nanostructures with multifarious catalytic activities for switchable/hybrid catalytic degradation processes. *J. Colloid. Interf. Sci.* **2018**, *509*, 502–514. [CrossRef]
16. Kuang, Q.; Yang, S. Template Synthesis of Single-Crystal-Like Porous $SrTiO_3$ Nanocube Assemblies and Their Enhanced Photocatalytic Hydrogen Evolution. *ACS Appl. Mater. Interfaces.* **2013**, *5*, 3683–3690. [CrossRef]
17. Chou, T.P.; Zhang, Q.; Russo, B.; Fryxell, G.E.; Cao, G. Titania Particle Size Effect on the Overall Performance of Dye-Sensitized Solar Cells. *J. Phys. Chem. C* **2007**, *111*, 6296–6302. [CrossRef]
18. Wang, J.; Inada, H.; Wu, L.; Zhu, Y.; Choi, Y.; Liu, P.; Zhou, W.-P.; Adzic, R.R. Oxygen Reduction on Well-Defined Core−Shell Nanocatalysts: Particle Size, Facet, and Pt Shell Thickness Effects. *J. Am. Chem. Soc.* **2009**, *131*, 17298–17302. [CrossRef]
19. Kango, S.; Kalia, S.; Celli, A.; Njuguna, J.; Habibi, Y.; Kumar, R. Surface modification of inorganic nanoparticles for development of organic–inorganic nanocomposites—A review. *Prog. Polym. Sci.* **2013**, *38*, 1232–1261. [CrossRef]
20. Verma, A.; Stellacci, F. Effect of Surface Properties on Nanoparticle–Cell Interactions. *Small* **2010**, *6*, 12–21. [CrossRef]
21. Huang, X.P.; Chen, Y.; Walter, E.; Zong, M.R.; Wang, Y.; Zhang, X.; Qafoku, O.; Wang, Z.M.; Rosso, K.M. Facet-Specific Photocatalytic Degradation of Organics by Heterogeneous Fenton Chemistry on Hematite Nanoparticles. *ACS Catal.* **2019**, *53*, 10197–10207. [CrossRef] [PubMed]
22. Chen, C.Q.; Duan, F.F.; Zhao, S.C.; Wang, W.K.; Yang, F.; Nuansingc, W.; Zhang, B.Y.; Qin, Y.; Knez, M. Porous Fe_2O_3 nanotubes with α-γ phase junction for enhanced charge separation and photocatalytic property produced by molecular layer deposition. *Appl. Catal. B Environ.* **2019**, *248*, 218–225. [CrossRef]
23. Huang, Y.; Han, C.; Liu, Y.Q.; Nadagouda, M.N.; Machala, L.; O'Shea, K.E.; Sharma, V.K.; Dionysiou, D.D. Degradation of atrazine by $Zn_xCu_{1-x}Fe_2O_4$ nanomaterial-catalyzed sulfite under UV–vis light irradiation: Green strategy to generate SO_4^{-}. *Appl. Catal. B Environ.* **2018**, *221*, 380–392. [CrossRef]
24. Rusevova, K.; Köferstein, R.; Rosell, M.; Richnow, H.H.; Kopinke, F.-D.; Georgi, A. $LaFeO_3$ and $BiFeO_3$ perovskites as nanocatalysts for contaminant degradation in heterogeneous Fenton-like reactions. *Chem. Eng. J.* **2014**, *239*, 322–331. [CrossRef]
25. Hu, Z.T.; Chen, Z.; Goei, R.; Wu, W.Y.; Lim, T.K. Magnetically recyclable Bi/Fe-based hierarchical nanostructures via self-assembly for environmental decontamination. *Nanoscale* **2016**, *8*, 12736–12746. [CrossRef]
26. Yao, Y.J.; Cai, Y.M.; Lu, F.; Wei, F.Y.; Wang, X.Y.; Wang, S.B. Magnetic recoverable $MnFe_2O_4$ and $MnFe_2O_4$-graphene hybrid as heterogeneous catalysts of peroxymonosulfate activation for efficient degradation of aqueous organic pollutants. *J. Hazard. Mater.* **2014**, *270*, 61–70. [CrossRef] [PubMed]
27. Jauhar, S.; Singhal, S.; Dhiman, M. Manganese substituted cobalt ferrites as efficient catalysts for H_2O_2 assisted degradation of cationic and anionic dyes: Their synthesis and characterization. *Appl. Catal. A Gen.* **2014**, *486*, 210–218. [CrossRef]
28. Chen, W.; Xiong, L.S.; Chen, F.X. Solvothermal synthesis of sub-200 nm Fe3O4 submicrospheres with enhanced catalytic performances by using acicular goethite as solid precursor. *Micro Nano Lett.* **2017**, *9*, 711–713. [CrossRef]
29. Zhu, J.N.; Zhu, X.Q.; Cheng, F.F.; Li, P.; Wang, F.; Xiao, Y.W.; Xiong, W.W. Preparing copper doped carbon nitride from melamine templated crystalline copper chloride for Fenton-like catalysis. *Appl. Catal. B Environ.* **2019**, *256*, 117830. [CrossRef]
30. Borthakur, S.; Saikia, L. $ZnFe_2O_4$@g-C_3N_4 nanocomposites: An efficient catalyst for Fenton-like photodegradation of environmentally pollutant Rhodamine B. *J. Environ. Chem. Eng.* **2019**, *7*, 103035. [CrossRef]
31. Huang, S.Q.; Zhang, Q.; Liu, P.Y.; Ma, S.J.; Xie, B.; Yang, K.; Zhao, Y.P. Novel up-conversion carbon quantum dots/α-FeOOH nanohybrids eliminate tetracycline and its related drug resistance in visible-light responsive Fenton system. *Appl. Catal. B Environ.* **2019**, *263*, 118336. [CrossRef]
32. Liu, Y.Y.; Jin, W.; Zhao, Y.P.; Zhang, G.S.; Zhang, W. Enhanced catalytic degradation of methylene blue by α-Fe_2O_3/graphene oxide via heterogeneous photo-Fenton reactions. *Appl. Catal. B Environ.* **2017**, *206*, 642–652. [CrossRef]

33. Wang, L.L.; Zhu, Y.; Yang, D.; Zhao, L.; Ding, H.M.; Wang, Z.H. The mixed marriage of copper and carbon ring-g-C_3N_4 nanosheet: A visible-light-driven heterogeneous Fenton-like catalyst. *Appl. Surf. Sci.* **2019**, *488*, 728–738. [CrossRef]
34. Yuan, H.M.; Li, H.; Zhang, T.S.; Li, G.H.; He, T.M.; Du, F.; Feng, S.H. A $K_2Fe_4O_7$ superionic conductor for all-solid-state potassium metal batteries. *J. Mater. Chem. A* **2018**, *6*, 8413–8418. [CrossRef]
35. Zhou, J.; Liu, X.; Hu, F.L.; Zou, H.H.; Li, X.J. A new 1-D extended vanadoborate containing triply bridged metal complex units. *Inorg. Chem. Commun* **2012**, *25*, 51–54. [CrossRef]
36. Guo, X.J.; Wang, K.B.; Li, D.; Jia, J.B. Heterogeneous photo-Fenton processes using graphite carbon coating hollow $CuFe_2O_4$ spheres for the degradation of methylene blue. *Appl. Surf. Sci.* **2017**, *420*, 792–801. [CrossRef]
37. Wongso, V.; Chen, C.J.; Razzaq, A.; Kamal, N.A.; Sambudi, N.S. Hybrid kaolin/TiO_2 composite: Effect of urea addition towards an efficient photocatalyst for dye abatement under visible light irradiation. *Appl. Clay Sci.* **2019**, *180*, 105158. [CrossRef]
38. Wolski, L.; Walkowiak, A.; Ziolek, M. Formation of reactive oxygen species upon interaction of Au/ZnO with H_2O_2 and their activity in methylene blue degradation. *Catal. Today* **2019**, *333*, 54–62. [CrossRef]
39. Wang, X.T.; Zhou, J.Q.; Zhao, S.; Chen, X.; Yu, Y. Synergistic effect of adsorption and visible-light photocatalysis for organic pollutant removal over $BiVO_4$/carbon sphere nanocomposites. *Appl. Surf. Sci.* **2018**, *453*, 394–404. [CrossRef]
40. Wang, D.J.; Shen, H.D.; Guo, L.; Wang, C.; Fu, F. Porous BiOBr/Bi_2MoO_6 Heterostructures for Highly Selective Adsorption of Methylene Blue. *ACS Omega.* **2016**, *1*, 566–577. [CrossRef]
41. Brown, I.D.; Altermatt, D. Bond-valence parameters obtained from a systematic analysis of the Inorganic Crystal Structure Database. *Acta Crystallogr. Sect. B Struct. Sci.* **1985**, *41*, 244–247. [CrossRef]
42. Keene, T.D.; D'Alessandro, D.M.; Krämer, K.W.; Price, J.R.; Price, D.J.; Decurtins, S.; Kepert, C.J. $[V_{16}O_{38}(CN)]^{9-}$: A Soluble Mixed-Valence Redox-Active Building Block with Strong Antiferromagnetic Coupling. *Inorg. Chem.* **2012**, *51*, 9192–9199. [CrossRef] [PubMed]
43. Sun, Y.J.; Yan, Y.; Wang, Y.Y.; Li, Y.; Li, J.Y.; Yu, J.H. High proton conduction in a new alkali metal-templated open-framework aluminophosphate. *Chem. Commun.* **2015**, *51*, 9317–9319. [CrossRef] [PubMed]
44. Gui, D.X.; Zheng, T.; Xie, J.; Cai, Y.W.; Wang, Y.X.; Chen, L.H.; Diwu, J.; Chai, Z.F.; Wang, S. Significantly Dense Two-Dimensional Hydrogen-Bond Network in a Layered Zirconium Phosphate Leading to High Proton Conductivities in Both Water-Assisted Low-Temperature and Anhydrous Intermediate-Temperature Regions. *Inorg. Chem.* **2016**, *55*, 12508–12511. [CrossRef] [PubMed]
45. Slade, R.C.T.; Hardwick, A.; Dickens, P.G. Investigation of H + motion in NAFION film by pulsed 1H NMR and A.C. conductivity measurements. *Solid State Ion.* **1983**, *9*, 1093–1098. [CrossRef]
46. Bernard, L.; Fitch, A.; Wright, A.F.; Fender, B.E.F.; Howe, A.T. Mechanisms of hydrogen diffusion and conduction in $DUO_2AsO_4{\cdot}4D_2O$ as inferred from neutron diffraction evidence. *Solid State Ion.* **1981**, *5*, 459–462. [CrossRef]
47. Dey, C.; Kundu, T.; Banerjee, R. Reversible phase transformation in proton conducting Strandberg-type POM based metal organic material. *Chem. Commun.* **2012**, *48*, 266–268. [CrossRef]
48. Yang, T.; Sun, J.L.; Eriksson, L.; Li, G.B.; Zou, X.D.; Liao, F.H.; Lin, J.H. $Na5[MB_{24}O_{34}(OH)_{12}]{\cdot}nH_2O$ ($M = Cr^{3+}$, Al^{3+}): Unprecedented Spherelike Polyborate Clusters from Boric Acid Flux Synthesis. *Inorg. Chem.* **2008**, *47*, 3228–3233. [CrossRef]

catalysts

MDPI

Article

Photocatalytic Oxidation of Methyl *Tert*-Butyl Ether in Presence of Various Phase Compositions of TiO_2

Marcel Šihor, Martin Reli *, Michal Vaštyl, Květoslava Hrádková, Lenka Matějová and Kamila Kočí

Institute of Environmental Technology, VŠB-Technical University of Ostrava, 17. listopadu 2172/15, 70800 Ostrava, Czech Republic; marcel.sihor@vsb.cz (M.Š.); michal.vastyl@vsb.cz (M.V.); kvetoslava.hradkova@vsb.cz (K.H.); lenka.matejova@vsb.cz (L.M.); kamila.koci@vsb.cz (K.K.)

* Correspondence: martin.reli@vsb.cz; Tel.: +420-597-327-304

Received: 21 November 2019; Accepted: 21 December 2019; Published: 26 December 2019

Abstract: MTBE (methyl *tert*-butyl ether) represents a rising threat to the environment, especially drinking water, and its effective removal (with all by-products) is necessary. Even a very low concentration of MTBE makes the water undrinkable; therefore, an effective treatment has to be developed. This work is focused on MTBE photocatalytic oxidation in presence of various TiO_2 photocatalysts with different phase composition prepared by different methods. It was confirmed the phase composition of TiO_2 had the most significant influence on the photocatalytic degradation of MTBE. The rutile phase more easily reduces adsorbed oxygen by photogenerated electrons to superoxide radical, supporting separation of charge carriers. The presence and concentrations of by-products have to be taken into account as well. The conversion of total organic carbon (TOC) was used for the comparison, 40% of TOC was removed after 1 h of irradiation in presence of TiO_2–ISOP–C/800 photocatalyst composed of anatase and rutile phase.

Keywords: photocatalysis; MTBE oxidation; TiO_2; anatase-rutile

1. Introduction

Environmental protection, especially water and air, is representing a serious challenge for current science. Transportation belongs among the pollution sources which negatively influence both air and water environment. The air pollution from transportation is well discussed, and there are various restrictions to decrease exhaust emissions as much as possible. However, the water pollution connected to transportation is not so apparent, especially to public knowledge. For example, methyl *tert*-butyl ether (MTBE), a gasoline additive, is a chemical compound used for increasing oxygen content in gasoline [1].

MTBE as fuel additive was used in 1979 for the first time. It was used to replace lead and as an octane enhancer [2]. The production of MTBE has increased ever since. The total consumption of MTBE reached 22.4 Mt annually in 2016, and this number is expected to increase to 26.5 Mt in 2021 [3]. Considering very high production levels of MTBE, it is expected to find this compound in the environment. Since the MTBE is rather easily dissolved in water, it is water sources where MTBE can be found. The problem is when drinking water is contaminated because even very low concentrations of MTBE can make drinking water undrinkable due to its offensive taste and odor. Since MTBE in drinking water has attracted attention quite recently, its harmful effects on human health after digestion are unclear [2]. However, higher concentrations of MTBE were reported to depress the nervous system, be genotoxic, irritate skin and eyes [1]. Water containing very low amount of MTBE, around 20 ppb and more, already smell like turpentine.

There are several ways how MTBE gets into the environment. Main sources of MTBE are accidental fuel leakages during transportation of storage containers and car accidents, but also unburned gasoline spilled from boats directly to surface waters [1]. As a result, MTBE is the second most commonly detected volatile organic compound in surface waters.

Currently, there are several methods for removing MTBE from water, for example, various adsorptions, air stripping, biodegradation, electrochemical oxidation, and advanced oxidation processes. All these methods are described and discussed in a review paper focused on technologies for removal of MTBE [1]. Advanced oxidation processes, especially photocatalysis proved to be a very promising method. It is the vision of very low costs for operating the technology that makes it so interesting. Photocatalysis found its way into basically all research fields during the last decade, and removal of MTBE is no exception. The photocatalytic decomposition of MTBE was studied using various semiconductor catalysts, which were modified or immobilized on the support, the catalysts were mostly based on TiO_2 [4–8] or ZnO [9–11].

This work is investigating the photocatalytic degradation of methyl *tert*-butyl ether in presence of various TiO_2 photocatalysts with different phase composition prepared by different methods.

2. Results and Discussion

The texture properties of TiO_2 photocatalysts prepared by various methods were evaluated using nitrogen physisorption (Table 1). The shape of the nitrogen adsorption/desorption isotherms of most of the photocatalysts can be categorized as IV type isotherms according the IUPAC classification [12], basically corresponding to the mesoporous materials. Based on the similarities of the shapes of isotherms, the TiO_2 photocatalysts may be divided to two groups of mesoporous materials; the first one includes TiO_2 photocatalysts prepared from titanium (IV) isopropoxide (TiO_2–ISOP–C/400, TiO_2–ISOP–PFC), the second one includes TiO_2 photocatalyst prepared from titanyl sulphate (TiO_2–TYS–C/450). TiO_2–ISOP–C/800 was not included in any of these two groups, since it is a nonporous material showing very low specific surface area (measured by Kr physisorption). The TiO_2–ISOP–C/400 and TiO_2–ISOP–PFC hysteresis loops were identified as the H2 type belonging to mesoporous adsorbents where the porous structure is complex and is not well-defined. It is evident the ISOP–based TiO_2 photocatalysts possess similarly smaller mesopores (pore width < 15 nm) (Figure 1b). However, TiO_2–ISOP–PFC shows significantly higher specific surface area and pore volume than TiO_2–ISOP–C/400 (Table 1), which may be attributed to the fact that TiO_2 nanocrystallites are less aggregated due to crystallization in pressurized fluids than under thermal treatment, and this different processing also results in different crystallinity of TiO_2–ISOP–PFC (i.e., bicrystalline anatase–brookite mixture). The TiO_2–TYS–C/450 hysteresis loop may be classified as the H3 type associated with aggregates of smaller TiO_2 nanocrystallites of broad crystallite size-distribution. Its porous structure comprises some macropores (Figure 1b). Concerning the effect of preparation parameters on TiO_2 textural properties, it is evident that the crystallization of ISOP–based TiO_2 in pressurized hot water and methanol led to lowered aggregation/sintration of TiO_2 crystallites reflected to enhanced specific surface area and well-developed porous structure of TiO_2 compared, e.g., to TiO_2–ISOP–C/400.

Table 1. Textural properties of investigated TiO_2 photocatalysts.

Photocatalyst Labeling	Physisorption		UV–Vis
	S_{BET} (m^2 g^{-1})	V_{ne} (cm^3_{liq} g^{-1})	Band Gap Energy (eV)
TiO_2–TYS–C/450	137	0.226	3.18
TiO_2–ISOP–C/400	80	0.120	3.04
TiO_2–ISOP–PFC	171	0.265	3.11
TiO_2–ISOP–C/800	0.99	—	2.90
TiO_2–P25	44	0.208	3.22

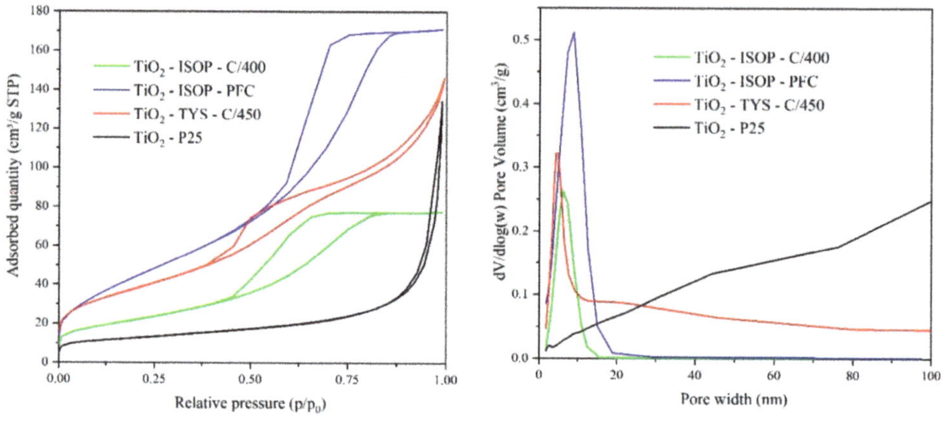

(**a**) (**b**)

Figure 1. Adsorption/desorption isotherms (**a**) and evaluated pore-size distributions (**b**) of investigated TiO_2 photocatalysts.

In order to evaluate the influence of preparation method of TiO_2 on phase composition, the XRD analysis has been conducted (Figure 2). It is clear the preparation method of TiO_2 significantly influences its phase composition (Table 2). The processing with pressurized hot fluids leads to a formation of bicrystalline phase anatase–brookite, and the calcination results in anatase phase or combination of anatase and rutile, depending on the calcination temperature.

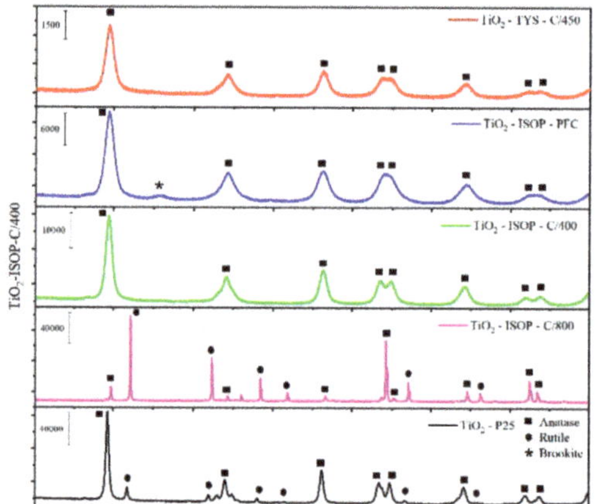

Figure 2. XRD patterns of investigated TiO_2 photocatalysts.

Optical properties were evaluated by UV–vis DRS technique, and the results are shown in Figure 3. The evaluation of band gaps of each photocatalyst was done from Tauc plots after the recalculation of reflectance according to the Kubelka–Munk function (Table 1). The indirect band gaps were evaluated for TiO_2 photocatalysts, therefore, the (K-M·hν) function has to be to power $\frac{1}{2}$ in order to obtain band gap energy. It is clear the preparation method significantly influences the band gap energy of resulting photocatalyst via its phase composition. The lowest band gap energy was 2.90 eV for TiO_2–ISOP–C/800, which contained the highest amount of rutile phase. The largest band gap energy was obtained for the commercial TiO_2–P25 (Evonik).

Table 2. Structural properties of investigated photocatalysts.

Photocatalyst Labeling	Phase Composition (wt.%)	Crystallite-Size (nm)	Facets (hkl)
TiO_2–TYS–C/450	Anatase	7.6	(101) (200)
TiO_2–ISOP–C/400	Anatase	10.3	(101) (200)
TiO_2–ISOP–PFC	79% Anatase	6.5	(101) (200)
	21% Brookite	5.2	(211)
TiO_2–ISOP–C/800	75% Anatase	112	(110) (101) (200)
	25% Rutile	356	(101) (200)
TiO_2–P25	85% Anatase	24	(110) (101) (200)
	15% Rutile	43	(101) (200)

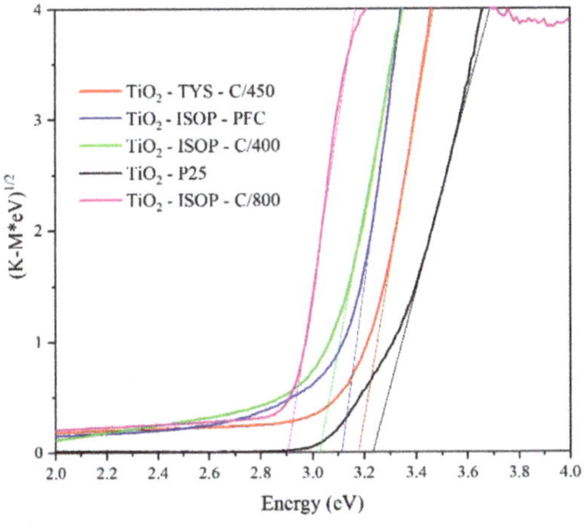

Figure 3. Tauc plots of investigated TiO_2 photocatalysts.

Photocatalysts TiO_2–TYS–C/450 and TiO_2–ISOP–C/400 were prepared by different methods, which leads to different S_{BET} and crystallite-size of anatase (Tables 1 and 2), but their morphology is almost the same (Figure 4). Nevertheless, both photocatalysts contained anatase phase only, and their photocatalytic activity toward removal of organic carbon was comparable. On the other hand, the presence of brookite phase beside anatase led to an increase of TOC removal. The increase in photocatalytic activity can be explained by formation of heterojunction between anatase and brookite phase, as was mentioned earlier [13,14]. TEM image confirmed clusters of very fine particles, which is the benefit of processing by pressurized hot fluids (Figure 4c) [15]. On the other hand, the larger crystallites and through that lower specific surface area was observed at TiO_2–P25 photocatalyst (Figure 4d).

The photocatalytic degradation of MTBE is presented in form of its decreasing concentration over time (Figure 5). MTBE belongs among the volatile organic compounds with a boiling temperature 55.2 °C, therefore time for creating equilibrium between gas and liquid phase is necessary. This is clearly evident from the blank test where the concentration of MTBE decreased during the first 75 min and then stayed more or less constant. Therefore, the 75 min time period was chosen to be sufficient for creating the equilibrium. On the other hand, when pure photolysis was conducted (no photocatalyst present) a significant decrease in MTBE concentration was detected, even after this 75 min dark time. It is clear that the strong UV irradiation itself decomposes the MTBE molecule. The highest decrease of MTBE concentration was observed in the presence of biphasic photocatalysts containing rutile phase (TiO_2–ISOP–C/800 and TiO_2–P25).

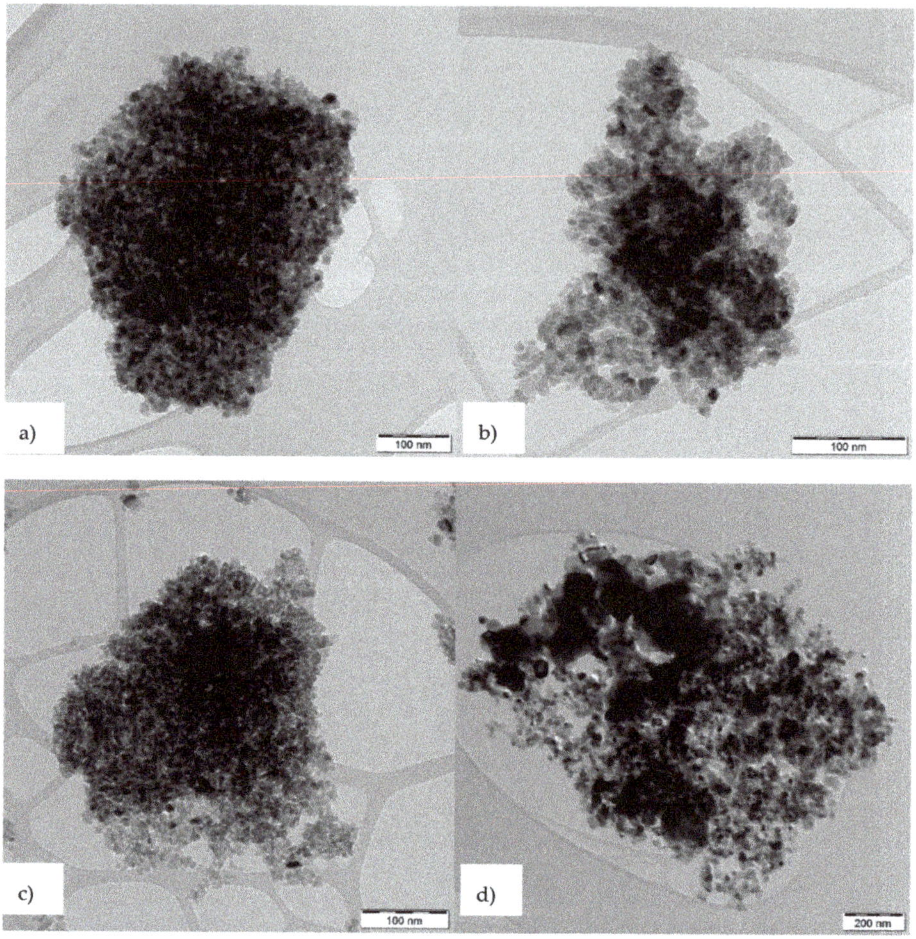

Figure 4. TEM images of (**a**) TiO_2–ISOP–C/400; (**b**) TiO_2–TYS–C/450; (**c**) TiO_2–ISOP–PFC; (**d**) TiO_2–P25.

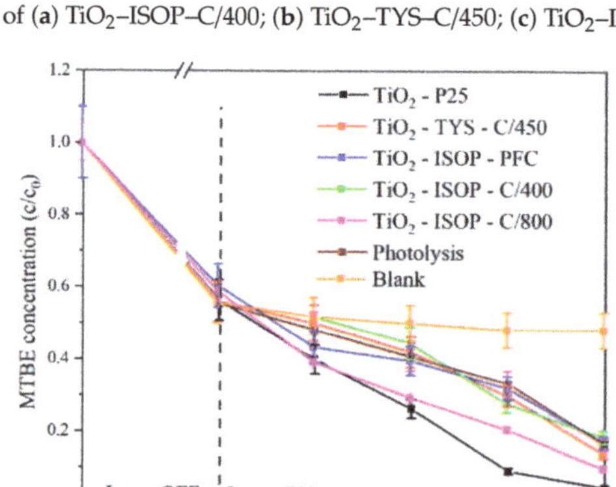

Figure 5. The dependence of MTBE (methyl *tert*-butyl ether) concentration on time over TiO_2 photocatalysts.

The degradation of organic compounds from the waste water is always a tricky one. Since the organic molecule can be oxidized in multiple ways, various by-products, sometimes even more dangerous than the original one, can be produced. There are several by-products reported during the photocatalytic degradation of MTBE, such as formic acid [16], acetaldehyde [16], acetone [17], *tert*-butyl alcohol [16–18], 2-methyl-1-propen [17], *tert*-butyl formate [17], 2-methyl propanoic acid [17], and CO_2 [16,17]. The by-products' production depends on experimental setup and can be different for various research groups.

For the above reasons, it is necessary to monitor not only the decrease of MTBE concentration, but also to analyze the formation of intermediates. Altogether, four different by-products were detected in our case. The presence of each by-product was confirmed by GC/MS and exact concentration in various time was determined the same way as MTBE concentration (SPME method coupled with GC/FID). Figure 6 shows the correlation between the concentration of each by-product and MTBE conversion for individual measurements after 60 min of irradiation. There were four different by-products recognized; acetone (AC), 2-methylprop-1-ene (MP), methyl acetate (MA), and *tert*-butyl formate (TBF). Only 2-methylprop-1-ene was detected in case of blank test, which suggests it is already present in the stock solution of MTBE. The detected by-products were also reported by other groups [17,19]. Since no change in pH was detected during the reaction, we can presume no acidic products, such as formic acid, acetic acid, or propanoic acid, were generated. The main by-product was *tert*-butyl formate. However, the concentration and ratios of by-products' concentrations varied for each photocatalyst, pointing toward the importance in photocatalyst preparation.

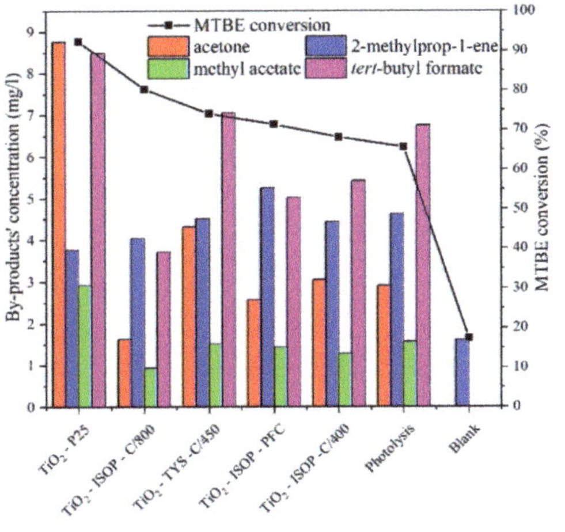

Figure 6. Correlation between by-products' concentrations and MTBE conversion for each photocatalyst, photolysis, and blank measurement after 60 min of irradiation (254 nm).

The formation of by-products in the presence of various photocatalysts is different, which is also evident from Figure 6. In order to compare the photocatalytic activity of each photocatalyst, the amount of total organic carbon (TOC) was calculated and depicted in Figure 7. All prepared photocatalysts performed higher efficiency in TOC removal in comparison with TiO_2–P25. From this point of view, the comparison of two anatase–rutile photocatalysts, i.e., TiO_2–ISOP–C/800 and TiO_2–P25, was very interesting. Even though the removal of MTBE is slightly lower in presence of TiO_2–ISOP–C/800 compared to TiO_2–P25, its photocatalytic activity is higher if we take into account removal of total organic carbon (Figure 7). It suggests that the TiO_2–ISOP–C/800 photocatalyst is more selective toward complete mineralization to CO_2 instead of partial oxidation to organic by-products.

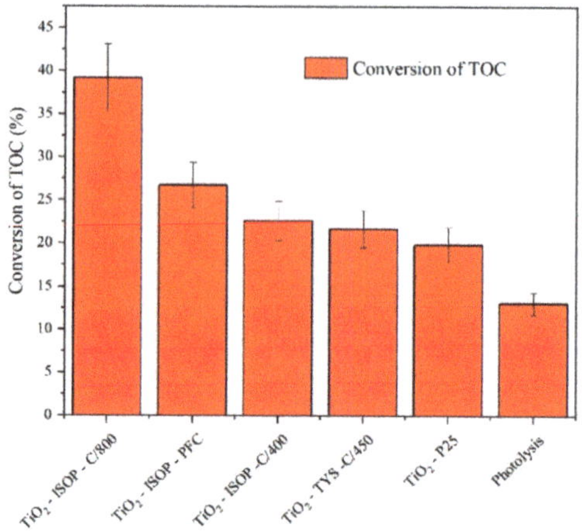

Figure 7. Conversion of total organic carbon (TOC) in presence of each photocatalyst and photolysis after 60 min of irradiation (254 nm).

The photocatalytic activity of photocatalysts depends on many factors such as phase composition, crystallite size, specific surface area, band gap energy, and so on. However, the specific reaction has to be also taken into an account. Prieto-Mahaney et al. [20] studied the correlation between structural and physical properties and photocatalytic activities for five different reactions and 35 TiO_2 samples. They studied six properties, specific surface area, density of lattice defects, primary and secondary particle size, and existence of anatase and rutile phases, to obtain intrinsic dependence of photocatalytic activities on the properties. The role of these properties was significantly depended on the type of photocatalytic reaction. Due to the fact that they did not study photocatalytic decomposition of MTBE, it is not possible compare these results with results obtained in this work. In case of MTBE oxidation, the specific surface area and the crystallite size did not play an important role. The photocatalyst with the highest photoactivity (TiO_2–ISOP–C/800) has the lowest specific surface area and the largest crystallite size. On the basis of the obtained results, we can say that the phase composition was the decisive parameter in the studied reaction. Titanium dioxide has different structures in anatase, rutile, and brookite phase. These differences strongly influence their physicochemical properties.

Our results also confirmed the importance of phase composition of TiO_2. Usually, the presence of two phases leads to a better photocatalytic activity due to the heterojunction between these two phases, which can promote the separation of photogenerated electrons and holes [21]. After absorption of photon, photogenerated electrons can migrate from phase with the higher conduction band (CB) to the phase with lower CB. At the same time, the holes from the phase with a lower valence band (VB) migrate to the phase with a higher VB. For this reason, electrons are separated on one phase of TiO_2 and holes on another one, and both, electrons and holes, can be more efficiently utilized for redox reactions [21,22].

Even though, Figure 6 shows a very slight increase in MTBE conversion in presence of photocatalysts composed solely of anatase or anatase/brookite phase compared to photolysis and photocatalysts containing anatase/rutile proved significantly higher photocatalytic activity toward MTBE oxidation (TiO_2–P25 and TiO_2–ISOP–C/800 removed 95% and 80% of MTBE, respectively), the TOC degradation is the more important factor, which must be considered (Figure 7). Why is the TOC degradation so significant? While TBF can be relatively rapidly degraded as the reaction proceeds, the AC and MA are not only persistent toward radical oxidation, but also a possible oxidation product of TBF [23,24] Due to this reason, the TiO_2–ISOP–C/800 photocatalyst may be considered as the most suitable photocatalyst for degradation of MTBE (Figure 7). This photocatalyst proved not only high conversion of MTBE, but especially the highest conversion of total organic carbon (the smallest amount of by-products). The TiO_2–P25 showed a very high activity toward removal of MTBE, however, only partial oxidation of organic pollutants was accomplished (Figure 7). In the presence of TiO_2–ISOP–C/800 photocatalyst, the smallest amount of AC (1.64 mg L^{-1}) and MA (0.95 mg L^{-1}) was formed in comparison with TiO_2–P25 (8.77 mg L^{-1} of AC and 2.93 mg L^{-1} of MA). Additionally, another of the bi-phasic photocatalysts (TiO_2–ISOP–PFC), which contained anatase and brookite phase, exhibited good selectivity to the total mineralization. In the presence of TiO_2–ISOP–PFC, 2.58 mg L^{-1} of AC and 1.44 mg L^{-1} of MA were formed. One of the reasons to explain lower photocatalytic activity of TiO_2–P25 toward TOC removal could be a smaller portion of rutile phase in comparison with TiO_2–ISOP–C/800 photocatalyst.

In addition to the formation of heterostructures and the associated reduction in the recombination of electrons and holes, the type of TiO_2 phase is important as well. The rutile phase is more stable, however, usually less photoactive than anatase. It is worth mentioning the charge recombination rate is lower in case of rutile [25]. The most significant difference between the anatase and rutile phases of TiO_2 is in concentration of adsorbed oxygen. The molecular oxygen more easily interacts with oxygen vacancies than with TiO_2 surface, and in addition to that, the interaction is much stronger for rutile than anatase [25]. This fact results in electron transfer from surface defects to an adsorbed oxygen, leading to a formation of superoxide anion (O_2^-). Superoxide anion is a reactive oxygen species, however, its utilization for oxidation of organic compounds is generally low. Nevertheless, $O_2^{\bullet-}$ is an important part in the oxidation process carried out in acidic or neutral environment. It is especially useful in degradation of phenolic compounds, hydroxyl radicals, on the other hand, better oxidize organic intermediates to CO_2 [25]. The above mentioned highlights the necessity of oxygen atmosphere (air) for the oxidation of MTBE. This was confirmed by Barreto et al. [26] who conducted the photocatalytic degradation of MTBE without oxygen as one of the blank tests, and almost no decrease in MTBE concentration was observed.

The adsorbed oxygen molecules reacted with the electron and created superoxide anion radicals (1), which can undergo a few steps, resulting in production of hydrogen peroxide (2) and (3) [16]. The hydrogen peroxide can be reduced by electrons, as well resulting in more hydroxyl radicals and, through that, increasing the conversion of MTBE (4).

$$O_2 + e^- \rightarrow O_2^{\bullet-} \tag{1}$$

$$O_2^{\bullet-} + H^+ \rightarrow HO_2^{\bullet} \tag{2}$$

$$HO_2^{\bullet} + HO_2^{\bullet} \rightarrow H_2O_2 + O_2 \tag{3}$$

$$H_2O_2 + e^- \rightarrow OH^{\bullet} + OH^- \tag{4}$$

$$OH^- + h^+ \rightarrow OH^- \tag{5}$$

Due to this fact, there are two sources of hydroxyl radical-hydrogen peroxide can be produced in the reaction mixture of two-electron transfer (Equation (6)) or water oxidation (Equation (7)) (its formation was found, etc. in [26]).

$$O_2 + 2\,e^- + 2\,H^+ \rightarrow H_2O_2 \tag{6}$$

$$2\,H_2O + 2\,h^+ \rightarrow H_2O_2 + 2\,H^+ \tag{7}$$

For the above reasons, the anatase rutile combination is more effective in oxidizing MTBE than the anatase brookite combination.

Since rutile has significantly higher ability to adsorb oxygen than anatase [25], it is clear why TiO_2–ISOP–C/800 sample has higher photocatalytic activity toward TOC conversion than all the other photocatalysts. There is the highest amount of rutile present and therefore, there is the highest separation of charge carriers, either through heterojunction or by easier reduction of adsorbed oxygen to superoxide anion radical.

Reaction Mechanism

Mohebali et al. [17] proposed two reaction path ways leading to either 2-methyl propanoic acid or acetic acid. The important thing to know is what initiates the oxidation of organic molecule; is it photogenerated hole itself or hydroxyl radical created from the water oxidation? It is well known electron–hole pair is generated after absorption of photon with sufficient energy by TiO_2 (1). In order to find out whether the reaction is initiated by hole or hydroxyl radical experiment with isopropyl alcohol was carried out (see Supplementary Materials). Isopropyl alcohol serves as hydroxyl radical scavenger, therefore, the reaction rate would decrease if hydroxyl radicals oxidize MTBE and stayed the same in case holes oxidize MTBE molecule [27]. The test was carried out repeatedly in presence of TiO_2–P25. Since the conversion of MTBE decreased from 92% (without isopropanol) to 37% (with 0.19 mL isopropanol) we can safely assume the oxidation of MTBE is initiated by hydroxyl radicals created from single hole water oxidation (2). These findings are in an agreement with Hwang et al. [27]. The most represented by-product, *tert*-butyl formate (TBF), is formed through two steps according (3).

The reduction of adsorbed oxygen by photogenerated electron helps to charge carriers' separation and lowers the recombination rate. Therefore, another experiment was conducted. This time, chloroform as superoxide anion radical scavenger was added (see Supplementary Materials). Chloroform willingly interacts with superoxide anion radicals shifting the equilibrium so more superoxide anion radicals are produced and the separation of charge carriers is even more enhanced. This was confirmed experimentally where the addition of chloroform (0.016 mL) increased the MTBE conversion to 100% from 92% in presence of TiO_2–P25.

3. Materials and Methods

3.1. Photocatalysts Preparation Method

Altogether, 4 different TiO_2 photocatalysts were prepared, using either titanyl sulphate or titanium (IV) isopropoxide as Ti-precursor depending on the used chemical method (thermal hydrolysis vs. sol-gel), and different processing (pressurized hot solvents crystallization vs. calcination). The preparation method and processing for individual TiO_2 photocatalysts is described in Table 3.

Table 3. Information about preparation of investigated TiO_2 photocatalysts.

Photocatalysts Labeling	Preparation		Processing	
	Method	Precursor	Method	Conditions
TiO_2–TYS–C/450	Thermal hydrolysis	Titanyl sulphate	Calcination	450 °C (2 h), 3 °C min^{-1}
TiO_2–ISOP–C/400	Sol-gel	Titanium (IV) isopropoxide	Calcination	400 °C (4 h), 10 °C min^{-1}
TiO_2–ISOP–PFC	Sol-gel	Titanium (IV) isopropoxide	Pressurized hot fluids crystallization	200 °C, 10 MPa, 1.5 L H_2O + 0.25 L CH_3OH + 0.1 L H_2O, 3.5–4.5 mL min^{-1}
TiO_2–ISOP–C/800	Sol-gel	Titanium (IV) isopropoxide	Calcination	800 °C (4 h), 5 °C min^{-1}

The preparations of individual TiO_2 were following:

TiO_2–TYS–C/450: The titanium precursor stock solution was prepared from titanyl sulphate monohydrate ($TiOSO_4$ H_2O). The solution was stirred with a spindle stirrer for 3 days until complete dissolution of titanyl sulphate monohydrate in water. The final concentration of stock solution was 100 g of TiO_2 L^{-1} of solution. Then the stock solution was mixed with 0.5 wt.% H_2SO_4 solution and stirred at temperature of 80 °C on a magnetic stirrer for 60 min. The solution was cooled at laboratory temperature. After cooling, a 20 wt.% NaOH solution was added until the pH of the solution rose to 7. The resulting suspension was filtered using a Buchner funnel. The collected precipitate was washed with demineralized water until the sulfates were removed. The presence of sulfates was verified using $BaCl_2$ solution. The obtained precipitate was dried in an oven at 50 °C for 24 h to constant weight. Afterwards, the precipitate was calcined in an oven at 450 °C for 2 h using a temperature ramp of 3 °C min^{-1}.

TiO_2–ISOP–C/400: Titania sol was prepared by mixing cyclohexane, Triton X-114, demineralized water, and titanium (IV) isopropoxide in molar ratio of 11:1:1:1. Firstly, cyclohexane, Triton X-114, and water were mixed and stirred for 15 min. Then titanium (IV) isopropoxide was added and the micellar sol was mixed for 30 min for homogenization. Afterwards, the sol was poured into a Petri dish and aged for 48 h on air. The gelation of sol took place. The gel was crashed to 3 × 3 mm small pieces, placed in crucibles, and calcined in an oven at 400 °C for 4 h using a temperature ramp of 10 °C min^{-1} to obtain TiO_2 powder. The powder sample was sieved to particle size < 0.160 mm.

TiO_2–ISOP–PFC: The titanium-based gel was prepared identically as mentioned in the case of TiO_2–ISOP–C/400, but the crashed gel was crystallized at 200 °C and 10 MPa, using the sequence of solvents: 1.5 L of demineralized water, 0.25 L of methanol, and 0.1 L of demineralized water with a flow rate of 3.5–4.5 mL min^{-1} to obtain TiO_2 powder. The powder sample was sieved to particle size < 0.160 mm.

TiO_2–ISOP–C/800: The titanium-based gel was prepared identically as mentioned in the case of TiO_2–ISOP–C/400, but the crashed gel was calcined in an oven at 800 °C for 4 h using a temperature ramp of 5 °C min^{-1} to obtain TiO_2 powder. The powder sample was sieved to particle size < 0.160 mm.

TiO_2–P25: Commercially available TiO_2 powder (particle size < 0.09 mm) used as a reference photocatalyst to be compared with investigated prepared photocatalysts.

3.2. *Characterization of TiO_2 Photocatalysts*

Each photocatalyst was characterized by several characterization techniques, such as nitrogen physisorption [28], powder X-ray diffraction [29], transmission electron microscopy [28], and diffuse reflectance UV-vis spectroscopy [30]. For more detail see Supplementary Materials.

3.3. *Photocatalytic Degradation of MTBE*

The annular batch photoreactor was used for the photocatalytic degradation of MTBE. The photoreactor was homemade from stainless steel and was fitted with stoppers around its inner periphery in order to achieve better mixing. Total volume of the photoreactor was 305 mL. The suspension of 0.1 g of photocatalyst in the 100 mL MTBE solution was used for the photocatalytic test. The concentration of MTBE was 70 μmol L^{-1} (52 mg L^{-1}). The mixture was stirred using magnetic stirrer at 600 rpm. The reactor was sealed in order to monitor composition of the gas phase, which was air at the beginning of the reaction. The gas sample was taken at 0 h and analyzed on GC/BID. The solution was left in the dark for 75 min in order to reach adsorption/desorption equilibrium. Reaction itself started by turning on the 8 W Hg lamp with peak intensity at 254 nm, which was placed in quartz glass tube in the axis of symmetry of the reactor. The photocatalytic test was carried out for 60 min, and samples were taken each 15 min. Liquid samples (2 mL) were taken through septum at the bottom of the reactor, and the photocatalyst was filtered out using syringe filters with quartz pre-filter. The composition of the liquid phase was analyzed using SPME method (Solid Phase Micro Extraction) and GC/FID.

4. Conclusions

MTBE represents a rising threat to the environment and its effective removal (with all the by-products) is necessary. This work investigates MTBE oxidation in presence of various TiO_2 photocatalysts prepared by different methods. The TiO_2–ISOP–C/800 photocatalyst showed significantly higher activity toward TOC removal and also high activity for MTBE conversion. Base on the results the phase composition of TiO_2 was found to be the key parameter in the photocatalytic degradation of

MTBE. Notably, the presence of the heterojunction in biphasic TiO_2 photocatalysts is profitable, because it enables better separation of electrons and holes. Furthermore, the rutile phase, unlike brookite phase, more easily reduce adsorbed oxygen by photogenerated electrons to superoxide radical, which also supports the separation of charge carriers leading to higher photocatalytic activity.

Supplementary Materials: The following are available online at http://www.mdpi.com/2073-4344/10/1/35/s1, Figure S1: Dependence of MTBE conversion on time in presence of TiO_2 P25 without and with scavengers of $OH^\bullet$ and $O_2^{\bullet -}$.

Author Contributions: Photocatalytic measurements, M.Š.; preparation of analytical method, K.H.; photocatalysts preparation, M.V. and L.M.; Writing—Original draft preparation, M.R.; Writing—Review and editing, K.K. and M.R. All authors have read and agreed to the published version of the manuscript.

Funding: This research was funded by EU structural funding in Operational Programme Research Development and Education "IET-ER", grant number CZ.02.1.01/0.0/0.0/16_019/0000853, by Ministry of Industry and Trade, grant number CZ.01.1.02/0.0/0.0/16_084/0010305 and Large Research Infrastructure "ENREGAT" supported by the Ministry of Education, Youth and Sports of the Czech Republic, grant number LM2018098.

Conflicts of Interest: The authors declare no conflict of interest.

References

1. Levchuk, I.; Bhatnagar, A.; Sillanpää, M. Overview of technologies for removal of methyl *tert*-butyl ether (MTBE) from water. *Sci. Total Environ.* **2014**, *476*, 415–433. [CrossRef] [PubMed]
2. U.S. Environmental Protection Agency (EPA). Methyl Tertiary Butyl Ether (MTBE)—Overview. Available online: https://archive.epa.gov/mtbe/web/html/faq.html (accessed on 16 October 2018).
3. Media, A. *Argus MTBE Annual 2017*; Argus Media: London, UK, 2017; pp. 1–6.
4. Park, S.E.; Joo, H.; Kang, J.W. Photodegradation of methyl tertiary butyl ether (MTBE) vapor with immobilized titanium dioxide. *Sol. Energy Mater. Sol. Cells* **2003**, *80*, 73–84. [CrossRef]
5. Zang, Y.; Farnood, R. Photocatalytic decomposition of methyl *tert*-butyl ether in aqueous slurry of titanium dioxide. *Appl. Catal. B* **2005**, *57*, 275–282. [CrossRef]
6. Orlov, A.; Jefferson, D.A.; Tikhov, M.; Lambert, R.M. Enhancement of MTBE photocatalytic degradation by modification of TiO_2 with gold nanoparticles. *Catal. Commun.* **2007**, *8*, 821–824. [CrossRef]
7. Rodríguez-González, V.; Zanella, R.; Del Angel, G.; Gómez, R. MTBE visible-light photocatalytic decomposition over Au/TiO_2 and Au/TiO_2–Al_2O_3 sol–gel prepared catalysts. *J. Mol. Catal. A Chem.* **2008**, *281*, 93–98. [CrossRef]
8. Mirhoseini, F.; Salabat, A. Removal of methyl *tert*-butyl ether as a water pollutant by photodegradation over a new type of poly(methyl methacrylate)/TiO_2 nanocomposite. *Polym. Compos.* **2018**, *39*, 1248–1254. [CrossRef]
9. Seddigi, Z.S.; Ahmed, S.A.; Bumajdad, A.; Danish, E.Y.; Shawky, A.M.; Gondal, M.A.; Soylak, M. The efficient photocatalytic degradation of methyl *tert*-butyl ether under Pd/ZnO and visible light irradiation. *Photochem. Photobiol.* **2015**, *91*, 265–271. [CrossRef] [PubMed]
10. Seddigi, Z.S.; Ahmed, S.A.; Bumajdad, A.; Gonadal, M.A.; Danish, E.Y.; Shawky, A.M.; Yarkandi, N.H. Photocatalytic degradation oftert-butyl alcohol andtert-butyl formate using palladium-doped zinc oxide nanoparticles with UV irradiation. *Desalin. Water Treat.* **2015**, 1–10. [CrossRef]
11. Klauson, D.; Gromyko, I.; Dedova, T.; Pronina, N.; Krichevskaya, M.; Budarnaja, O.; Oja Acik, I.; Volobujeva, O.; Sildos, I.; Utt, K. Study on photocatalytic activity of ZnO nanoneedles, nanorods, pyramids and hierarchical structures obtained by spray pyrolysis method. *Mater. Sci. Semicond. Process.* **2015**, *31*, 315–324. [CrossRef]
12. Sing, K.S.W.; Everett, D.H.; Haul, R.A.W.; Moscou, L.; Pierotti, R.A.; Rouquerol, J.; Siemieniewska, T. Reporting Physisorption Data for Gas/Solid Systems. *Pure Appl. Chem.* **1985**, *57*, 603–619. [CrossRef]
13. Reli, M.; Kobielusz, M.; Matějová, L.; Daniš, S.; Macyk, W.; Obalová, L.; Kuśtrowski, P.; Rokicińska, A.; Kočí, K. TiO_2 Processed by pressurized hot solvents as a novel photocatalyst for photocatalytic reduction of carbon dioxide. *Appl. Surf. Sci.* **2017**, *391*, 282–287. [CrossRef]
14. Matějová, L.; Šihor, M.; Lang, J.; Troppová, I.; Ambrožová, N.; Reli, M.; Brunátová, T.; Čapek, L.; Kotarba, A.; Kočí, K. Investigation of low Ce amount doped-TiO_2 prepared by using pressurized fluids in photocatalytic N_2O decomposition and CO_2 reduction. *J. Sol-Gel Sci. Technol.* **2017**, *84*, 158–168. [CrossRef]
15. Matějová, L.; Matěj, Z.; Fajgar, R.; Cajthaml, T.; Šolcová, O. TiO_2 powders synthesized by pressurized fluid extraction and supercritical drying: Effect of water and methanol on structural properties and purity. *Mater. Res. Bull.* **2012**, *47*, 3573–3579. [CrossRef]

16. Araña, J.; Peña Alonso, A.; Doña Rodríguez, J.M.; Herrera Melián, J.A.; González Díaz, O.; Pérez Peña, J. Comparative study of MTBE photocatalytic degradation with TiO_2 and Cu-TiO_2. *Appl. Catal. B* **2008**, *78*, 355–363. [CrossRef]
17. Mohebali, S. Degradation of methyl *t*-butyl ether (MTBE) by photochemical process in nanocrystalline TiO_2 slurry: Mechanism, by-products and carbonate ion effect. *J. Environ. Chem. Eng.* **2013**, *1*, 1070–1078. [CrossRef]
18. Hu, Q.; Zhang, C.; Wang, Z.; Chen, Y.; Mao, K.; Zhang, X.; Xiong, Y.; Zhu, M. Photodegradation of methyl *tert*-butyl ether (MTBE) by UV/H_2O_2 and UV/TiO_2. *J. Hazard. Mater.* **2008**, *154*, 795–803. [CrossRef]
19. Selli, E.; Letizia Bianchi, C.; Pirola, C.; Bertelli, M. Degradation of methyl *tert*-butyl ether in water: Effects of the combined use of sonolysis and photocatalysis. *Ultrason. Sonochem.* **2005**, *12*, 395–400. [CrossRef]
20. Prieto-Mahaney, O.-O.; Murakami, N.; Abe, R.; Ohtani, B. Correlation between Photocatalytic Activities and Structural and Physical Properties of Titanium(IV) Oxide Powders. *Chem. Lett.* **2009**, *38*, 238–239. [CrossRef]
21. Bai, S.; Jiang, J.; Zhang, Q.; Xiong, Y. Steering charge kinetics in photocatalysis: Intersection of materials syntheses, characterization techniques and theoretical simulations. *Chem. Soc. Rev.* **2015**, *44*, 2893–2939. [CrossRef]
22. Zhang, Y.; Gan, H.; Zhang, G. A novel mixed-phase TiO_2/kaolinite composites and their photocatalytic activity for degradation of organic contaminants. *Chem. Eng. J.* **2011**, *172*, 936–943. [CrossRef]
23. Huang, K.C.; Couttenye, R.A.; Hoag, G.E. Kinetics of heat-assisted persulfate oxidation of methyl *tert*-butyl ether (MTBE). *Chemosphere* **2002**, *49*, 413–420. [CrossRef]
24. Hetflejš, J.; Šabata, S.; Kuncová, G. *terc*-butylmethylether a jeho degradace oxidačními procesy. *Chem. Listy* **2007**, *101*, 1011–1019.
25. Buchalska, M.; Kobielusz, M.; Matuszek, A.; Pacia, M.; Wojtyla, S.; Macyk, W. On Oxygen Activation at Rutile- and Anatase-TiO_2. *ACS Catal.* **2015**, *5*, 7424–7431. [CrossRef]
26. Barreto, R.D.; Gray, K.A.; Anders, K. Photocatalytic degradation of methyl-*tert*-butyl ether in TiO_2 slurries: A proposed reaction scheme. *Water Res.* **1995**, *29*, 1243–1248. [CrossRef]
27. Hwang, S.; Huling, S.G.; Ko, S. Fenton-like degradation of MTBE: Effects of iron counter anion and radical scavengers. *Chemosphere* **2010**, *78*, 563–568. [CrossRef]
28. Troppová, I.; Šihor, M.; Reli, M.; Ritz, M.; Praus, P.; Kočí, K. Unconventionally prepared TiO_2/g-C_3N_4 photocatalysts for photocatalytic decomposition of nitrous oxide. *Appl. Surf. Sci.* **2018**, *430*, 335–347. [CrossRef]
29. Reli, M.; Troppová, I.; Šihor, M.; Pavlovský, J.; Praus, P.; Kočí, K. Photocatalytic decomposition of N_2O over g-C_3N_4/$BiVO_4$ composite. *Appl. Surf. Sci.* **2019**, *469*, 181–191. [CrossRef]
30. Reli, M.; Svoboda, L.; Šihor, M.; Troppová, I.; Pavlovský, J.; Praus, P.; Kočí, K. Photocatalytic decomposition of N_2O over g-C_3N_4/WO_3 photocatalysts. *Environ. Sci. Pollut. Res. Int.* **2017**, 1–12. [CrossRef]

catalysts

MDPI

Article

Enhanced Photocatalytic Ozonation of Phenol by Ag/ZnO Nanocomposites

Junmin Peng [1], Tong Lu [1], Hongbo Ming [1], Zhengxin Ding [1], Zhiyang Yu [1,2], Jinshui Zhang [1] and Yidong Hou [1,*]

1 State Key Laboratory of Photocatalysis on Energy and Environment, College of Chemistry, Fuzhou University, Fuzhou 350002, China; junminpeng@163.com (J.P.); tlutlutlu@163.com (T.L.); minghongbo@126.com (H.M.); zxding@fzu.edu.cn (Z.D.); yuzyemlab@gmail.com (Z.Y.); jinshui.zhang@fzu.edu.cn (J.Z.)

2 Xiamen Tungsten Co., LTD. Xiamen 361126, China

* Correspondence: ydhou@fzu.edu.cn; Tel.: +86-591-22865857

Received: 4 November 2019; Accepted: 27 November 2019; Published: 30 November 2019

Abstract: Ag/ZnO nanocomposites were synthesized and applied in the photocatalytic ozonation of phenol. Their crystal, textural, morphological, optical, and electrochemical properties were investigated by XRD, Raman, SEM, TEM, UV–Vis diffuse reflectance spectroscopy (DRS), X-ray photoemission spectroscopy (XPS), and photoluminescence (PL) techniques in detail. The results indicated that silver nanoparticles were well dispersed on the surface of porous ZnO and the intimate contacts were formed at the Ag/ZnO interfaces. This prominently favored the separation and transfer of photoinduced electrons from ZnO to Ag nanoparticles for the activation of ozone to produce •OH and $•O_2^-$. As a result, a significant enhancement in photocatalytic ozonation of phenol was achieved over Ag/ZnO catalysts. It also showed a synergistic effect between photocatalysis and ozonation.

Keywords: ZnO; silver; photocatalytic ozonation; phenol; synergistic effect

1. Introduction

The combination of ozone and heterogeneous photocatalysis is one of the promising advanced oxidation technologies for the decomposition of organic contaminants [1–3]. In this process, ozone can effectively capture the photogenerated electrons from the semiconductor to form ozonide radicals, which not only obviously inhibits the recombination of the electron-hole couples, but also remarkably increases the utilization efficiency of the dissolved ozone [4]. Thus, a significant enhancement in the decomposition and mineralization of organic pollutants should be realized in the process of photocatalytic ozonation. To advance the practical application of photocatalytic ozonation technology, it is highly desirable to develop high-performance catalysts that can initiate a great synergism in the combination of photocatalysis and ozonation [5–12].

As a typical semiconductor, ZnO is a promising candidate for photocatalytic ozonation reactions, owing to its good photocatalytic properties and unique function for ozonation activation [13–17]. In addition, easy synthesis, low cost, and environmental friendliness also enable the large-scale application of ZnO in the photocatalytic ozonation of organic pollutants [18]. However, owing to the rapid recombination of the photoexcited electron-hole pairs, the development of ZnO for photocatalytic ozonation reactions is greatly hindered and only moderate activity is achieved on pure ZnO for wastewater treatment [19].

Surface decoration of a semiconductor with metal nanoparticles is a facile and effective strategy to advance photocatalytic performances because the formation of heterostructure between metal and the semiconductor can greatly promote the separation and transfer of photogenerated electron–hole pairs [20–23]. In addition, the supported metal nanoparticles can work with the semiconductor to cocatalyze chemical reactions [24–26]. In this study, Ag nanoparticles were employed to decorate ZnO for photocatalytic ozonation of organic pollutants by means of their unique properties. Experimentally, Ag nanoparticles were deposited onto the ZnO surface via a photochemical approach and the resultant Ag/ZnO catalysts were well investigated by XRD, Brunauer-Emmett-Teller (BET), TEM, SEM, X-ray photoemission spectroscopy (XPS), and electrochemical measurement. Phenol was chosen as a model contaminant to assess the catalytic performance. Electron paramagnetic resonance (EPR), along with quenching experiments, was conducted to probe the role of active species in the phenol removal and mineralization. Furthermore, the possible reaction mechanism over Ag/ZnO nanocomposites was proposed.

2. Results and Discussion

2.1. Catalysts Characterization

Figure 1a displays the XRD patterns of pure ZnO and Ag/ZnO samples. The strong and sharp diffraction peaks marked with "♣" can be assigned to the hexagonal wurtzite ZnO (JCPDS File No. 36-1451). The loading of Ag on ZnO does not change its crystal structure. When the Ag content increases to a certain amount (1.5 wt%) the characteristic diffraction peaks of Ag appear at 38.1° and 44.3° (JCPDS File No. 04-0783). The peak intensity becomes stronger with the increase of Ag content, indicating that the Ag/ZnO composite is successfully prepared. Furthermore, no shift in the ZnO diffraction peak is found among the samples, implying that Ag is deposited on the surface of ZnO rather than incorporated into the lattice of ZnO. Figure 1b presents the Raman spectra of ZnO and 1.5 wt%Ag/ZnO. The pure ZnO displays several bands at 433, 377, and 327 cm^{-1}, corresponding to the E_2, A_1(TO), and A_1 modes of hexagonal wurtzite ZnO, respectively [27]. The strong E_2 peak implies the good crystallinity of the ZnO nanoparticle. After loading the Ag nanoparticles, the intensity of the characteristic bands for ZnO is greatly increased, this is probably due to the local field of the metallic Ag on the surface of ZnO [28]. The new bands between 500 and 900 cm^{-1} could originate from the enhanced Raman scattering by coupled LO-phonon-plasmon modes of ZnO with the presence of Ag nanoparticles [29].

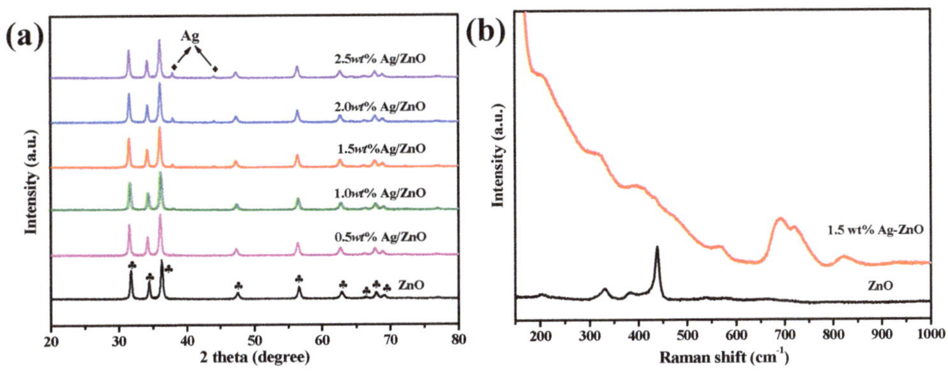

Figure 1. (**a**) XRD patterns and (**b**) Raman spectra of the samples.

SEM was taken to observe the morphology of pure ZnO and 1.5 wt% Ag/ZnO. In Figure 2, pure ZnO has a porous flake-like morphology composed of numerous aggregated nanoparticles and no obvious change in morphology is observed when ZnO is loaded with Ag. Ag/ZnO was further characterized by TEM. The porous structure is from the aggregation of ZnO nanoparticles (Figure 3a), which is in good agreement with the observation in the SEM image. The HR-TEM image (Figure 3b) clearly presents the lattice distance of 0.247 nm and 0.235 nm, corresponding to the (101) plane of

ZnO and the (111) plane of Ag [26], respectively. This means that the intimate contact between ZnO and Ag is formed, which will facilitate the separation for the photogenerated carriers of the catalyst. Meanwhile, the energy-dispersive X-ray spectroscopy (EDX) analysis was done to study the chemical constitution of Ag/ZnO. In Figure 3, Ag/ZnO comprises of three elements (O, Zn, and Ag) and the Ag element is well dispersed on the surface of ZnO. Additionally, the surface area and pore structure of ZnO and Ag/ZnO were studied by N_2-sorption analysis (Figure S1). The surface area of Ag/ZnO (20 m^2/g) is slightly lower than that of ZnO (25 m^2/g). This could be attributed to the pore structure blocking by the Ag nanoparticles.

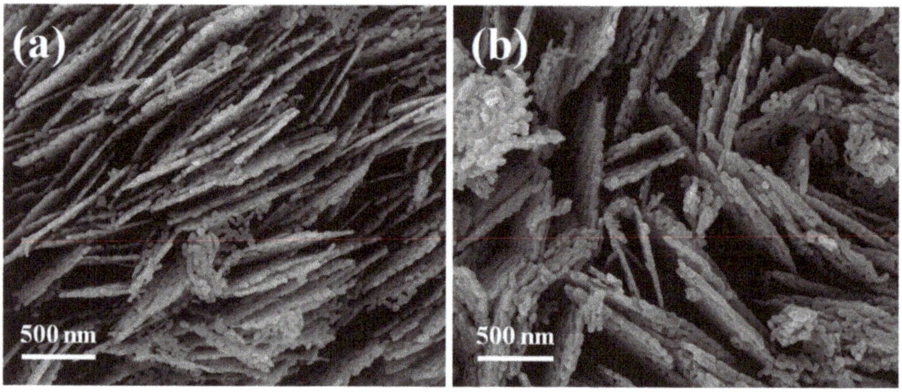

Figure 2. SEM images of (**a**) ZnO and (**b**) 1.5 wt% Ag/ZnO.

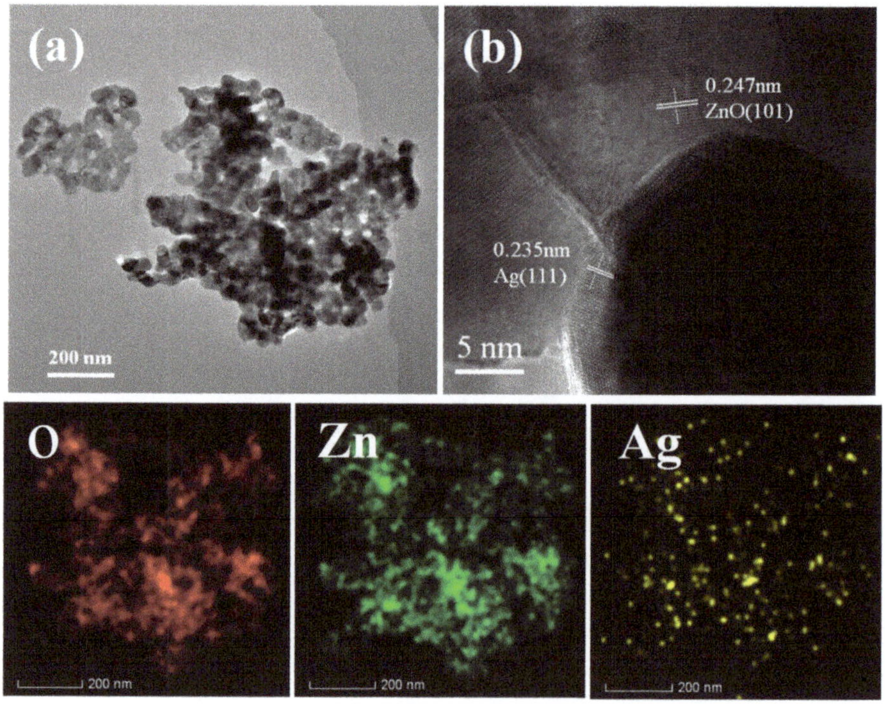

Figure 3. (**a**,**b**) TEM and energy-dispersive X-ray spectroscopy (EDX) mapping images of 1.5 wt% Ag/ZnO.

The chemical states of ZnO and Ag/ZnO were carefully examined by X-ray photoemission spectroscopy (XPS). In the survey spectra (Figure 4a), Zn, O, and C are present. ZnO, Zn, O, Ag, and C exist in the Ag/ZnO, and no other elements are observed. In Figure 4b, the Zn 2p spectra has two symmetrical peaks at 1021.4 and 1044.6 eV, corresponding to Zn 2p3/2 and Zn 2p1/2, respectively. This is indicative of the presence of Zn^{2+} in the samples, according to the reported data [30,31]. In Figure 4c, both samples present similar O1s spectra, which can be fitted into two peaks (531.7 eV and 530.2 eV), corresponding to the surface adsorbed oxygen and the lattice oxygen of ZnO [32], respectively. Figure 4d shows the Ag 3d XPS spectrum of Ag/ZnO. The two peaks centered at 373.4 and 367.4 eV correspond to Ag 3d3/2 and Ag 3d5/2, respectively. The binding energies are much lower than those of pure metallic Ag (374.2 eV and 368.2 eV), indicative of the strong interaction and charge transfer between Ag and ZnO [33–35], which will favor the separation of the photogenerated charges and suppress their recombination.

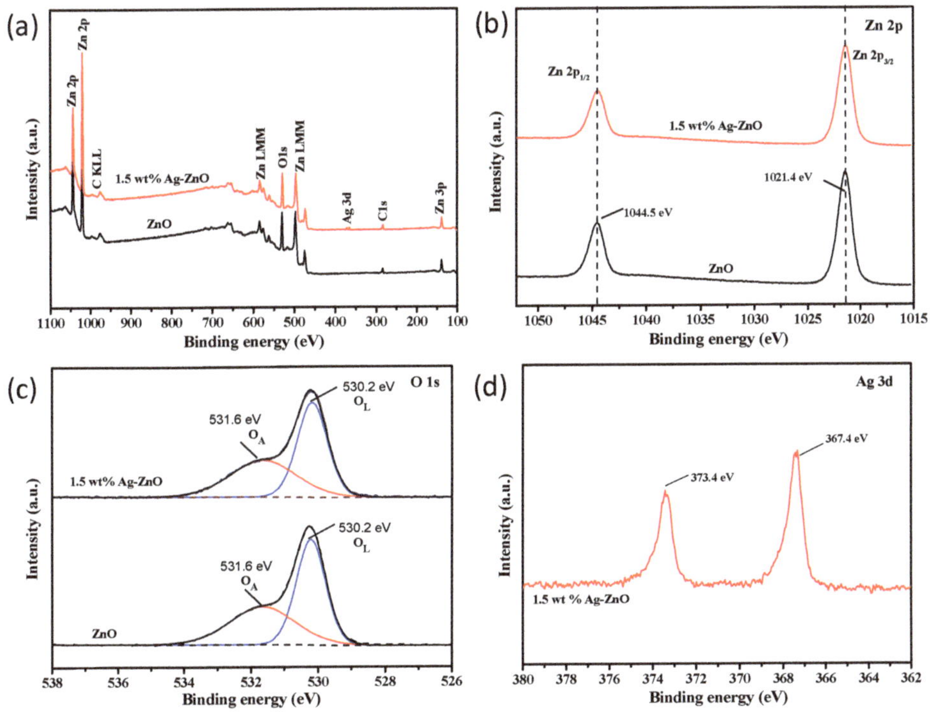

Figure 4. The X-ray photoemission spectroscopy (XPS) spectra of ZnO and 1.5 wt% Ag/ZnO, (**a**) survey, (**b**) Zn 2p, (**c**) O 1s, and (**d**) Ag 3d.

Figure 5a displays the UV-Vis diffuse reflectance spectra of ZnO and Ag/ZnO. Pure ZnO has a strong absorption in the ultraviolet region (400 nm), in accordance with the wide band-gap feature of ZnO semiconductor. In comparison to ZnO, Ag/ZnO exhibits an extra absorption band in the visible region, which can be assigned to the strong surface plasmon resonance of the metallic Ag nanoparticle [36]. This also confirms that the as-synthesized sample is composed of metallic Ag and ZnO. Photoluminescence (PL) is a useful tool to get insightful evidence about the recombination of the photogenerated electron-hole pairs, thus the PL spectra for ZnO and Ag/ZnO were carried out. As shown in Figure 5b, there is a visible emission band around 600 nm in the PL spectra. The emission band may be due to the electron transition in various kinds of defect states [37]. It can be obviously seen that Ag/ZnO has a much lower emission intensity than bare ZnO, suggesting that the recombination of photogenerated electron-hole couples is significantly suppressed [36]. This will facilitate the process involved in the participation of the photogenerated carriers in the reaction.

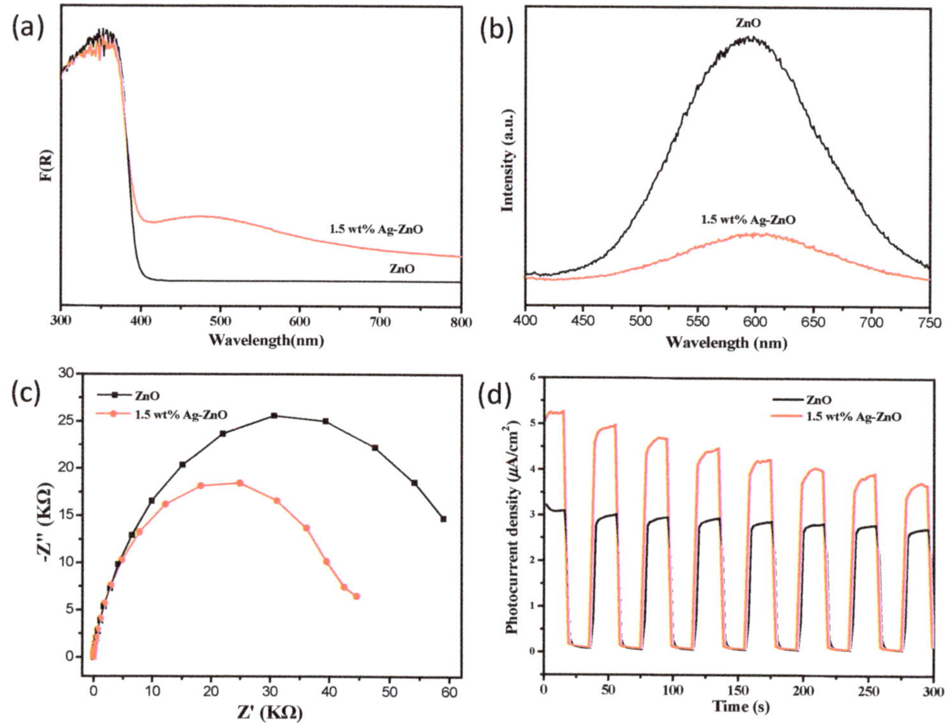

Figure 5. (**a**) The UV-Vis diffuse reflectance spectra, (**b**) the photoluminescence spectra, (**c**) electrochemical impedance spectroscopy (EIS) Nyquist plots, and (**d**) the photocurrent response curves of ZnO and 1.5 wt% Ag/ZnO.

Electrochemical impedance spectroscopy (EIS) has been commonly conducted to explore the charge transfer capability of the catalysts. The EIS Nyquist plots of ZnO and Ag/ZnO (Figure 5c) display semicircles related to the charge transfer resistance (R_{ct}) across the catalyst-electrolyte interface. A smaller semicircle arc means a lower R_{ct} and faster charge transfer. Ag/ZnO shows a much smaller radius than bare ZnO, indicating that Ag deposition can promote the interfacial electron transfer [38]. The transient photocurrent response is closely associated with the separation efficiency of the photogenerated carriers. Generally, a higher photocurrent implies a higher separation efficiency of photogenerated electron-hole pairs. Figure 5d displays the photocurrent-time curves for ZnO and Ag/ZnO during repeated ON/OFF irradiation cycles. Ag/ZnO exhibits an enhanced photocurrent response compared to ZnO, suggesting that Ag/ZnO shows an enhanced efficiency in the photogenerated charges separation [39]. This will be advantageous to boost its activity in the photocatalytic ozonation reaction.

2.2. Activity

Phenol is a typical prevalent contaminant in wastewater streams and is often resistant to conventional water treatments. Thus, phenol was used as a model compound to evaluate the catalytic activities of ZnO and Ag/ZnO. Figure 6 shows the change of the normalized concentration of phenol and total organic carbon (TOC) as a function of the reaction time over different catalysts. The phenol and TOC removals are determined to 86% and 10% in the UV-O_3 process, respectively. When ZnO was used as the catalyst in the UV-O_3 process, the phenol and TOC removals increased to 100% and 52%. Compared to pure ZnO, Ag/ZnO exhibited an enhanced performance and 1.5 wt% Ag/ZnO showed the best activity with the TOC removal of 62% in 60 min. Thus, we conducted the comprehensive comparison between 1.5 wt% Ag/ZnO and ZnO.

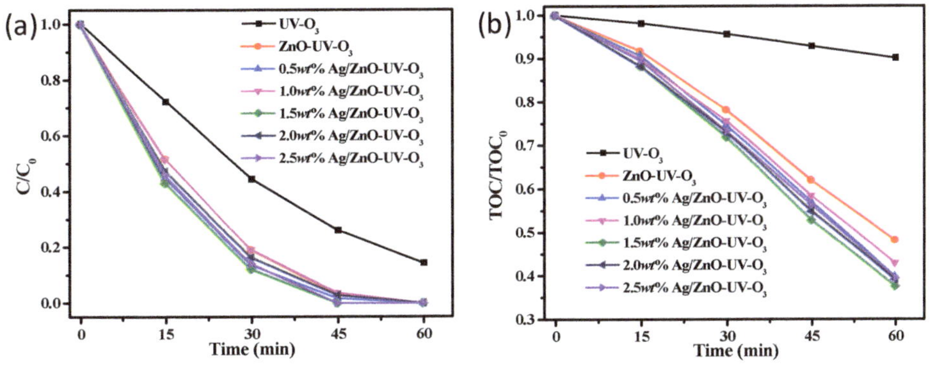

Figure 6. (**a**) The degradation and (**b**) mineralization of phenol in the photocatalytic ozonation process over ZnO and Ag/ZnO samples. Gaseous [O_3]: 10 mg/L; O_3 flow rate: 50 mL/min; catalyst dosage: 0.5 g/L; phenol concentration: 50 mg/L.

To better understand the process of photocatalytic ozonation, we have studied the activities of ZnO and Ag/ZnO in different oxidation processes (Figure 7). There is no observable removal of phenol under UV irradiation in the absence of a catalyst. However, in the presence of ZnO, a rapid degradation of phenol occurs under UV irradiation. The introduction of ozone into the photocatalytic system can further increase the degradation rate of phenol (Figure 7a). The phenol degradation data can be expressed in a linear pattern using a pseudo-first kinetics model and the deduced kinetics constants are shown in Figure 7b. In the case of the ozonation process, ZnO behaves as a good catalyst for the phenol degradation efficiency and the corresponding kinetics constant is 0.05 min^{-1}, about two times as high as that in ozone alone. The kinetics constant becomes slightly lower when ZnO is deposited with silver nanoparticles. It has been reported that Ag was an effective catalyst in decomposing ozone [40], thus the presence of silver nanoparticles might impair the utilization of ozone in the ozonation process. Interestingly, Ag/ZnO exhibits much higher photocatalytic activity than bare ZnO, which can be attributed to the higher photoinduced carrier separation by the formation of Ag/ZnO heterostructure. In the case of the combination of photocatalysis and ozonation, the kinetics constant of 0.07 min^{-1} is achieved over Ag/ZnO, which is higher than that in the ZnO-UV-O_3 system (0.055 min^{-1}). Furthermore, the kinetics constant is greatly larger than the sum of those in Ag/ZnO-UV (0.014 min^{-1}) and O_3 (0.022 min^{-1}). This enhancement in catalytic activity could be ascribed to the synergistic effects between photocatalysis and ozonation with Ag/ZnO. In addition, the Ag/Zn-UV-O_3 process exhibited better performance for phenol removal than the reported oxidation process with the ZnO-based catalyst (Table S1).

It is a vital issue whether or not the organic molecules can be completely oxidized into nontoxic CO_2 in oxidation reactions. Thus, the total of carbon (TOC)removal in different oxidation processes is also investigated in detail and the result is shown in Figure 7c,d. It exhibits a similar trend to the phenol removal in the different oxidation processes. In the system of Ag/ZnO-UV-O_3, the TOC removal rate can reach 0.4 mg/L/min, which is higher than those in Ag/ZnO-UV (0.14 mg/L/min) and O_3 (0.03 mg/L/min). These results also illustrate the synergy which occurred in the photocatalytic ozonation process. Among the different processes, Ag/ZnO-UV-O_3 exhibits the best activity for the phenol degradation and mineralization. Additionally, the stability of Ag/ZnO in the photocatalytic ozonation of phenol was investigated. In Figure 8, the removal of phenol remained at nearly 100% and the mineralization rate was kept at around 61% in the cyclic runs. Meanwhile, we also studied the Ag^+ leaching in the photocatalytic ozonation process and the leaching amount of the Ag ion was 0.004, 0.006, 0.009, and 0.004 ppm in the four cycles, respectively. No obvious change is observed in the SEM image and the EDX analysis for Ag/ZnO after the reaction (Figure S2). This indicates that Ag/ZnO exhibits robust stability in the photocatalytic ozonation of phenol. Furthermore, this catalyst can be extensively used for the removal of tetracycline hydrochloride (Figure S3).

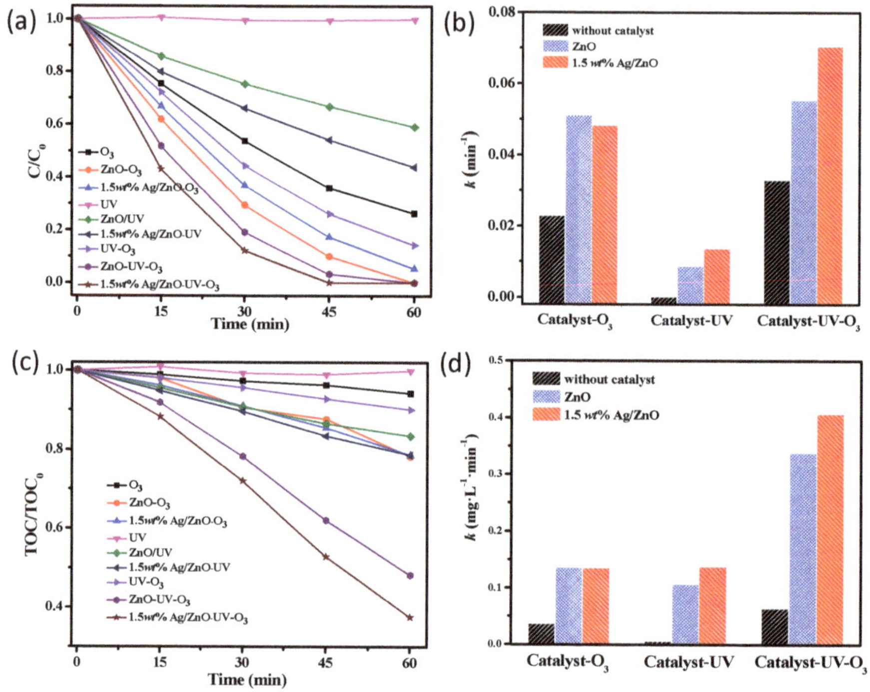

Figure 7. (**a**) The degradation and (**c**) mineralization of phenol in different oxidation processes, (**b**) the fitted kinetics degradation constants, and (**d**) TOC removal rates. Gaseous O_3 concentration: 10 mg/L; O_3 flow rate: 50 mL/min; catalyst dosage: 0.5 g/L; phenol concentration: 50 mg/L.

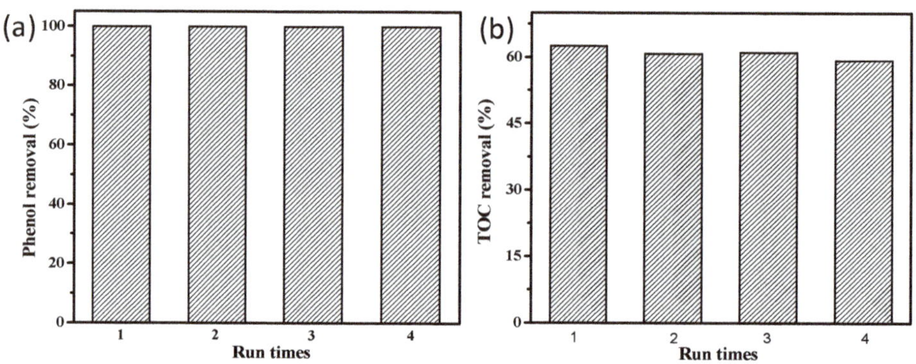

Figure 8. (**a**) The degradation and (**b**) mineralization of phenol in 60 min by a 1.5 wt% Ag/ZnO photocatalytic ozonation reaction for four cycles. Gaseous [O_3]: 10 mg/L; O_3 flow rate: 50 mL/min; catalyst dosage: 0.5 g/L; phenol concentration: 50 mg/L.

2.3. Reaction Mechanism

To monitor the active species (•OH and •O_2^-) generated in the photocatalytic ozonation process, EPR experiments were carried out. •OH detection was done in the medium of ultrapure water. In Figure 9a, a four-line spectra with the intensity ratio of 1:2:2:1, characteristic of DMPO/•OH [41,42], is observed over ZnO-UV-O_3 and Ag/ZnO-UV-O_3, and no obvious DMPO/•OH signal is found in the UV-O_3 process. It is well know that •OH has strong oxidation ability and it can deeply mineralize the organic contaminants. Moreover, the •OH produced in the Ag/ZnO-UV-O_3 system will result in a higher efficiency in the TOC removal. Methanol was used as the reaction medium instead of ultrapure water to detect •O_2^-. Figure 9b clearly displays the EPR signals of the four-line spectra with the intensity ratio of 1:1:1:1, indicative of DMPO/•O_2^- adducts [43–45]. When Ag/ZnO is employed as the catalyst, the intensity ratio is 1:1:1:1, indicative of DMPO/•O_2^- adducts [43–45]. When Ag/ZnO is employed as the catalyst, the intensity of the DMPO/•O_2^- signal is significantly increased. The produced active species also accelerate the removal and mineralization of phenol. To further verify the role of the active species in the phenol degradation over Ag/ZnO, the radical trapping experiments were carried out by using Tert-butanol (TBA) and p-benzoquinone (*p*-BQ) as the scavengers for •OH and •O_2^- [46,47], respectively. In Figure S4, the kinetics constants for phenol degradation dramatically decrease to 0.03 min^{-1} and 0.005 min^{-1} in the presence of 10 mM TBA and 10 mM p-BQ, respectively. This illustrates that both •OH and •O_2^- contribute to the decomposition of phenol, which is consistent with the EPR results.

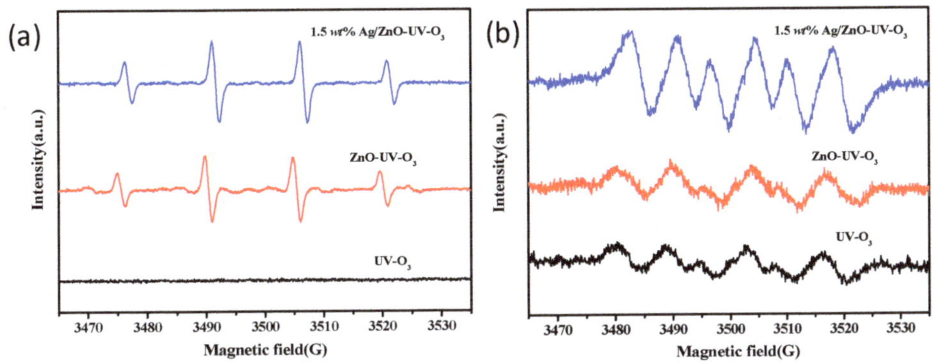

Figure 9. The electron paramagnetic resonance (EPR) spectra of DMPO/•OH (**a**) and DMPO/•O_2^- (**b**) over the UV-O_3 systems with different catalysts.

On the basis of the above results and analysis, the possible reaction mechanism for the photocatalytic ozonation of phenol over the Ag/ZnO catalyst was proposed. In Figure 10, when Ag/ZnO is irradiated with UV light, electrons are excited to the conduction band (CB) and holes are generated simultaneously in the valence band (VB). The photo-induced charges will migrate to the surface of ZnO, the holes oxidize phenol or react with H_2O to produce •OH, and the photogenerated electrons in the conduction band of ZnO will quickly transfer to Ag nanoparticles by virtue of the intimate contact between ZnO and Ag. This will significantly reduce the recombination of the photogenerated electron-hole pairs. Meanwhile, the electrons will be trapped by O_3 to produce active species such as •OH and •O_2^-, and these active species will efficiently degrade the organic pollutants and mineralize them into CO_2.

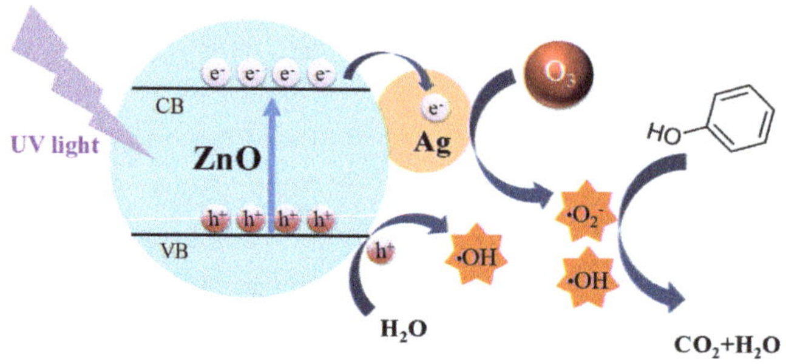

Figure 10. The proposed mechanism for the photocatalytic ozonation of phenol over Ag/ZnO.

3. Materials and Methods

3.1. Catalyst Preparation and Characterization

$Zn(CH_3COO)_2{\cdot}2H_2O$ (4.46 g) and $CO(NH_2)_2$ (1.80 g) were dissolved in 40 mL deionized water, which was transferred into a 100 mL Teflon-lined stainless steel autoclave and then heated at 120 °C for 2 h. After the autoclave naturally cooled down to room temperature, the precipitation was filtered and washed with water/ethanol and then dried at 60 °C. ZnO was obtained by heating the dry precipitation at 400 °C for 2 h. Ag/ZnO was prepared by the photodeposition method. Typically, 0.60 g of ZnO powder was dispersed in 100 mL of the mixture of methanol and water (10%/90%, *v*/*v*) containing different amounts of $AgNO_3$ under stirring. The mixed suspension was irradiated with UV light for 30 min. The precipitate was then filtered, dried, and finally heated at 300 °C for 2 h. The sample was denoted as "*xwt*% Ag/ZnO", where *x* is the theoretical value. The actual Ag loading amount was also determined by an inductively coupled plasma mass spectrometry (ICP) measurement (Table S2).

The crystal phase of the samples was identified by an X-ray Diffraction (XRD, Bruker D8 Advance, Germany) with Cu Ka radiation (λ = 1.5406 Å). The morphology and microstructure was characterized by scanning electron microscopy (SEM, Hitachi S4800, Japan) and transmission electron microscopy (TEM, FEI Talos F200S, Netherlands). Brunauer-Emmett-Teller specific surface area was determined on a Micromeritics ASAP2020 apparatus (USA). The UV-Vis diffuse reflectance spectra (DRS) of the samples were carried out on a Varian Cary 500 Scan. The X-ray photoelectron spectrum (XPS) was obtained by a PHI Quantum 2000 XPS (USA) with the C1s peak (284.6 eV) as a reference. The photoluminescence (PL) spectrum was measured on a Hitachi F900 fluorescence spectrophotometer (Japan). The electron spin resonance (ESR) spectrum was recorded using a Bruker model A300 spectrometer (Germany) to detect reactive species, with 5, 5-Dimethyl-1-pyrroline-N-oxide (DMPO) as the spin-trapping agent. The Ag leaching in the catalyst was investigated by inductively coupled plasma mass spectrometry (Avio 200, Perkin Elmer) in the photocatalytic ozonation process.

3.2. Photocatalytic Ozonation Experiments

To evaluate catalytic activity of the catalysts, all experiments were conducted in the customized tubular quartz container (diameter: 50 mm; height: 250 mm) equipped with a sampling aperture, a long straight quartz tube (φ6*1 mm) inserted into the bottom as ozone inlet. In a typical process for phenol removal, the tubular reactor contained a 160 mL suspension with 50 mg/L of phenol (pH = 6.8) and 0.5 g/L of the catalyst, surrounded by four portable UV lamps (Philips, BL 6W/10, 365 nm, Netherlands) and the gaseous ozone was continually bubbled into the reactor via the ozone inlet at a flow rate of 50 mL/min. During the run, a certain volume of the suspension was extracted from the reactor at each regular interval and then the supernate was obtained by using a 0.22 μm microfilter for the phenol and the TOC analysis. The phenol concentration was determined by high performance liquid chromatography (HPLC) with a UV detector at 270 nm. The mixture of acetonitrile and water (30%/70%, *v*/*v*) was used as the mobile phase with a flow rate of 1.0 mL/min. The total organic carbon (TOC) of the phenol solution was analyzed with a Shimadzu TOC-VCPH analyzer. Additionally, single ozonation, photocatalysis, and adsorption experiments were carried out as contrast experiments under similar conditions. To achieve the adsorption-desorption equilibrium of the phenol/catalyst, the suspension was stirred for 30 min in the dark before the reaction for all experiments.

4. Conclusions

Ag/ZnO nanocomposites were successfully synthesized by a simple photodeposition approach and the catalysts were applied in the photocatalytic ozonation of organic wastewater. In comparison to bare ZnO, Ag/ZnO showed enhanced catalytic performance in the phenol removal and its mineralization, and the highest activity was achieved over 1.5 wt%Ag/ZnO. The synergy between photocatalysis and ozonation for phenol mineralization was also observed over Ag/ZnO, and it exhibited robust stability in the cyclic runs. Mechanism studies revealed, Ag loading promotes the separation and transfer efficiency of electron-holes on the surface of ZnO by virtue of the formed heterojunction between Ag and ZnO, thus more active species (•OH and $•O_2^-$) are produced to drive the oxidation reaction for the advanced treatment of the organic pollutants. This investigation demonstrates an approach in developing efficient catalysts to advance photocatalytic ozonation for water decontamination.

Supplementary Materials: The following are available online at http://www.mdpi.com/2073-4344/9/12/1006/s1, Figure S1: The N_2 adsorption-desorption isotherms for ZnO and 1.5 wt% Ag-ZnO, Figure S2: the SEM image and EDX analysis for 1.5 wt% Ag-ZnO after the reaction, Figure S3: The mineralization of tetracycline hydrochloride in different oxidation processes, Figure S4: The radical trapping experiments in the photocatalytic ozonation process using 1.5 wt% Ag/ZnO as a catalyst, Table S1: The recent reports on ZnO-based catalysts for phenol removal, Table S2: The ICP analysis of the Ag/ZnO samples.

Author Contributions: Conceptualization, Y.H.; investigation, J.P., T.L. and H.M.; writing—original draft preparation, J.P., Z.D., J.Z. and Y.H.; writing—review and editing, Z.Y. and Y.H.

Funding: This research was funded by National Key Technologies R&D Program of China (2014BAC13B03), NSFC (U1805255 and 21972022), Natural Science Foundation of Fujian Province (2018J01681), the Award Program for Outstanding Young Innovative Talents of Fujian province (00332111), and the Independent Research Project of State Key Laboratory of Photocatalysis on Energy and Environment (SKLPEE-2017A03).

Conflicts of Interest: The authors declare no conflict of interest.

References

1. Agustina, T.E.; Ang, H.M.; Vareek, V.K. A review of synergistic effect of photocatalysis and ozonation on wastewater treatment. *J. Photochem. Photobiol. C: Photochem. Rev.* **2015**, *6*, 264–273. [CrossRef]
2. Xiao, J.D.; Xie, Y.B.; Cao, H.B. Organic pollutants removal in wastewater by heterogeneous photocatalytic ozonation. *Chemosphere* **2015**, *121*, 1–17. [CrossRef] [PubMed]
3. Mehrjouei, M.; Muller, S.; Moller, D. A review on photocatalytic ozonation used for the treatment of water and wastewater. *Chem. Eng. J.* **2015**, *263*, 209–219. [CrossRef]
4. Pan, Z.H.; Cai, Q.H.; Luo, Q.; Li, X.W. Mechanism and kinetics of H-acid degradation in TiO_2/O_3/UV process. *Can. J. Chem. Eng.* **2014**, *92*, 851–860. [CrossRef]
5. Marque, G.; Rodriguez, E.M.; Maldonado, M.I.; Alvarez, P.M. Integration of ozone and solar TiO_2-photocatalytic oxidation for the degradation of selected pharmaceutical compounds in water and wastewater. *Sep. Purif. Technol.* **2014**, *136*, 18–26. [CrossRef]
6. Liu, X.L.; Guo, Z.; Zhou, L.B.; Yang, J.; Cao, H.B.; Xiong, M.; Xie, Y.B.; Jia, G.R. Hierarchical biomimetic $BiVO_4$ for the treatment of pharmaceutical wastewater in visible-light photocatalytic ozonation. *Chemosphere* **2019**, *222*, 38–45. [CrossRef] [PubMed]
7. Yang, J.; Xiao, J.D.; Cao, H.B.; Guo, Z.; Rabeah, J.; Bruckner, A.; Xie, Y.B. The role of ozone and influence of band structure in WO_3 photocatalysis and ozone integrated process for pharmaceutical wastewater treatment. *J. Hazard. Mater.* **2018**, *360*, 481–489. [CrossRef]
8. Rey, A.; Mena, E.; Chavez, A.M.; Beltran, F.J.; Medina, F. Photocatalytic water treatment over WO_3 under visible light irradiation combined with ozonation. *Chem. Phys. Lett.* **2010**, *500*, 86–89.

9. Mano, T.; Nishimoto, S.; Kameshima, Y.; Miyake, M. Water treatment efficacy of various metal oxide semiconductors for photocatalytic ozonation under UV and visible light irradiation. *Chem. Eng. J.* **2015**, *264*, 221–229. [CrossRef]
10. Ling, Y.; Liao, G.Z.; Xu, P.; Li, L.S. Fast mineralization of acetaminophen by highly dispersed Ag/g-C_3N_4 hybrid assisted photocatalytic ozonation. *Sep. Purif. Technol.* **2019**, *216*, 1–8. [CrossRef]
11. Yin, J.; Liao, G.Z.; Zhou, J.L.; Huang, C.M.; Ling, Y.; Lu, P.; Li, L.S. High performance of magnetic $BiFeO_3$ nanoparticle-mediated photocatalytic ozonation for wastewater decontamination. *Sep. Purif. Technol.* **2016**, *168*, 134–140. [CrossRef]
12. Lee, K.M.; Lai, C.W.; Ngai, K.S.; Juan, J.C. Recent developments of zinc oxide based photocatalyst in water treatment technology: A review. *Water Res.* **2016**, *88*, 428–448. [CrossRef] [PubMed]
13. Orge, C.A.; Soares, O.; Ramalho, P.S.F.; Pereira, M.F.R.; Faria, J.L.; Salomé, G.P. Magnetic nanoparticles for photocatalytic ozonation of organic pollutants. *Catalysts* **2019**, *9*, 703. [CrossRef]
14. Zhai, X.; Chen, Z.L.; Zhao, S.Q.; He, W.; Yang, L. Enhanced ozonation of dichloroacetic acid in aqueous solution using nanometer ZnO powders. *J. Environ. Sci.* **2010**, *22*, 1527–1533. [CrossRef]
15. Yuan, X.J.; Yan, X.; Xu, H.M.; Li, D.Y.; Sun, L.; Cao, G.; Xia, D.S. Enhanced ozonation degradation of atrazine in the presence of nano-ZnO: Performance, kinetics and effects. *J. Environ. Sci.* **2017**, *61*, 3–13. [CrossRef] [PubMed]
16. Wu, J.F.; Su, T.M.; Jiang, Y.X.; Xie, X.L.; Qin, Z.Z.; Ji, H.B. In situ DRIFTS study of O_3 adsorption on CaO, gamma-Al_2O_3, CuO, alpha-Fe_2O_3 and ZnO at room temperature for the catalytic ozonation of cinnamaldehyde. *Appl. Surf. Sci.* **2019**, *412*, 290–305. [CrossRef]
17. Chen, W.H.; Liu, Q.F.; Tian, S.Q.; Zhao, X.J. Exposed facet dependent stability of ZnO micro/nano crystals as a photocatalyst. *Appl. Surf. Sci.* **2019**, *470*, 807–816. [CrossRef]
18. Lam, S.M.; Sin, J.C.; Abdullah, A.Z.; Mohamed, A.R. Degradation of wastewaters containing organic dyes photocatalysed by zinc oxide: A review. *Desalin. Water Treat.* **2012**, *41*, 131–169. [CrossRef]
19. Jing, L.Q.; Zhou, W.; Tian, G.H.; Fu, H.G. Surface tuning for oxide-based nanomaterials as efficient photocatalysts. *Chem. Soc. Rev.* **2013**, *42*, 9509–9549. [CrossRef]
20. Jing, L.Q.; Wang, D.J.; Wang, B.Q.; Li, S.D.; Xin, B.F.; Fu, H.G.; Sun, J.Z. Effects of noble metal modification on surface oxygen composition, charge separation and photocatalytic activity of ZnO nanoparticles. *J. Mol. Catal. A: Chem.* **2006**, *244*, 193–200.
21. Qi, K.Z.; Cheng, B.; Yu, J.G.; Ho, W.K. Review on the improvement of the photocatalytic and antibacterial activities of ZnO. *J. Alloys Compd.* **2017**, *727*, 792–820. [CrossRef]
22. Liu, X.Q.; Iocozzia, J.; Wang, Y.; Cui, X.; Chen, Y.H.; Zhao, S.Q.; Li, Z.; Lin, Z.Q. Noble metal-metal oxide nanohybrids with tailored nanostructures for efficient solar energy conversion, photocatalysis and environmental remediation. *Energy Environ. Sci.* **2017**, *10*, 402–434. [CrossRef]
23. Wu, A.P.; Tian, C.G.; Yan, H.J.; Hong, Y.; Jiang, B.J.; Fu, H.G. Intermittent microwave heating-promoted rapid fabrication of sheet-like Ag assemblies and small-sized Ag particles and their use as co-catalyst of ZnO for enhanced photocatalysis. *J. Mater. Chem. A* **2014**, *2*, 3015–3023. [CrossRef]
24. Xie, W.; Li, Y.Z.; Sun, W.; Huang, J.C.; Xie, H.; Zhao, X.J. Surface modification of ZnO with Ag improves its photocatalytic efficiency and photostability. *J. Photochem. Photobiol. A* **2010**, *216*, 149–155. [CrossRef]
25. Raji, R.; Sibi, K.S.; Gopchandran, K.G. ZnO: Ag nanorods as efficient photocatalysts: Sunlight driven photocatalytic degradation of sulforhodamine B. *Appl. Surf. Sci.* **2019**, *427*, 863–875.
26. Zhua, X.L.; Liang, X.H.; Wang, P.; Dai, Y.; Huang, B.B. Porous Ag-ZnO microspheres as efficient photocatalyst for methane and ethylene oxidation: Insight into the role of Ag particles. *Appl. Surf. Sci.* **2018**, *456*, 493–500. [CrossRef]
27. Wang, R.P.; Xu, G.; Jin, P. Size dependence of electron-phonon coupling in ZnO nanowires. *Phys. Rev. B* **2004**, *69*, 13303–13304. [CrossRef]
28. Zhu, G.; Liu, Y.; Xu, H.; Chen, Y.; Shen, X.; Xu, Z. Photochemical deposition of Ag nanocrystals on hierarchical ZnO microspheres and their enhanced gas-sensing properties. *CrystEngComm* **2012**, *14*, 719–725. [CrossRef]
29. Cacciato, G.; Bayle, M.; Pugliara, A.; Bonafos, C.; Zimbone, M.; Privitera, V.; Grimaldi, M.G.; Carles, R. Enhancing carrier generation in TiO_2 by a synergistic effect between plasmon resonance in Ag nanoparticles and optical interference. *Nanoscale* **2015**, *7*, 13468–13476. [CrossRef]

30. Mou, H.; Song, C.; Zhou, Y.; Zhang, B.; Wang, D. Design and synthesis of porous Ag/ZnO nanosheets assemblies as super photocatalysts for enhanced visible-light degradation of 4-nitrophenol and hydrogen evolution. *Appl. Catal. B Environ.* **2018**, *221*, 565–573. [CrossRef]
31. Liu, Y.T.; Zhanga, Q.P.; Xua, M.; Yuan, H.; Chen, Y.; Zhang, J.; Luo, K.; Zhang, J.Q.; You, B. Novel and efficient synthesis of Ag-ZnO nanoparticles for the sunlight induced photocatalytic degradation. *Appl. Surf. Sci.* **2019**, *476*, 632–640. [CrossRef]
32. Zheng, Y.H.; Zheng, L.R.; Zhan, Y.Y.; Lin, X.Y.; Zheng, Q.; Wei, K.M. Ag/ZnO heterostructure nanocrystals: Synthesis, characterization, and photocatalysis. *Inorg. Chem.* **2007**, *46*, 6980–6986. [CrossRef] [PubMed]
33. Lu, W.; Gao, S.; Wang, J. One-pot synthesis of Ag/ZnO self-assembled 3D hollow microspheres with enhanced photocatalytic performance. *J. Phys. Chem. C* **2008**, *112*, 16792–16800. [CrossRef]
34. Liu, X.; Li, W.; Chen, N.; Xing, X.X.; Dong, C.J.; Wang, Y.D. Ag-ZnO heterostructure nanoparticles with plasmon-enhanced catalytic degradation for Congo red under visible light. *RSC Adv.* **2015**, *5*, 34456–34465. [CrossRef]
35. Lai, Y.; Meng, M.; Yu, Y. One-step synthesis, characterizations and mechanistic study of nanosheets-constructed fluffy ZnO and Ag/ZnO spheres used for Rhodamine B photodegradation. *Appl. Catal. B: Environ.* **2010**, *100*, 491–501. [CrossRef]
36. Vaiano, V.; Matarangolo, M.; Murcia, J.J.; Rojas, H.; Navío, J.A.; Hidalgo, M.C. Enhanced photocatalytic removal of phenol from aqueous solutions using ZnO modified with Ag. *Appl. Catal. B: Environ.* **2018**, *225*, 197–206. [CrossRef]
37. Hasabeldaim, E.H.H.; Ntwaeaborwa, O.M.; Kroon, R.E.; Coetsee, E.; Swart, H.C. Enhanced green luminescence from ZnO nanorods. *J. Vac. Sci. Technol. B* **2019**, *37*, 011201. [CrossRef]
38. Nouri, H.; Habibi-Yangjeh, A.; Azadi, M. Preparation of Ag/ZnMgO nanocomposites as novel highly efficient photocatalysts by one-pot method under microwave irradiation. *J. Photochem. Photobiol. A* **2014**, *281*, 59–67. [CrossRef]
39. Huang, M.L.; Weng, S.X.; Wang, B.; Hu, J.; Fu, X.Z.; Liu, P. Various facet tunable ZnO crystals by a scalable solvothermal synthesis and their facet-dependent photocatalytic activities. *J. Phys. Chem. C* **2014**, *18*, 25434–25440. [CrossRef]
40. Imamur, S.; Ikebata, M.; Ito, T.; Ogita, T. Decomposition of ozone on a silver catalyst. *Ind. Eng. Chem. Res.* **1991**, *30*, 217–221. [CrossRef]
41. Wang, X.Y.; Wang, A.Q.; Ma, J. Visible-light-driven photocatalytic removal of antibiotics by newly designed C_3N_4@$MnFe_2O_4$-graphene nanocomposites. *J. Hazard. Mater.* **2017**, *336*, 81–92. [CrossRef] [PubMed]
42. Ding, J.; Dai, Z.; Qin, F.; Zhao, H.P.; Zhao, S.; Chen, R. Z-scheme $BiO_{1-x}Br/Bi_2O_2CO_3$ photocatalyst with rich oxygen vacancy as electron mediator for highly efficient degradation of antibiotics. *Appl. Catal. B: Environ.* **2017**, *205*, 281–291. [CrossRef]
43. Zheng, Y.; Yu, Z.H.; Lin, F.; Guo, F.S.; Alamry, K.A.; Taib, L.A.; Asiri, A.M.; Wang, X.C. Sulfur-doped carbon nitride polymers for photocatalytic degradation of organic pollutant and reduction of Cr(VI). *Molecules* **2017**, *22*, 572. [CrossRef] [PubMed]
44. Yu, L.H.; Zhang, X.Y.; Li, G.W.; Cao, Y.T.; Shao, Y.; Li, D.Z. Highly efficient $Bi_2O_2CO_3$/BiOCl photocatalyst based on heterojunction with enhanced dye-sensitization under visible light. *Appl. Catal. B: Environ.* **2016**, *187*, 301–309. [CrossRef]
45. Dong, Y.M.; Yang, H.X.; He, K.; Wu, X.; Zhang, A.M. Catalytic activity and stability of Y zeolite for phenol degradation in the presence of ozone. *Appl. Catal. B: Environ.* **2008**, *82*, 163–168. [CrossRef]
46. Yang, T.T.; Peng, J.M.; Zheng, Y.; He, X.; Hou, Y.; Wu, L.; Fu, X. Enhanced photocatalytic ozonation degradation of organic pollutants by ZnO modified TiO_2 nanocomposites. *Appl. Catal. B: Environ.* **2018**, *221*, 223–234. [CrossRef]
47. Li, W.J.; Li, D.Z.; Lin, Y.M.; Wang, P.X.; Chen, W.; Fu, X.Z.; Shao, Y. Evidence for the active species involved in the photodegradation process of methyl orange on TiO_2. *J. Phys. Chem. C* **2012**, *116*, 3552–3560. [CrossRef]

catalysts MDPI

Review

The Role of Catalytic Ozonation Processes on the Elimination of DBPs and Their Precursors in Drinking Water Treatment

Fernando J. Beltrán *, Ana Rey and Olga Gimeno

Departamento de Ingeniería Química y Química Física, Instituto Universitario de Investigación del Agua, Cambio Climático y Sostenibilidad, Universidad de Extremadura, 06006 Badajoz, Spain; anarey@unex.es (A.R.); ogimeno@unex.es (O.G.)

* Correspondence: fbeltran@unex.es

Abstract: Formation of disinfection byproducts (DBPs) in drinking water treatment (DWT) as a result of pathogen removal has always been an issue of special attention in the preparation of safe water. DBPs are formed by the action of oxidant-disinfectant chemicals, mainly chlorine derivatives (chlorine, hypochlorous acid, chloramines, etc.), that react with natural organic matter (NOM), mainly humic substances. DBPs are usually refractory to oxidation, mainly due to the presence of halogen compounds so that advanced oxidation processes (AOPs) are a recommended option to deal with their removal. In this work, the application of catalytic ozonation processes (with and without the simultaneous presence of radiation), moderately recent AOPs, for the removal of humic substances (NOM), also called DBPs precursors, and DBPs themselves is reviewed. First, a short history about the use of disinfectants in DWT, DBPs formation discovery and alternative oxidants used is presented. Then, sections are dedicated to conventional AOPs applied to remove DBPs and their precursors to finalize with the description of principal research achievements found in the literature about application of catalytic ozonation processes. In this sense, aspects such as operating conditions, reactors used, radiation sources applied in their case, kinetics and mechanisms are reviewed.

Keywords: ozone; catalytic ozonation; photocatalytic ozonation; disinfection by-products; natural organic matter; humic acids; drinking water; chlorination

1. Introduction

Safety of drinking water has always been a concern for humankind from at least 500 BC when natural materials were used to purify water [1]. However, it was not until the end of the Middle Age when significant steps in the treatment of water were noticed. In fact, the discovery of the microscope at the end of the 16th century was, years later in 1850, the milestone to know the reason of pandemic problems associated to the use of water. With the aid of microscopes the presence of pathogens in drinking water was discovered and the use of disinfectants would become soon later. Specifically, in 1854 chlorine was first used in London to remove bad odors coming from sewers. Although at that time the disinfectant power of chlorine was not yet well known, the role of contaminated water to spread pandemic illnesses had already been confirmed with the cholera epidemic of London [2]. In 1879 chlorine was first used as a disinfectant also for sewage and in 1903 in drinking water treatment plants [3]. Since then, chlorine has been widely used for some other operations in drinking water treatment such as taste and odor removal, keeping safe water distribution systems, biofouling control, and color removal, to cite a few. For more than 70 years, chlorine was used with great success as a water disinfectant. However, water chlorination began to be questioned after the US National Organics Reconnaissance Survey [4] that revealed the presence of halogenated compounds in 80 USA drinking water plants in 1975. This survey was based on previous studies of Rook [5] and Bellar et al. [6] in 1974 sponsored by the United States Environmental Protection

Agency (USEPA) that initiated a work on the analysis of contaminants in rivers such as the Mississippi River at New Orleans [7]. In these works, six main halogenated compounds were identified: Four trihalomethanes ($CHCl_3$, $CHCl_2Br$, $CHClBr_2$, and $CHBr_3$), CCl_4 and 1,2-dichloroethane [8]. Soon after, many more halogenated compounds, both of volatile and non-volatile nature, were detected in chlorine disinfected water and municipal wastewater secondary effluents. These compounds were not previously present in the untreated waters. For instance, Glaze and Henderson, in 1975, [9] detected in a municipal secondary effluent 39 halogenated compounds of aromatic and aliphatic type with one to at least four chlorine atoms in their molecules to give a total 3 to 4 mg L^{-1} of Cl. These halogenated compounds took the name of disinfection byproducts (DBPs) and the concern about their possible toxic character prompt the development of four questions and subsequent research lines: Which were the organic precursors of DBPs? what kind of alternative disinfectants could be applied? which different groups of DBPs were formed? and what toxicity these compounds have? A brief description about the findings of these research lines is presented below.

2. Nature of Chlorinated DBPs Precursors

There are multiple and variable natural substances present in surface waters that mainly come from plant degradation and animal wastes. Most of them are constituted by macromolecules containing numerous aromatic and aliphatic structures to which other simpler compounds, both of natural or anthropogenic origin, can be linked such as sugars, amino acids, metals, or pesticides [10,11].

These macromolecules are called humic substances that are mainly formed by three fractions: Humic and fulvic acids and humin [12,13]. Humin is not soluble in water and the acid fractions can be separated after precipitation of humic acid at pH = 1. They are mostly colored from yellow to black and of high molecular weight. Up to 90% of total dissolved organic carbon (DOC) in surface waters is due to humic substances with fulvic acid as the major contributing fraction with about 80%. In surface waters, DOC can reach up to 60 mg L^{-1} but the usual concentration ranges from 1 to 6 mg L^{-1} [14]. Figure 1 shows an example of this type of macromolecules where it can be seen the presence of polyphenol structures.

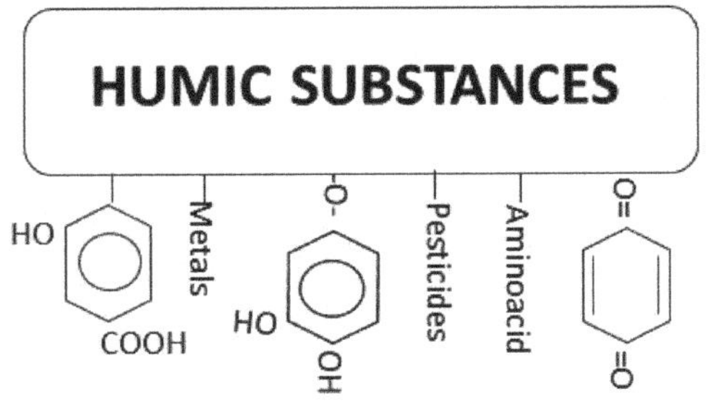

Figure 1. General structure of humic acids.

These polyphenol structures contain nucleophilic points where electrophilic substances such as chlorine can react. These reactions are responsible of halogenated compounds formation [15]. For instance, the molecule of resorcinol that can be found forming part of some humic macromolecule reacts with chlorine as shown in Scheme 1:

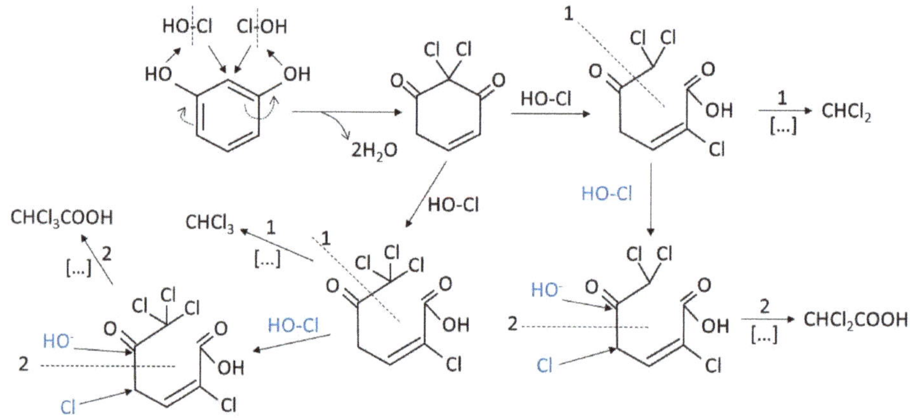

Scheme 1. Mechanism of halogenated compounds formation from the reaction of hypochlorous acid with resorcinol in water: (Way 1): Volatile Chloroform and dichloroethane and (way 2): Non-volatile dichloro and trichloroacetic acids.

As it is seen from Scheme 1 the electrophilic agent (chlorine or hypochlorous acid) attacks the ortho position in the benzene molecule with respect to both hydroxyl groups. This position is strongly activated by the hydroxyl groups becoming a nucleophilic point. Also, chlorine reacts with simpler molecules such as ketones through the well-known haloform reaction [16] shown in Scheme 2.

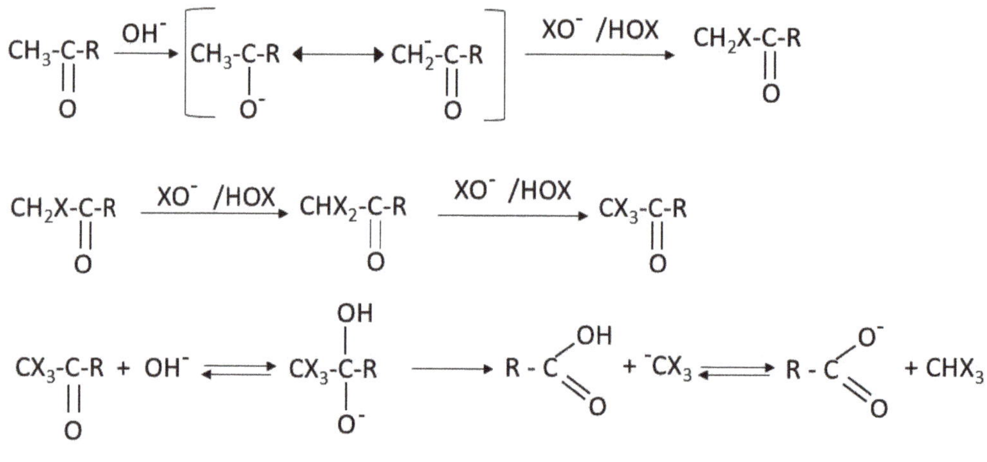

Scheme 2. Reaction of haloform formation (from [16]).

It should be noted that there are other compounds formed during the treatment of water when oxidants or disinfectants (chlorine, ozone, advanced oxidation processes, etc.) are applied. Precursors of these compounds are other pollutants, most of them of anthropogenic origin, such as pesticides, polynuclear aromatics, pharmaceutical compounds, and products of personal care [17–19]. These reaction products are also called DBPs because the reactive agent is a disinfectant. In this work, however, only DBPs coming from disinfection of drinking waters originated in reactions of disinfectants with humic and fulvic acids or generally called natural organic matter (NOM) are considered [20].

3. Alternative Disinfectants to Chlorine

The first possible disinfectant agents to be applied in water treatment alternatives to chlorine were ozone ([21]) and UV radiation. Table 1 shows the oxidation potential of main oxidants that can be used in drinking water treatment and their relative value with respect to that of ozone.

Table 1. Oxidation power of some oxidizing-disinfectants of drinking water [1].

Oxidant-Disinfectant	Oxidation Potential, V	Relative Oxidation Power [2]
Ozone	2.07	1.00
Hydrogen peroxide	1.77	0.86
Potassium permanganate	1.49	0.72
Hypochlorous acid	1.49	0.72
Chlorine	1.36	0.66
Hypobromous acid	1.33	0.64
Chlorine dioxide	1.28	0.62
Monochloramine	1.16	0.56

[1] At 25 °C, relative to hydrogen electrode. [2] Based on ozone.

It can be seen that ozone has the highest oxidizing power among these disinfectants and its oxidation potential is 1.39, 1.52, 1.62, and 1.78 times more oxidant than hypochlorous acid, chlorine, chlorine dioxide, and monochloramine, respectively [22]. These values are in accordance with the CT parameter that gives a measure of the disinfectant power. CT is the product of the disinfectant concentration and the detention time [23]. This reaction time is that needed for the 10% of the water flow through the disinfection contactor be in contact with the indicated disinfectant concentration. CT varies with the nature of pathogen, disinfectant, and disinfection level to be reached. Accordingly, the lower the concentration or the detention time the higher the disinfection power. Table 2 shows, as an example, CT values for the four drinking water chemical oxidants to remove from water two logs or 99% of some microorganisms. It can be seen that in all cases ozone presents the lowest CT values with monochloramine as the weakest disinfectant [23]. Below some comments about chlorine alternative oxidant-disinfectants are given with special emphasis on ozonation.

Table 2. CT values of drinking water disinfectants to inactivate 99% of some pathogens [1].

Pathogen	Chlorine	Chlorine Dioxide	Chloramine	Ozone
E. Coli	0.034–0.05	0.4–0.75	95–180	0.02
Rotavirus	0.01–0.05	0.2–2.1	3810–6480	0.006–0.05
G. Lambia Cyst	47–150	-	-	0.5–0.6
G. Muris	30–630	7.2–18.5	1400	1.8–2.0

[1] pH between 6 and 7 except for chloramines with pH between 8 and 9 [23]. CT values in mg L^{-1}min.

3.1. Ozone

Ozone, similar to chlorine, was first used as bactericide at the end of the 19th century. The first drinking water treatment plant (DWTP) using ozone as disinfectant was in the small city of Oudshorrn in Holland in 1893. After that, ozone was applied, always for disinfection purposes in other cities, such as Wiesbaden, Germany, in 1901, but the first important city using ozone was Nice, France, in 1906. This plant, called Bon Voyage, has been using this oxidant-disinfectant since then [21]. Application of ozone in DWTP, however, experienced a great boost after trihalomethanes (THMs) discovery in chlorinated drinking water and, also, years later, from 1987, when Glaze et al. defined the advanced oxidation processes (AOPs) after studying the combination of ozone, hydrogen peroxide and UVC radiation to yield hydroxyl free radicals [24]. During the second part of the 20th century most of the fundamentals of ozone chemistry in water had already been uncovered [25]; and consequently, its possible application for water treatment [23]. Ozone can react with DOM through two possible routes: Direct reactions on nucleophilic points or addition reactions on unsaturated moieties (double and triple carbon bonds) that form H_2O_2:

$$O_3 + \text{Nucleophilic points}/\text{–C} = \text{C–} \xrightarrow{\text{Direct reactions}} H_2O_2 + \text{Intermediates} \quad (1)$$

At high pH it predominates the indirect reaction or reaction with hydroxyl radicals formed in ozone decomposition [26].

$$O_3 + OH^- \rightarrow \text{free radicals} \rightarrow HO^\bullet \Rightarrow HO^\bullet + \text{Organics} \xrightarrow{\text{Indirect reactions}} \text{Intermediates} \quad (2)$$

Many ozone reactions with humic substances are of the electrophilic substitution type. These reactions are faster than those with chlorine because the highest oxidizing power and oxidation kinetics of/with ozone [27–29]. In Schemes 3 and 4, the Criegge mechanism, the reaction of ozone with a double carbon bond, and an electrophilic aromatic substitution reaction of ozone with the carbon hydrogen bond in ortho position with respect to the hydroxyl group of a phenol molecule are, respectively, shown [30].

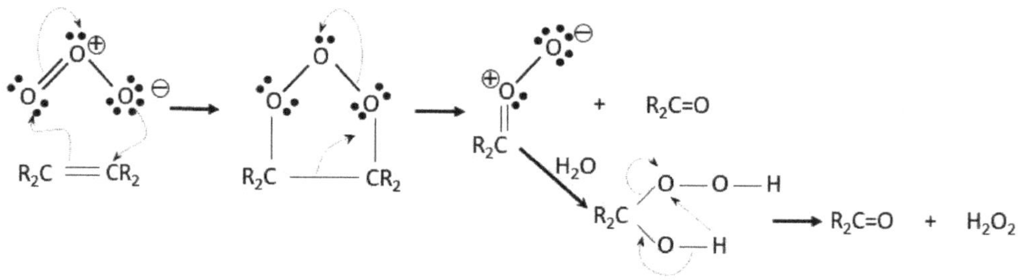

Scheme 3. Cycloaddition of ozone to a carbon double bond.

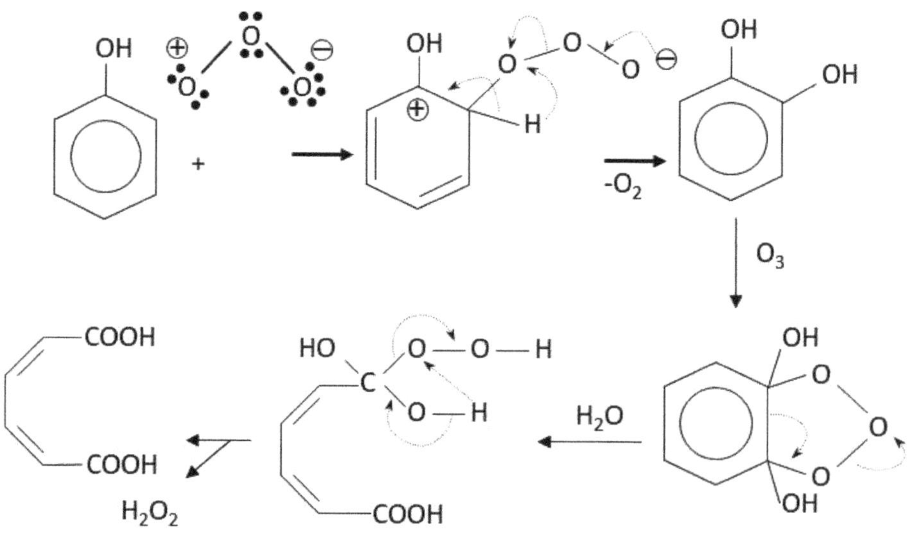

Scheme 4. Electrophilic aromatic substitution reaction of O_3 with phenol.

These are selective fast reactions of ozone with unsaturated compounds and nucleophilic points of organics. Furthermore, in these reactions hydrogen peroxide is formed. The reaction of the ionic form of this oxidant with ozone constitutes the main initiation step of hydroxyl radical formation mechanism in ozone involving advanced oxidation processes [27]. Regarding Scheme 4, after hydroxyl substitution in the phenol molecule, aromatic ring breaking is followed due to the cycloaddition of another ozone molecule and, finally, other ozone addition reactions to unsaturated carboxylic acids formed lead to low molecular weight saturated carboxylic acids. These acids, aldehydes and, in some cases, ketones are usually the end products of ozone reactions which are not precursors of halogenated compounds with the exception of some ketones. Then, previous utilization of ozone to chlorination allows significant reduction of trihalomethane formation potential (THMFP) or halogenated compounds formation potential (TOXFP). In case of high alkalinity, hydroxyl radical oxidation takes place to also yield similar byproducts that direct ozone reactions. Another advantage of ozone is its high capacity for being combined with other agents (oxidants, catalysts, and/or radiation) to yield more hydroxyl radicals, what are called ozone AOPs. This has implications even on THMs or TOX removal since ozone alone is not reactive with many halogenated compounds. On the contrary, hydroxyl radicals generated in ozone AOPs can react with THMs or TOX. However, some ozone DBPs can also be harmful, for instance, when water contains bromide ion. In this case, bromoform and hypobromous acid are formed and, eventually, depending on the conditions, bromate ion, an important priority pollutant. Then, the levels of bromide in surface water needs surveillance if ozone is used for THMFP and TOXFP control. Some considerations about bromate control have been specifically reviewed in Section 9. Also, the ozone dose has to be controlled because if added very low, it could increase the precursors concentration due to partial oxidation of first TOX precursors. For instance, water samples containing 1 mg L^{-1} 1,3-cyclohexanedione, a DBP precursor, treated with an ozone dose of 30 mg L^{-1} during 15 and 30 min and then diluted 100 times and chlorinated (1 mg L^{-1} Cl_2 dose) led to $CHCl_3$ concentrations of 77 and 15 µg L^{-1}, respectively, while without preozonation, $CHCl_3$ formation was 50 µg L^{-1} [31]. Ozone added at very high dose could form bromoform and other brominated organics such as bromohydrines [32]. In spite of its faster kinetics through two possible routes and high disinfectant and oxidizing power application of ozone also present another drawback: It cannot be used as residual disinfectant and chlorine or another residual disinfectant has to be used at the end of a DWTP. In any case, the residual disinfectant dose needed in a DWTP with a pre-ozonation step is much lower than in a conventional process without pre-ozonation, especially when a biological activated carbon step is placed after ozonation. For example, Langlais et al. [23] reported that chlorine demand was reduced from 4.5 mg L^{-1} to 3.5 and 2.7 mg L^{-1} when water from St. Rose treatment plant in Laval (Canada) was treated with rapid sand filtration, rapid sand filtration and ozonation, and rapid sand filtration plus ozonation and biological activated carbon, respectively. The savings in chlorine dose were in some case comparable to the cost of ozonation [23].

3.2. Chlorine Dioxide

Chlorine dioxide started to be used in DWTP in 1944, that is, much later than chlorine or ozone. The first application of chlorine dioxide in DWTP was for taste and odor control in Niagara Falls, State of New York. Then, it was most commonly used for bleaching in textile industries [21]. Depending on the way of synthesis, concentration and presence or not of chlorine, chlorine dioxide may lead to THMs or TOX formation while reacting with humic substances. Chlorine dioxide when used alone no THMs or TOX is formed but accompanied by chlorine, which could be formed in its preparation method, halogenated compounds can appear in the treated water. In a detailed studied, Rav-Acha [33] already described the main reactions and mechanisms of chlorine dioxide with many organics including humic substances. He concluded that since the main way of chlorine dioxide reaction with these substances is as one electron acceptor, its reactions a more selective and mainly yield quinones, hydroquinones and low molecular weight aldehydes and carboxylic acids. For instance, Scheme 5 shows a proposed mechanism of reaction of chlorine dioxide and phenolate, also a precursor molecule of TOX [33].

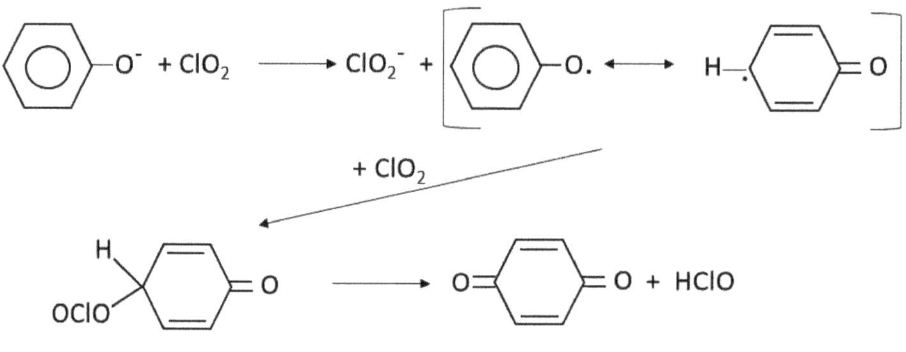

Scheme 5. Reaction of chlorine dioxide and phenol, from [23].

It is shown that the final products are p-benzoquinone and hypochlorous acid but some amount of chlorite is also formed. The presence of hypochlorous acid is likely responsible of TOX formation when precursor molecules are in excess [34]. More information about chlorine dioxide DBPs and toxicity is given in Sections 4 and 5.

3.3. Chloramines

Chloramines are formed in the reaction of chlorine with amine in a three consecutive steps where mono, di and trichloramine are formed:

$$HClO + NH_3 \rightarrow NH_2Cl + H_2O \tag{3}$$

$$HClO + NH_2Cl \rightarrow NHCl_2 + H_2O \tag{4}$$

$$HClO + NHCl_2 \rightarrow NCl_3 + H_2O \tag{5}$$

For practical cases, it is monochloramine the species used as disinfectant. The way of chloramine formation, from addition of chlorine to amine or vice versa, already present in water or by adding a previously prepared monochloramine solution to the water, highly affects the appearing of chlorinated byproducts because of the free chlorine remaining in the first two cases [35]. However, even in the case of monochloramine solution application to water, hypochlorite (at alkaline pH) or hypochlorous acid (at acid pH) eventually will appear in water due to monochloramine hydrolysis. Then, using monochloramine as disinfectant always leads to chlorinated organics. Monochloramine is the less effective disinfectant and oxidant species among the four reviewed in this work but its capability of forming chlorinated organics is also the lowest. As a consequence, chloramines, depending on the TOXFP of the water, can be recommended as secondary disinfectant for the control of DBPs because their residual disinfectant property. A different question arises when talking about nitrogen containing disinfection byproducts (NDBPs) such as haloacetamides (HAcAm). In this case, chloramination may result in a higher formation potential of these compounds than chlorination.

3.4. UV Disinfection

Although chemical disinfectants are used in DWTPs because of the need of a residual disinfectant and DBPs control, the use of UV disinfection of water is highly extended due to some advantages over chemical oxidants as reported by Song et al. [36]. These advantages are the absence of DBPs, low health risk, easy operation and maintenance, minimum reaction time, and high disinfection power. However, problems associated to UV disinfection arise when treating natural waters containing humic substances. Efficiency of disinfection highly depends on the transparency of water and the presence of colored water due to humics reduces the UV absorption and lower pathogen removal rates. This negative effect could be reduced by increasing the energy dose and lowering the wavelength of emitted radiation but at a higher cost.

3.5. Other Disinfection Processes

Literature also gives examples of water disinfection processes different from the use of chemical oxidants or radiation though some of them are intimately related with them. This happens, for example, with the simultaneous use of radiation and a semiconductor catalyst, that is, photocatalytic oxidation [37]. This process has the attractive possibility of using solar light to excite the catalyst [38,39]. In photocatalytic oxidation, an AOP process, hydroxyl radicals generated can inactivate microorganisms, including viruses, bacteria, spores, and protozoa. However, since hydroxyl radicals are strong oxidants, DBPs similar to those found in ozone processes can be produced and cytotoxicity of catalyst particles requires further studies. Other examples of disinfection processes are based on membranes [40], hydrodynamic cavitation [41], nanoparticles such as in photocatalysis or in membranes [42], or with magnetic materials [43]. However, more work is needed so that these processes can be put into practice.

4. Nature of DBPs from Classical Oxidant-Disinfectants Agents Used in DWTPs

Soon after the uncover of THMs in chlorinated drinking water many works were carried out to ascertain the nature of generated products not only from chlorine but also from the application of the other alternative oxidants commented before. One of the first studies on this matter was carried out by Coleman et al. in 1980 [44] who identified 400 compounds from a chlorinated fulvic acid extract. Among these compounds, polynuclear chlorinated aromatic hydrocarbons and polychlorinated biphenyls represented an important fraction. Also, regarding DBPs of alternative oxidants, Glaze in 1986 [45] published a review on ozonation DBPs, highlighting the formation of hydroperoxidic byproducts, unsaturated aldehydes and hypobromous acid or even permanganate ions, the two later from the oxidation of bromide and manganous ions, respectively. Regarding chlorine dioxide DBPs, at that time, Werdehoff and Singer in 1987 [46] reported some TOX and THMs formation from chlorine dioxide oxidation of humics. They also checked that these chlorinated organic compounds were mainly due to residual chlorine formed accompanying the synthesis of chlorine dioxide. They also reported the formation of chlorite ion as main inorganic byproduct. Throughout these last three decades, a lot of work has been done on DBPs identification [32] and some reviews on this matter have been published [47]. Also, published studies report works not only on DBPs identified from chlorination [48], but also from chloramination [49,50], ozonation [32], and chlorine dioxide NOM oxidation [51,52]. Most of identified DBPs come from water chlorination and mainly involve organochlorine compounds but some treat the presence of brominated (Br-DBPs) and iodinated (I-DBPs) DBPs formed from chlorine, ozone or chlorine dioxide when natural waters contain bromide or iodide, respectively [32,53]. In these works, oxidation of bromide and iodide yields hypobromous and hypoiodous acids, respectively, which in turn oxidize NOM to form Br-DBPs and I-DBPs. Also, from the oxidation of chlorine, ozone or chlorine dioxide, bromite, and iodite are formed that in a fast step are oxidized to bromate and iodate, respectively.

DBPs are classified as regulated (R-DBPs) and non-regulated (NR-DBPs) whether or not some maximum contaminant level (MCL) has been imposed from official government environmental organizations such as USEPA or the European Chemical Agency (ECA), (called standard values in this case). Regulated DBPs are among the most halogen organics identified. These are four out of ten THMs: Those containing chlorine and bromine and five out of nine haloacetic acids (HAAs) (only with chlorine and bromine atoms): Mono, di and trichloroacetic acids and mono and dibromoacetic acids. For total THM, USEPA, and ECA have imposed MCL (or standard values) of 80 and 100 μg L^{-1}, respectively, while for total HAAs, MCL are 60 μg L^{-1} imposed for both official organisms. Regarding inorganic DBPs, bromate has 10 μg L^{-1} and chlorite 1000 and 250 μg L^{-1}, also from these organizations, USEPA and ECA, respectively. The rest of the identified DBPs are non-regulated and main family groups are listed in Table 3 with some representative example compound. Many of these DBPs are formed from chlorination but also from chloramination, ozonation and chlorine dioxide application. Richardson et al. [54] in an extensive and detailed review gives numerous examples of DBPs from the oxidation with the four mentioned disinfectants and from sequential oxidations such as ozonation or chlorine dioxide treatment followed by chlorination or chloramination. New DBPs are continuously being identified as Pan et al. [55] did in 2017. These authors reported 13 new polar phenolic chlorinated and brominated DBPs or How et al., also in 2017 [56], about organic chloramines. Literature shows detailed explanations about the formation mechanisms of these DBPs which are out of the scope of this work.

Table 3. Family groups of unregulated DBPs identified in water during disinfection processes, representative compounds and average detection concentration levels (ADC) [1].

NR-DBP Family Group	Representative Compound	ADC, μg L^{-1}
Halogenated compounds		
THMs [4]	Iodoform	0.2 [2]
HAAs [4]	Bromochloracetic acid	<1 [1]
Halonitromethanes	chloropicrin (trichloronitromethane)	0.5 [2]
Haloacetonitriles	dichloroacetontrile	3.08 [2]
Haloacetamides	dichloroacetamide	1.62 [2]
Haloamines	N-chloroaminoacetic acid [3]	-
Haloaldehydes	trichloroacetaldehyde	3.67 [2]
Haloketones	trichloropropanone	3.55 [2]
Halofuranones	MX: 3-chloro-4-(dichloromethyl)-5-hydroxy-2(5H)-furanone	<1.0 [1]
Haloquinones	2,6-Dichloro-1,4-benzoquinone	<1.0 [1]
Iodinated DBP (other than THM and HAACs)	Iodoacetaldehyde	<1.0 [1]
Halogencyanide	Cyanogen chloride	1.54 [2]
Non halogenated compounds		
Aldehydes	Formaldehyde	3.46 [2]
Ketones	Dimethylglyoxal	<1.0 [1]
N-Nitrosamines	N-nitrosodimethylamine	0.01 [2]

[1] Richardson et al. [54] also shows other less important family groups. [2] From [57]. [3] From [58]. [4] Regulated DBPs.

5. Issues Related to DBPs Toxicity

As it was reported in the 1970s the reason for applying alternative disinfectants to chlorine was the potential toxic character of THMs and other halogenated organics found in finished chlorinated drinking water [59]. Specifically, the US National Cancer Institute in 1976 published that chloroform was carcinogenic in rodents [60]. Soon after, epidemiological studies suggested some relation between chlorinated drinking water and the occurrence of bladder, colon, and rectal cancer [61,62]. Since then, many studies on the evaluation of safety and hazard of DBPs have been reported [63]. According to DeMarini [64], at present, 20 out of 22 DBPs are rodent carcinogens, more than 100 genotoxic and 1000 water samples have been found to be mutagenic. It has been shown that brominated DBP are more carcinogenic than the chlorinated ones [54] and genotoxicity and cytotoxicity decrease in the following order for halogenated DBPs: Iodinated > brominated > chlorinated [65]. Generally, every DBP evaluated is genotoxic [64]. Comparing chlorinated and ozonated DBPs, the former are more genotoxic than the latter, at least, with *Salmonella* [66]. Regarding the way of DBPs exposure, some works [67] have reported higher cancer risk with the inhalation/dermal way than with oral intake. From inhalation or dermal way, volatile DBPs go directly to the blood stream, bypassing the liver, and once in the bladder they can be activated by some mutagen. From oral intake, DBPs go first to the liver where they could be detoxified. This is particularly important in swimming pool water and bath spa water where, in addition, many nitrogen containing DBPs (N-DBPs) have been found likely due to urine present in these waters. N-DBPs like nitrosamines have been found even more cytotoxic and genotoxic than their corresponding halogenated organics [68].

6. The Role of AOPs in the Removal of Precursors and DBPs

Different AOPs have been applied to eliminate NOM or formed DBPs in a refinement stage focused on drinking water production. AOPs are capable to completely mineralize organic contaminants at mild operating conditions of pressure and temperature though the balance between cost and mineralization requirements is a key issue. At moderate conditions, partial oxidation of NOM components into more reactive compounds to form DBPs can occur. In contrast, a very high mineralization is usually expensive and may not be economically feasible. This section reviews some of the main issues concerning the application of different AOPs for NOM and DBPs removal.

6.1. Elimination of Precursors or DBPs Formation Potential

Advanced oxidation processes have been studied for DBPs precursor removal due to their oxidation ability through hydroxyl radical (HO$^\bullet$) fast and non-selective reactions with organic compounds [69,70]. However, research efforts are needed in terms of improved efficiency, development of less expensive installation and operating conditions [70].

DBP formation potential (DBPFP) represents the level of the formation of different DBPs (or total TOX) in the worst scenario using high excess of disinfectant. This a useful tool to assess the effectiveness of a specific treatment to remove the DBPs precursors, mainly NOM. Since 1980 the number of publications related to AOPs applied to NOM or surrogate removal to reduce DBPFP has progressively increased. The search presented in Figure 2 shows the number of publications in JCR (source WOS) between 1980–2020 and includes different combinations of the keywords: Advanced oxidation, natural organic matter, disinfection by-products, trihalomethanes, haloacetic acids, halonitromethanes, haloketones, haloacetamides, ozone; and their abbreviations, for drinking water treatment.

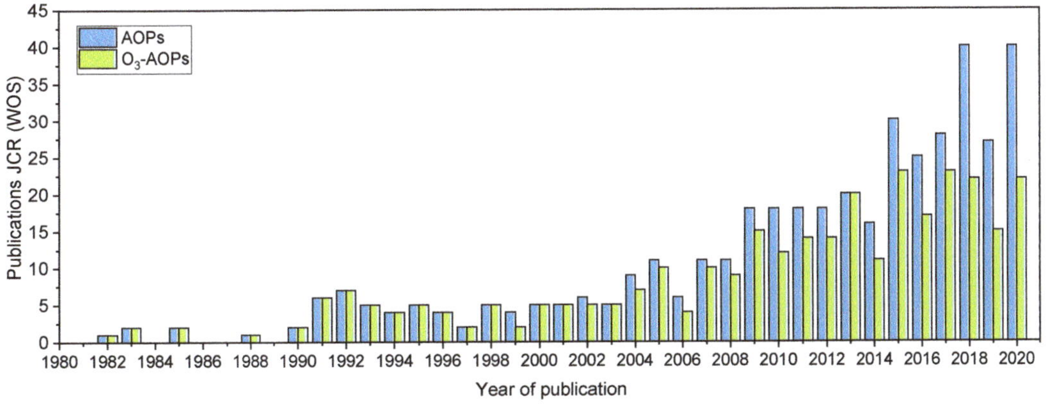

Figure 2. Reports in AOPs and ozone-AOPs for DBPFP or NOM removal for drinking water in the last 40 years.

Matilainen and Sillanpää [71] and Sillanpää et al. [70] comprehensively reviewed the application of AOPs for the removal of NOM in the periods between 2006–2009 and 2010–2016, respectively. This section is focused on the abatement DBPs precursors, mainly NOM, highlighting the formation of DBPs and some key factors in the application of AOPs for DBPFP removal without the aim of compiling all the research found. The application of these processes to the elimination of other microcontaminants in surface water such as pharmaceuticals or personal care products without subsequent DBPs analyses are not reviewed here.

6.1.1. Classic Ozonation Processes

The first study published by Glaze et al. [72] was focused on the degradation of natural trihalomethane precursors by O_3 and O_3/UVC processes in raw waters with high DOC levels. They found that secondary precursors of greater refractivity were produced by ozonation but they could be simultaneously destroyed by UVC combined with ozone. This was due to the formation of high concentration of hydroxyl radicals in the combined process, less selective than ozone, favoring indirect reactions to some extent. Since then, the works in the last two decades of the 20th century have been mainly related to classical ozone based-AOPs, with the main aim of increasing the production of hydroxyl radicals compared to O_3 alone, O_3/H_2O_2, O_3/UVC, O_3/H_2O_2/UVC and also the process H_2O_2/UVC [73]. These processes are well established and have been deeply investigated at lab scale, pilot scale and even at full-scale [74]. In general, the oxidation changes the nature of NOM structures into more hydrophilic compounds, being hydrophobic fractions (humic and fulvic structures) the main responsible for THMs and trihaloacetic acid formation, whereas the more hydrophilic ones are related to dichloroacetic acid generation and the formation of higher concentration of brominated DBPs [75–77]. In this line, even with the highest THMs and HAAs formation potential removal by combined O_3/UV treatment, bromine-containing DBPs present increased toxicity being the conditions between oxidant and UV dose crucial to minimize the risks [78]. Also, the particular characteristics of the NOM play a key role in the suitability of a specific treatment [70,79].

In addition to the classical ozonation and ozone-AOPs, homogeneous and heterogeneous catalytic treatments such as Fenton-related, photocatalytic oxidation or persulfate processes, have been studied for NOM or surrogates elimination in drinking water [69–71].

6.1.2. Fenton Related Processes

Fenton reaction involves the combination of hydrogen peroxide and a metal salt or oxide catalyst, commonly iron, to produce hydroxyl radicals responsible for NOM oxidation and mineralization through the following main reactions:

$$Fe^{2+} + H_2O_2 \rightarrow Fe^{3+} + OH^- + HO^\bullet \quad (6)$$

$$Fe^{3+} + H_2O_2 \rightarrow Fe^{2+} + H^+ + HO_2^\bullet \quad (7)$$

The presence of radiation in photo-Fenton process accelerates the reduction of Fe^{3+} improving the yield of $HO^\bullet$ generation:

$$Fe^{3+} + H_2O \xrightarrow{h\nu} Fe^{2+} + H^+ + HO^\bullet \quad (8)$$

Murray and Parsons [80] studied the elimination of NOM through Fenton and photo-Fenton using UVC as radiation source. They found that the three treatments at optimum conditions reached 90% removal of DOC (DOC_0 = 7.5 mg L^{-1}), dropping THMFP from 140 to below 10 μg L^{-1}. The economic assessment disfavored these processes compared to conventional treatments but neither the use of a different radiation source nor the possibility of a more restrictive regulation (not only THMs), were considered. In this line, Moncayo-Lasso et al. [81] treated surface water by solar photo-Fenton in a CPC reactor (compound parabolic collector) using natural sunlight. They demonstrated the performance of the process to mineralize up to 80% of NOM (TOC_0 = 7.44–7.81 and 4.12–5.02 mg L^{-1}) from surface water containing dissolved iron or with Fe^{2+} supplement at pH = 5. They found a beneficial effect of photoactive natural components present in the natural water on the process leading to a more efficient mineralization than using dihydroxy-benzene as a model compound. Unfortunately, the DBPFP before and after chlorine disinfection was not analyzed in this work either the formation of other oxidation by-products. Moncayo-Lasso et al. [82] also studied the photo-Fenton process using simulated solar radiation to eliminate NOM at pH = 6.5, which introduces an advantage with respect to Fenton related treatments usually carried out at pH near 3. They used Fe^{3+} at low concentration for the treatment of surface water with TOC_0 = 7.1 mg L^{-1}, demonstrating the effectiveness of the system for the transformation and partial mineralization of NOM previously treated with a sand filter system. The THMFP dropped from 160 to 20 μg L^{-1} at the optimum conditions. However, the transformation of NOM can lead to different DBPs precursors that have not been considered in these works.

6.1.3. Photocatalytic Oxidation

Heterogeneous photocatalysis based in the interaction of a semiconductor, mainly *TiO_2*, with radiation to produce reactive oxidizing species (ROS) is a complex mechanism that can be generally described by the following reactions:

$$TiO_2 \xrightarrow{h\nu} h^+ + e^- \quad (9)$$

$$h^+ + (H_2O, OH^-, Ti-OH) \rightarrow HO^\bullet \quad (10)$$

$$e^- + O_2 \xrightarrow{[...]} HO^\bullet \quad (11)$$

where organic molecules can react both in the catalyst surface with photogenerated holes (h^+) or in the liquid phase with hydroxyl radicals or other ROS.

Photocatalytic oxidation using suspended TiO_2 for NOM removal has been studied by different authors. Lee and Ohgaki [83] observed an initial increase in THMFP at short treatment times with the system TiO_2/UVC and different surface waters with DOC_0 = 1.1–3.5 mg L^{-1}. This effect has been subsequently observed by different authors with different surface waters using TiO_2/UVA [84,85]), TiO_2/UVC [86] and even TiO_2/solar radiation [87]; concluding, in general, that sufficient radiation doses are needed to achieve the desired elimination, depending on raw water characteristics, and an initial negative impact of photocatalysis on DBPFP is expected at low UV doses.

Apart from suspended TiO_2, Murray and Parsons [88] studied different supported TiO_2 demonstrating that can be effective combined with UVC for NOM removal. Later, Murray et al. [89] studied pelletized TiO_2 to adsorb NOM from different surface waters (DOC_0 = 17.0–5.4 mg L^{-1}) at bench scale and then used UVC to regenerate the TiO_2 pellets by oxidation. They observed a reduction of DOC concentration of source water by 70% with the subsequent THMFP reduction in a multistage process. Also Kent et al. [90] compared the efficiency of suspended and fixed TiO_2 using nanostructured thin films and UVC with river water (DOC_0 = 5.34 mg L^{-1}). No complete DOC elimination was reached but they obtained removals of total trihalomethane formation potential (TTHMFP) and total haloacetic acid formation potential (THAAFP) of approximately 20% and 90%, respectively, being fixed TiO_2 configuration less effective than suspended TiO_2.

6.1.4. Sulfate-Radical Processes

Sulfate radical-based AOPs are based on the generation of $SO_4^{\bullet-}$ radicals from persulfate (PS) or peroxymonosulfate (PMS) activation by temperature, metal ions, solid catalysts, radiation, ultrasound, etc. For NOM removal and DBPFP reduction, mainly homogeneous metal ions or UV activation have been used:

$$S_2O_8^{2-}+M^{n+} \rightarrow M^{(n+1)}+SO_4^{\bullet-}+SO_4^{2-}\ (M = \text{metal ion}) \quad (12)$$

$$HSO_5^{-}+M^{n+} \rightarrow M^{(n+1)}+SO_4^{\bullet-}+HO^{-}\ (M = \text{metal ion}) \quad (13)$$

$$S_2O_8^{2-}+h\nu \rightarrow 2SO_4^{\bullet-} \quad (14)$$

$$HSO_5^{-}+h\nu \rightarrow SO_4^{\bullet-}+HO^{\bullet} \quad (15)$$

$$SO_4^{\bullet-}+H_2O \rightarrow SO_4^{2-}+HO^{\bullet}+H^{+} \quad (16)$$

Lu et al. [91] studied the system Co^{2+}-PMS and reported the NOM reconfiguration upon exposure to sulfate radicals with a great increase in the DBPFP, concretely chloroform, trichloroacetic acid, and dichloroacetic acid. Also, the formation of reactive bromine species through $SO_4^{\bullet-}$ and Br^- reaction was demonstrated.

Hua et al. [92] studied the DBP alteration from NOM and model compounds after UV/PS treatment and subsequent chlorination. They found that the yields of THMs and dichloroacetonitrile (DCAN) from NOM decreased by 50% and 54%, respectively, after UV/PS followed with chlorination, whereas those of chloral hydrate (CH), 1,1,1-trichloropropanone and trichloronitromethane (TCNM) increased by 217%, 136%, and 153%, respectively. They demonstrated the impact of different structures of the precursors on DBP formation through different surrogates (benzoic acid, resorcinol or methylamines). Wang et al. [93] also studied Br-DBPs during UV/PS oxidation demonstrating the potential negative effects of Br^- on sulfate radical-AOPs, which need to be considered if this technology is applied in practice. In the same line, Wang et al. [94] studied the formation of iodinated DBPs during heat activated PS using phenol as model compound in the presence of iodide. They concluded that I^- was transformed into free iodine leading to iodinated DBPs that can further be degraded by additional $SO_4^{\bullet-}$ and transformed to iodate. All these works highlight the crucial impact of sulfate radicals in the presence of bromide, iodide or chloride.

6.1.5. Chlorine/UV Process

The combination of chlorine/UV has been recently studied for NOM degradation. With radiation of sufficient energy, the photolysis of hypochlorous acid and hypochlorite can lead to hydroxyl and chlorine radicals as follows:

$$HOCl \rightleftarrows H^{+}+ClO^{-} \quad pKa = 7.58 \quad (17)$$

$$HOCl + h\nu \rightarrow HO^{\bullet}+Cl^{\bullet} \quad (18)$$

$$ClO^{-}+h\nu \rightarrow O^{\bullet-}+Cl^{\bullet} \quad (19)$$

$$O^{\bullet-}+H_2O \rightarrow HO^{\bullet}+OH^{-} \quad (20)$$

Pisarenko et al. [95] investigated the generation of DBPs during NOM degradation by combination of chlorine (electrochemically generated or from hypochlorite solution) with UVC and UVA radiation. They found that $Cl^{\bullet}$ and $HO^{\bullet}$ radicals generated oxidized NOM from river water (DOC_0 = 2.6 mg L^{-1}) and different surrogates (o-methoxybenzoic acid and 4,6-dioxoheptanoic acid) increasing HAA formation but with no negative impact on THM formation. UVC resulted in lower overall DBP formation than UVA. Also, Wang et al. [96] studied the process at pilot scale with river water (DOC_0 = 1.5–3.5 mg L^{-1}). They concluded that the treatment was effective and the DBPs formed were comparable to that from H_2O_2/UV also tested. The main drawback was the formation of chlorate during the treatment up to 17% of the total chlorine used. Liu et al. [97] reported the formation of DBPs in the system chlorine-chloramine/UVC. They found that the mass ratio of HOCl to NH_2Cl played a key role in the control of DBPs formation and that the presence of Br^- increased DBPFP but it can be reduced with higher radiation doses (Br^- transformed to bromate).

6.1.6. Electrochemical AOPs

Electrochemical AOPs (EAOPs) have been mainly applied for wastewater more than for drinking water treatment due to technical and economic aspects [98]. Trellu et al. [99] studied the removal of humic acids from drinking water by anodic oxidation and electro-Fenton processes. They concluded that EAOPs are promising processes for the production of high quality organic-free water, particularly when BDD anode is used but no disinfection by-products were analyzed. Mao et al. [100] compared conventional ozonation with electro-peroxone (E-peroxone, E-H_2O_2/O_3) in the removal of NOM (DOC_0 = 3.2 mg L^{-1}). They found that E-peroxone was less efficient at reducing chlorination DBPs. However, the formation of bromate was significantly reduced in the combined treatment. More research is needed in these processes and their combinations to boost the application in real conditions.

6.1.7. Some Considerations on AOPs for DBPFP Removal

Many studies have determined that the application of AOPs operating with limited energy or chemical inputs can lead to an increase in the DBPFP as compiled Mayer et al. [101]. Therefore, intensified treatments with high enough oxidant concentration or exposure time are required. In this scenario, the effectiveness of ozone-based AOPs for DBPFP depletion for drinking water treatment has led to the investigation of other catalytic and photocatalytic ozone treatments with the aim of improving the use of ozone. These processes are deeply reviewed in Sections 7 and 8.

6.2. DBPs Removal

Although DBPs can be controlled through removing NOM in the drinking water treatment process, it is not only difficult to remove NOM completely, but DBPs are inevitably produced when the remaining chlorine is released to rivers and reacts with NOM. Because of variability in DBPs characteristics, complete elimination from drinking water by single technique is impossible. A quick search in database Scopus with key words "disinfection by-products removal AND oxidation processes" give more than two hundred entries.

6.2.1. Ozone Based Processes

Ozone-based processes, other than catalytic/photocatytic ozonation, have been used for removing DBPs from water supplies. As it can be inferred from Table 4, N-nitrosamines and chloroacetic acids (CAAs) are the main disinfection by-products investigated. N-nitrosamines are highly toxic disinfection by-products formed in drinking water treatment process. More than 80% of N-nitrosamines are carcinogenic, teratogenic and mutagenic. USEPA has classified N-nitrosodimethylamine (NDMA) as a probable human carcinogen [102]. N-nitrosamines cannot be sufficiently removed during conventional water and wastewater treatment processes and could permeate through reverse osmosis (RO) membrane filtration in water reclamation systems. In this sense, NDMA degradation by ozone-based AOPs has been investigated. However, these works indicated that conventional ozonation had very limited effect on NDMA oxidation [102]. Zhang et al. [102] found that, compared with UV irradiation, ozone was less effective for the abatement of N-nitrosamines. Less than 20% of N-nitrosamines were oxidized after ozonation. However, when 2 mg L^{-1} of H_2O_2 was added into ozonation system, the degradation efficiencies of N-nitrosamines were significantly improved. Xu et al. [103] investigated the removal efficiency of NDMA using UV/O_3 and evaluated the ability of UV/O_3 to diminish the regeneration of NDMA after degradation. It was showed that 99% NDMA removal was achieved both in UV and UV/O_3 system, while the oxidized fraction of NDMA was only 10% at pH = 6 by ozone alone ($[O_3]_0$ = 6.6 mg L^{-1} (0.14 mM), $[NDMA]_0$ = 0.1 mM). The introduction of ozone during the UV process had little influence on NDMA removal, but had a large impact on NDMA degradation products. Lee et al. [104] compared the efficiency of conventional ozonation and the AOP (O_3/H_2O_2) for NDMA oxidation. The degradation effectiveness of NDMA was only 13% ($[NDMA]_0$ = 1 mM) at pH = 7 and 12 min contact time, by 40 mM ozone alone. In contrast, 85% of NDMA was removed by the AOP (O_3/H_2O_2, $[O_3]_0$ = 40 mM, $[O_3]_0/[H_2O_2]_0$ = 2) under the same reaction conditions. It appeared that the reaction with $HO^\bullet$ radicals dominated NDMA ozonation, and methylamine (MA) was the primary amino product of this NDMA oxidation. In this sense, Lv et al. [105] investigated the degradation of NDMA by ozonation. The effects of initial NDMA concentration (50–800 ng L^{-1}), ozone dosage (0.02–0.2 mM), and pH (5–8) were studied in detail in the context of NDMA degradation. The amount of removed NDMA increased as the initial NDMA concentration rose ($[O_3]_0$ = 4.8 mg L^{-1}, pH = 7.6). Higher ozone dosage ($[NDMA]_0$ = 400 ng L^{-1}, pH = 7.6) and pH enhanced removal efficiency. Increasing pH was favorable for NDMA oxidation ($[O_3]_0$ = 0.1 mM, $[NDMA]_0$ = 400 ng L^{-1}). Inhibition of NDMA degradation was observed when a hydroxyl radical scavenger, t-butanol, was added during the ozonation process. Hydroxyl radicals generated from ozone played a critical role in the degradation of NDMA.

Table 4. Works on DBPs removal by ozone-based processes.

Target DBPs	Processes Applied	Reactor and Experimental Conditions	Main Results	Ref.
9 N-nitrosamines	O_3 O_3/H_2O_2	1 L glass beaker with magnetic stirring. 30 min degradation experiments; room temperature (21 °C); [N-nitrosamines] = 100 ng L^{-1}; $[H_2O_2]$ = 2 mg L^{-1}; $[O_3]$ = 1.5 mg L^{-1}	O_3/H_2O_2 process got the highest removal efficiency of 36.6% (NDMA)—91.4% (NDBA) among all the investigated methods. In comparison, ozonation merely removed less than 20% of N-nitrosamines (except NDphA for 29.3%).	[102]
Chloroacetic acids MCA DCA	O_3/H_2O_2 O_3 (pH = 11)	Planar falling film reactor. Seven UVA lamps (15W, 360 nm) fixed inside the reactor. Intensity UV light: 1 mW cm^{-2}; water flow rate: 1 L min^{-1}, volume 0.5 L; [CAA] = 1 mM; pH 3, pure gaseous oxygen rate of 10 L h^{-1}, ozone gas with 130 ± 5 mg L^{-1} ozone at a power of 30 W.	Chloroacetic acids are highly resistant towards direct ozonation in the darkness as only about 2% degradation was observed after 90 min treatment (pH 3). However, increasing the pH of the solution to 11 shows a dramatic improvement in the degradation efficiency and by the combination of O_3 with H_2O_2.	[106]
NDMA in ultrapure water and natural water (River Sanhaowu)	O_3 (pH = 7.6)	Batch and continuous experiments. A sealed cylindrical reactor with a volume of 5 L. The reactor was stirred mildly and set in the dark at room temperature (24 ± 1 °C). Buffer solution (5 mM phosphate and 1 mM carbonate, prepared with ultrapure water) and pH 7.6). $[O3]_0$ = 0.1 mM (4.8 mg L^{-1}), $[NDMA]_0$ = 400 ng L^{-1}.	Ozonation was an efficient process for NDMA degradation. The removal efficiency was affected by initial NDMA concentration; higher NDMA dosing required higher ozone utilization. NDMA oxidation was favored at high ozone dosage and high pH. NDMA ozonation under various pH as well as hydroxyl radical ($HO^\bullet$) inhibition experiments verified that $HO^\bullet$ generated from ozone dominated NDMA oxidation.	[105]

Table 4. *Cont.*

Target DBPs	Processes Applied	Reactor and Experimental Conditions	Main Results	Ref.
DCA in aqueous media.	O_3 O_3/UVC	The photo-reactor was a cylinder made of Teflon™ closed at both ends with two demountable, flat, circular windows made of quartz. Reactor length 5.2 cm, and the inner diameter 5.2 cm (V_{Rirra} = 110.4 cm^3). Dissolved ozone concentrations of 1.46 to 2.1 × 10^{-7} mol cm^{-3}. 15 and 40 W lamps. $[DCA]_o$ = 20, 40 and 50 ppm; pH 3.5 ± 0.1.	O_3 or UVC by themselves did not result in appreciable decomposition of DCA. Conversely, the O_3/UV combination can be considered a suitable process for degrading DCA in water. The combination of ozone and UVC radiation produces a significant amount of hydrogen peroxide as an important reaction by-product.	[107]
NDMA in distilled water.	O_3/UVC	A cylindrical glass reactor with 700 mL valid sample bulk. Low-pressure Hg lamp (8W, emission at 253.7 nm). $[NDMA]_0$ = 0.1 mmol L^{-1}, pH 6.0, irradiation 1000 W cm^{-2}, $[O_3]_0$ = 6.6 mg L^{-1}.	UV irradiation and the UV/O_3 combination are effective methods for NDMA removal from drinking water. The introduction of ozone into the UV process had little influence on the effectiveness of NDMA removal. However, it had a great influence on the formation of degradation products from NDMA. As the main products, DMA and NO^{2-} decreased markedly in the UV/O_3 process compared with UV irradiation.	[103]
DCA TCA	O_3 O_3/UVC O_3/H_2O_2 O_3/H_2O_2/UVC	A cylindrical stainless steel column (2 L vol.). Its diameter and height are 100 and 300 mm, respectively. Inside the reaction column is a quartz well containing a UV lamp with a diameter of 30 mm and a height of 300 mm—a 15-W low pressure mercury vapor lamp (254 nm). Ozone adding (mg min^{-1}): 0.3 ± 0.06; $[H_2O_2]_o$ = 2.5 mg L^{-1}; wavelength of UV lamp (nm): 254; power input of UV lamp (W): 15; reaction volume (L): 2; Initial DCA and TCA concentration (mg L^{-1}): 2.0.	O_3/UV showed to be more suitable for the decomposition of DCA and TCA in water among the six methods of oxidation. Decomposition of DCA was easier than TCA by AOPs.	[108]
NDMA in buffered deionized water and natural waters.	O_3 O_3/H_2O_2	500 mL glass bottle equipped with a dispenser. ($[NDMA]_0$ = 1 μM, $[O_3]_0$ = 40 μM, ratio of $[O_3]_0/[H_2O_2]_0$ = 2. 10 mM phosphate buffer.	In experiments with natural waters, NDMA could not be significantly oxidized during conventional ozonation. In the AOP O_3/H_2O_2, ozone doses of 160–320 mM ($[O_3]_0/[H_2O_2]_0$ = 2:1) were necessary for > 50% NDMA oxidation depending on the $HO^\bullet$ scavenging rates of the natural waters. Bromate formation may be the limiting factor for NDMA oxidation during ozonation and ozone-based AOPs in bromide containing waters.	[104]

On the other hand, it is well known that CAAs are widely present in water treated by chlorination processes and are resistant against ozonation in the darkness. The results obtained by Hama Aziz [106] showed that single ozonation was an inefficient method for the destruction of the CAAs as only about 2% degradation was observed after 90 min treatment (pH = 3). However, increasing the pH of the solution to 11 showed a dramatic improvement in the degradation efficiency. The fast decomposition of ozone molecule in alkaline solution (pH > 9) to generate powerful and non-selective hydroxyl radicals resulted in efficient degradation of organic pollutants. The addition of hydrogen peroxide to ozonation process, also improved monochloroacetic acid (MCA) and dichloroacetic acid (DCA) degradation. Lovato et al. [107] studied DCA decomposition in aqueous media employing ozone and UVC radiation. The influence of various parameters, such as pollutant initial concentration (20–50 mg L^{-1}), radiation photon fluency rate at the reactor window and ozone dissolved concentration (1.46 to 2.1 × 10^{-7} mol cm^{-3}) was studied. Oozne or UVC by themselves did not result in appreciable decomposition of DCA within the studied reaction time (180 min). Conversely, the O_3/UV combination can be considered a suitable process for degrading DCA in water. Also, the combination of ozone and UVC radiation produces a significant amount of hydrogen peroxide as an important reaction by-product. Finally, Wang et al. [108] studied the decomposition of DCA and trichloroacetic acid (TCA) from water by means of single oxidants: Ozone, UV radiation; and by the AOPs constituted by combinations of O_3/UV radiation, O_3/H_2O_2, O_3/H_2O_2/UV radiation. Single O_3 or UV did not result in perceptible decomposition of HAAs within the applied reaction time (30 min). O_3/UV showed to be more suitable for the decomposition of DCA and TCA in water among the methods of oxidation tested.

6.2.2. Ozone Free Processes

Regarding other than classic ozonation treatments [109], Table 5 shows the most illustrative works dealing with advanced oxidation systems used for removal of DBPs in water supplies in the last 25 years.

Table 5. Works on DBPs removal by ozone free processes.

Target DBPs	Processes Applied	Reactor Configuration	Experimental Conditions	Ref.
THMs HAAs HANs HKs	Heterogeneous Fenton-like reaction + GAC filtration.	Pilot plant that includes mixing tank, mechanical flocculation tank, tube settler, sand filter column, intermediate water tank, heterogeneous Fenton column, active carbon column, water-producing water tank and automatic control device.	The design flow of single-set process is 2 m^3 h^{-1}. Poly aluminum chloride (PACL): 30 mg L^{-1}; dosage of H_2O_2 is 0.15 mM (5 mg L^{-1}).	[110]
MCA DCA TCA	UVA/TiO_2; UVA/BiOCl.	50 mL quartz reactor with a recycling water glass jacket. Light source a 500 W xenon lamp (simulated sunlight, containing 4% UV light).	T = 298 K, 40 mg BiOCl/TiO_2 (P25) (1g L^{-1}) dispersed in 40 mL model pollutant (20 mg L^{-1}). Reaction time: 60 min.	[111]
TCA	Advanced oxidation/reduction processes (AORPs): the vacuum UV (VUV; 185 nm + 254 nm).	A tubular glass photoreactor. Internal diameter of 25 mm and a height of 400 mm. Working vol of 100 mL. 5.7 W dichromatic low-pressure mercury UV lamp emitting UV at two distinct wavelengths (185 and 254 nm).	50 mg L^{-1} TCA, pH = 7; reaction time = 20 min.	[112]

Table 5. *Cont.*

Target DBPs	Processes Applied	Reactor Configuration	Experimental Conditions	Ref.
THMs	Photo-Fenton process: $UV/Fe/H_2O_2$	Photochemical reactor equipped with a 254 nm low-pressure mercury UV-C lamp. Stirred with magnetic bar.	The average fluence rate was 0.93 mW cm^{-2}. T (25–31 °C). Reaction time: 60 min. CFe(II) = 2 mg L^{-1}. C_{H2O2} = 20 mg L^{-1}. pH = 7.2–7.4; $CHCl_3$=163.6 µg L^{-1}, CH_2BrCl = 145.3 µg L^{-1}, $CHBr_2Cl$ = 131.8 µg L^{-1}, CHBr3 = 124.6 µg L^{-1}.	[113]
DCA TCA	A combination of Ferrate [Fe(VI)] and UV irradiation.	Self-made photoreactor (effective volume 0.5 L) equipped with a cooling system and placed on a magnetic stirrer. A 75-W low-pressure mercury UV lamp (λmax = 365, 310, and 254 nm, respectively) with 0.225 $mWcm^{-2}$ intensity placed in the center of the reactor.	Experimental design: HAAs 100–1000 µg L^{-1}; pH: 3–9; Fe (VI) 10–40 mg L^{-1}; time 5–60 min.	[114]
4HANs 9THMs 4 HAcA 4 HAcAm Chloropicrin 500 nmol each	Reductive electrolysis.	Norit GAC (0.4 g) placed in a 1.5 cm × 4 cm cylinder constructed from sheet graphite to serve as the working cathode. It was then transferred to the cathodic chamber of an electrolysis cell.	A constant potential of −1000 mV vs. SHE was applied to the cathode, while cathodic chamber was continuously stirred with a Teflon-lined magnetic stir bar.	[115]
TCA	TiO_2 photocatalytic process combined with Fenton reagent.	Self-made cylindrical reactor with cool water recycling cloth and a UV light tube. The UV light (kmax = 254 nm) 50 W low pressure mercury lamp placed in the center of the reactor and equipped with a protective quartz tube.	Irradiation intensity was about 35 Mw cm^{-2}; TCAA aqueous solution (150 mL) containing TiO_2. Flow rate of 40 mL min^{-1} of O_2. [TCA] = 0.01 mmol L^{-1}, UV irradiation intensity= 35 mW cm^{-2}, $[TiO_2]$ = 1.0 g L^{-1}, $[Fe^{2+}]$ = 0.1 mmol L^{-1}, $[H_2O_2]$ = 1.8 mmol L^{-1}, natural pH = 6.	[116]
TCA	Ferrate(VI) a multipurpose chemical, is used as coagulant and oxidant.	Self-made photoreactor (effective volume 0.5 L) equipped with a cooling system and placed on a magnetic stirrer.	Initial pH of solution (3–9), ferrate (VI) dosage (1–10 mg L^{-1}), contact time (5–60 min), trichloroacetic acid (100–1000 µg L^{-1}), and initial turbidity (1–10 NTU).	[117]
TCA	Fenton with TiO_2 photocatalytic oxidation.	Self-made photoreactor equipped with cool water recycling and a UV light tube. 50 W low-pressure mercury lamp placed in the center of the reactor and equipped with a protective quartz tube.	Irradiation intensity was about 35 mW cm^{-2}. TCA aqueous solution (150 mL, 2 mg L^{-1}). $[TiO_2]$ = 1.0 g L^{-1}, $[Fe^{2+}]_0$ = 5.6 mg L^{-1}, $m(Fe^{2+}):m(H_2O_2)$ = 1:10, intensity (UV) = 35 mW cm^{-2}, flow rate (O_2) = 40 mL min^{-1}. pH = 5.8 (without adjustment).	[118]
THMs EPA method 551/1	$UV/ZnO/H_2O_2$	A photoreactor equipped with 4 UV lamps in the reactor corners.	5 mL of concentrated solution (30%) H_2O_2; 0.5 g of ZnO in 100 mL of drinking water samples with constant reaction time (1hr) and UV irradiation.	[119]

Table 5. *Cont.*

Target DBPs	Processes Applied	Reactor Configuration	Experimental Conditions	Ref.
4THMs	Sonophotolytic degradation.	Rectangular shape of stainless-steel reactor (L100, W100, H250) and ultraviolet lamps (4). Transducer was located in a bottom of the reactor.	The applied ultrasonic frequency was 500 kHz and the electrical powers were 0–52.55 W. Electrical power of each lamp was 10.5 W. 10 mg L^{-1} of THMs mixture (1.5 L).	[120]
TCA	A sequential Fe^0 (zero valent iron) and BAC column system.	Fe^0 column and a BAC column made of glass, 30 cm in length and 3 cm in id.	1.2 μM TCA; pH of the feed water was approximately 6.0. The feed was initially treated by the Fe^0 column (BET was 1.3 m^2 g^{-1}) and then the BAC column.	[121]
17 DBPs (i.e., halomethanes, haloacetonitriles, halopropanones, chloral hydrate, and trichloronitromethane) at low concentration (μg L^{-1})	Electrochemical reduction using a resin impregnated graphite cathode.	Flow-through electrochemical reactor consisting of two polycarbonate frames (internal dimensions of 20 × 5 × 1.2 cm).	The cathode potential from −700 to −900 mV vs. SHE.	[122]
TCA	High-frequency sonoelectrochemical methods.	A sonoreactor (0.5 L of volume) consisting of a cylindrical flask equipped with a cooling jacket where the electrodes (18 cm^2 on each side) were placed.	0.5 mM TCAA aqueous solution; 850 kHz ultrasonic irradiation, titanium cathode, and a platinized titanium anode.	[123]
MCA DCA TCA	Iron-based bimetallic particles: Two kinds of dry Pd/Fe nanoparticles (Pd/Fe^{-1} and Pd/Fe^{-2}).	A series of glass vials (60 mL). The vials were sealed with Teflon-lined rubber septa and aluminium cap.	Pd content = 0.1 wt%, Pd/Fe loading = 3 g L^{-1}, initial concentration of chloroacetic acid = 20 mg L^{-1}, and reaction time = 180 min.	[124]
4THMs	Photocatalysis with TiO^2 (slurry).	Photo-Cat Lab consists of an air compressor for oxygenation of the system; eight 75-watt, low-pressure, mercury arc bulbs in series; and a submicron-pore-size ceramic membrane filter that produces TiO2-free effluent.	Batch configuration. Process flow rate of 25 L min^{-1}; initial system volume of approximately 16 L. the average intensity of the UV bulbs was approximately 7.0 Mw cm^{-2}. 400 mg L^{-1} TiO_2.	[86]
MCA DCA TCA	Photocatalytic degradation over various bare and silver-deposited Degussa P25 TiO^2 particles.	A medium scale (V = 2.5 dm^3) photochemical reactor.	Irradiation performed under anaerobic and aerobic conditions. Flow rate of gases (air and Ar) was 40 dm^3 h^{-1}; light source (40 W, λmax = 350 nm), initial concentration of TiO_2 (rutile, anatase, P25) was 1 g dm^{-3}. initial concentration of MCA, DCA, and TCA was adjusted to be 1 mM.	[125]
DCA TCA	Electrochemical treatment (dehalogenation).	Packed-bed flow reactor. The reactor was composed of two glass compartments separated by a cationic exchange membrane.	Initial concentration of HAAs 10.5 mM. Flow rate: 1 mL min^{-1}; electrolysis potential: −0.200–−0.400 V (vs. SCE).	[126]

Table 5. *Cont.*

Target DBPs	Processes Applied	Reactor Configuration	Experimental Conditions	Ref.
4THMs	Ultrasonic (US) irradiation, hydrogen peroxide (H_2O_2), Fenton's oxidation, US/H_2O_2 and $US/H_2O_2/Fe^2$	200 mL conical closed glass reactor kept in a temperature-controlled bath	THMs solute mixture, at 10 mg L^{-1} for each compound, in deionized water. Initial pH adjusted to 3.5 by 1N H_2SO_4. 250–500 mg L^{-1} H_2O_2 and 20–40 mg L^{-1} $Fe^{(2+)}$.90 min reaction.	[127]
4THMs Iodoform (CHI_3)	Sonodegradation	200 mL conical closed glass reactor kept in a temperature-controlled bath	100 mL aqueous solution at 25 °C. frequency of 20 kHz; acoustic intensity 3.75 W cm^{-2}; initial pH 5.4–5.8.	[128]
4 THMs CHI_3	Sonodegradation	200 mL conical closed glass reactor kept in a temperature-controlled bath.	Ultrasonic frequency of 20 kHz. presence of inorganic components. increase the ultrasonic intensity, from 0.9 to 7.0 W cm^{-2}. the power density, from 0.123 to 0.368 W mL^{-1}.	[129]
4THMs CHI_3	Ultrasonic irradiation	200-mL conical closed glass reactor kept in a temperature-controlled bath.	Ultrasonic irradiation 20 kHz. Acoustic intensity 3.75 W cm^{-2}, and the power density 0.184 W mL^{-1}. initial pH 5.4–5.8 without buffer addition. Initial conc. of THMs 10 mg L^{-1}.	[130]
MCA DCA TCA	Photodegradation in the presence of titanium dioxide suspensions.	Reaction vessel 350-mm long quick fit condenser tube. Connections to a thermostatic water bath. Two mercury vapor greenhouse lamps were used to mimic sunlight radiation ($\lambda >$ 400 nm).	250 cm^3 of approximately 0.1 mol dm^{-3} solutions of the HAAs. 0.1 g of finely powdered titanium dioxide was added as a photocatalyst.	[131]
MCA DCA TCA MBA (bromoacetic acid)	Iron Fe (0) particles.	Batch experiments. 125 mL serum bottles.	Aqueous HAA solution buffered at pH 7.5 with 50 mM deoxygenated MOPS. 0.3 g of iron and rotated at 45 rpm. Different Initial HAA Concentrations: 15–405 μM.	[132]
TCA	Ultraviolet (UV) photolysis, ultrasound (US) sonolysis and their combination.	Quarz reactor immersed in a thermostathic bath. Two UV lamps located on either side of the sample reactor.	TCA (2.89×10^{-4} M). Vreacción 36 mL. pH 3.5; T = 30 °C.	[133]
HAAs: TCA TBA (tribromoacetic acid) CDBA (chlorodibromoacetic acid) BDCA (bromodichloroacetic acid)	Reduction with zerovalent iron (Fe^0).	Batch experiments (glass serum bottle).	0.5 g of Fe(0) in 36 mL of DI water. Desired conc. (100–200 μM). RT up to 94 h. No pH buffer employed. Initial pH values ranged 3.62 to 4.14 and final rose to 5.60 to 6.23.	[134]

Table 5. *Cont.*

Target DBPs	Processes Applied	Reactor Configuration	Experimental Conditions	Ref.
4THMs	Fentons Reagent.	The reactors consisted of 54 mL Pyrex test tubes filled to capacity with stock THM solution, plus Fenton's reagent, and sealed with septum-fitted screw caps.	pH = 3.5 $[THM]_o$ = 37.2 µg L^{-1} each $[H_20_2]$ = 3.7 Mm; $[Fe^{2+}]$ = 0.19 mM.	[135]
TCA	Photocatalysis with TiO_2 and thermal decomposition.	Photocatalysis: The reactor equipped with a suprasil-glass adapter for the UV-lamp. Reaction vessel has inlet and outlet ports of Teflon tubes for bubbling air and N_2. The apparatus was closed. Thermal: A double-walled thermostated jar of glass was used.	0.5 g TiO_2 were suspended in 500 mL aqueous solution containing 10 mmol (20 mM) of trichloroacetic acid. 20 W-Hg low pressure lamp.	[136]

As it can be seen from Table 5 among the various harmful DBPs removed from water supplies, HAAs are the most investigated. As it is known, they are the second most important DBPs after THMs and are highly stable and non-volatile compounds. Trichloroacetic acid (TCA) and dichloroacetic acid (DCA), the two main fractions of HAAs, are important because of their potential risks to human, aquatic, and plant life. Thus, realizing the danger of chloroacetic acids, different advanced oxidation techniques or electrochemical methods are proposed to eliminate them from water. A great deal of effort has been devoted to HAAs removal by biological activated carbon (BAC) processes. However, it has generally been found that while mono- and di-HAAs are relatively easily biodegradable, tri-HAAs are recalcitrant to biodegradation [121].

The breaking of C-Cl bond can result in the degradation of chloroacetic acids. TCA and degradation by-products can be reductively removed by zero valent iron (ZVI) or bimetallic particles. Reduction of HAAs through dehalogenation by using zero valent iron (Fe^0) or other element-doped Fe^0 (such as the bimetallic Pd/Fe, Cu/Fe and Si/Fe) have been reported [134]. In contrast to the order of HAAs biodegradability, the susceptibility of HAAs to Fe^0 reduction normally follows the order of tri-HAAs, di-HAAs, and lastly mono-HAAs. Di- and mono-HAAs were usually observed as the dehalogenation intermediates or end-products [121]. However, the potential release of metal ions affects water quality, leading to secondary pollution in long term. Electrochemical reductive dechlorination is efficient for TCA, and the stepwise dechlorination from TCA and DCA to MCA is achieved [126]. Because the degradation of TCA, DCA, or MCA strictly depends on current density, electrodes are easily corroded by high current in this electrochemical process. To increase the degradation efficiency of TCA and simultaneously protect the electrodes, Esclapez et al. [123] have developed a hybrid sonoelectrochemical process at a low current density and obtained satisfying dechlorination efficiency. Summarily, the break of C-Cl bond is an efficient degradation channel for TCA in ultrasonic (and sono-photolytic degradation [133]) or electrochemical or special reductive reagents reaction, and Cl^-, DCA, MCA and CO_2 are the major by-products in aqueous solution [116].

In addition, the breakage of C-C bond can also produce the removal of chloroacetic acids. In this sense, decomposition of mono-, di-, and trichloroacetic acid in aqueous titania suspensions was studied by several researchers [111]. It has been concluded that two types of reactions may be responsible for the TiO_2-mediated photodegradation of chloroacetic acids: (1) Direct reactions between the photogenerated charge carriers and the organic molecules and (2) reactions of hydroxyl radicals or other oxygen containing radicals with the organic molecules [125]. MCA and DCA are readily decomposed over UV-irradiated TiO_2 catalyst in aqueous media to form CO_2 and HCl. On the other hand, using bare TiO_2, TCA is degraded with a very low efficiency [111]. TCA has no C–H bond, and such a molecule has been found to be hardly reactive in TiO_2-based photocatalytic systems. For that reason, an efficient Fenton assisted TiO_2 photocatalytic hybrid process have developed [116,118]. These works report that exists a noticeable synergetic effect dominantly caused by the oxidation– reduction recycling reaction of $Fe^{3+} \leftrightarrow Fe^{2+}$ between Fenton and TiO_2 photocatalytic oxidation. Also, an AOP based on combination of Ferrate with UV illumination was investigated in terms of kinetics, by-products analysis and application of surface methodology for degradation of CAAs [114].

On the other hand, the most significant group of DBPs formed during chlorination is THMs. Compounds of this group: Chloroform, bromodichloromethane, chlorodibromomethane, and bromoform, were recognized as potential human or animal carcinogens [119], as detailed before. AOPs are promising techniques to efficiently and effectively convert these compounds into better biodegradable or less harmful substances. Several types of AOPs have been developed using Fenton Reagent [129], photo-Fenton process [113], UV/ZnO nanocatalyst/H_2O_2 [119], photocatalysis with titanium dioxide [86], ultrasonic degradation [127–130], and sonophotolytic process [120].

Finally, from Table 5, it can be concluded that major ozone-free processes tested for DBPs removal have been photocatalysis and sonodegradation/sonophotolytic degradation. From an applied point of view, Gerrity et al. [86] studied different scenarios using a pilot-scale TiO_2 photocatalysis reactor. DBPs may form in clear wells prior to distribution that could also be destroyed before the water is released into the distribution system. Since photocatalysis will destroy a chlorine residual, the water would have to be rechlorinated prior to discharge. The authors concluded that when focusing on photocatalytic destruction of DBPs rather than precursors and formation potential, several problems still exist. Optimization of this process should be considered in future studies to address some relatively high energy requirements and the need of rechlorination. In general, this technology has a number of promising aspects but should be considered on a case-by-case basis. Regarding to the high-frequency sonoelectrochemical methods for DPBs mitigation have been applied in a laboratory scale, Esclapez et al. [123] claim that the use of the electricity as unique reactant and the mineralization of the trichloroacetic acids and its by-products make the sonoelectrochemical technology a serious alternative to current technologies.

7. Elimination of DBP Precursors by Catalytic Ozonation Processes

In this section after description of some features about proposed mechanisms of catalytic ozonation, literature about removal of DBP precursors or what is called DPBFP is reviewed.

7.1. Catalytic Ozonation

Several strategies have been proposed to efficiently increase the generation of hydroxyl radicals during ozonation favoring indirect reactions. Many groups have started to investigate a new ozone advanced oxidation process named catalytic ozonation, that is, the simultaneous use of ozone and substances (catalysts) that activate its decomposition to increase the formation of hydroxyl radicals and also to increase not only the removal of pollutants from water but also the mineralization of the total organic carbon content [137]. Catalytic ozonation was starting, in fact, in the middle of 20th century, in 1949, with the use of soluble catalysts, that is, with metal cations such as Co^{2+} [138]. However, this kind of process was not feasible from health and environmental aspects since these catalysts, as heavy metals, are contaminants. Hence, studies were oriented to the use of solid materials of different nature, what it is called heterogeneous catalytic ozonation. It has to be highlighted; however, that also what is called homogeneous catalytic ozonation is a heterogeneous process since gaseous ozone is transferred to the water to react. Then, the catalytic ozonation process where the catalyst is a solid was named the heterogeneous catalytic ozonation process. According to WOS database in the last 10 years about 400 publications have been reported in literature about heterogeneous catalytic ozonation. Three main families of catalyst types have mainly been used since then: Metal oxide catalysts, carbonaceous materials and ceramic materials [139]. More recently, however, a new type of potential catalyst has appeared: Metal based organic frameworks (MOFs) of high specific surface area and tunable pore structure [140,141]. In addition to this number of publications some review works [142–145] have also appeared highlighting the main aspects concerning this process.

As a summary, it can be said that works already published deal with synthesis and characterization of catalysts, application of the new catalyst to remove some given organics taken as model compounds, mechanism of the process, a few of them study the kinetics and some also give data on toxicity, activity, and stability of the catalysts. In addition to the diversity of results obtained which is the consequence of the different methodologies, equipment, analytical procedures, etc., applied, the most relevant fact is the high variation in the reaction mechanism proposed which is also due to the variability of methods of catalyst preparation, analytical techniques, etc., and the lack of knowledge about catalyst stability, by-products formed and ecotoxicity. In fact, the mechanism of catalytic ozonation can be due to direct reactions of ozone or to reactions of hydroxyl radicals coming from the decomposition of ozone on the catalyst surface. Usually, it is expected that the mechanisms of catalytic ozonation fulfilled some of the following requirements:

- Ozone adsorbed on the surface of the catalysts is decomposed into reactive species.
- Organic molecules are adsorbed on the surface of the catalysts with subsequent ozone attacks.
- Both ozone and organic molecules are adsorbed on the catalyst surface and surface reactions take place.

Steps of these mechanisms follow the well-known LHHW or ER mechanisms [146]. These mechanisms, as indicated above, may imply both the formation of hydroxyl radicals or direct reactions between adsorbed ozone and target compounds. For instance, Liu et al. [147] proposed the following simple mechanism for the catalytic ozonation of nitrobenzene with a Zn/SiO_2 catalyst:

$$ZnO/SiO_2{+}O_3 \rightleftarrows S\text{–}HO \quad (21)$$

$$S\text{–}HO \rightarrow Fe_3O_4/CoO/Al_2O_3{+}HO^{\bullet} \quad (22)$$

$$HO^{\bullet}{+}pCNB \rightleftarrows \text{Products} \quad (23)$$

where it is seen that hydroxyl radicals are formed as adsorbed species, then they are release to water bulk to oxidize the organic compound. On the contrary, Andreozzi et al. [148] for the catalytic ozonation of oxalic acid with a manganese oxide catalyst reported the adsorption of the target compound (oxalic acid) to form a manganese oxalate complex that then reacts with dissolved ozone, that is, as a surface direct reaction:

$$Mn(III)OH_2^{+}{+}C_2O_4^{2-} \rightleftarrows MnC_2O_4^{-}{+}H_2O \quad (24)$$

$$MnC_2O_4^{-} \rightleftarrows Mn^{2+}{+}CO_2^{\bullet-}{+}CO_2 \quad (25)$$

$$MnC_2O_4^{-}{+}O_3{+}2H^{+} \rightarrow 2CO_2{+}O_2{+}MnOH_2^{+} \quad (26)$$

Other examples of catalytic ozonation mechanisms have been reported with hydroxyl radicals or ozone (adsorbed or in solution) as main oxidants. Martins and Quinta-

Ferreira [149] proposed the following LHHW mechanism for the Mn-Ce-O catalytic ozonation of phenolic acids:

$$\text{Mn–Ce–O} + \text{Phenolic acids(P)} \rightleftarrows \text{P–S} \tag{27}$$

$$O_3 + S \rightleftarrows O_3\text{–}S \tag{28}$$

$$\text{P–S} + O_3\text{–S} \rightarrow \text{Products} + \text{Mn–Ce–O}. \tag{29}$$

Mechanism proposals are not limited to metal oxide catalysts. For example, Ikhlaq et al. [150] studied the catalytic ozonation of some pharmaceuticals, volatile organic compounds and carboxylic acids with a ZSM-5 zeolite constituted by $xSiO_2/yAl_2O_3$. They proposed a LHHW mechanism as shown in Scheme 6 where ozonation also goes through a direct surface reaction between ozone and organic adsorbed species.

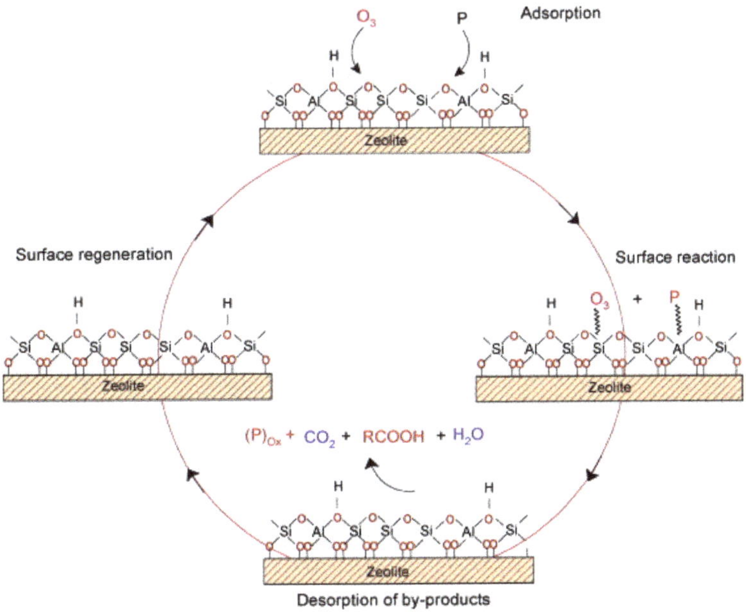

Scheme 6. Mechanism of catalytic ozonation over zeolite ZSM-5 [150] (Reprinted with permission of Appl. Catal. B: Environ, 154-155, Ikhlaq, A.; Brown, D.R.; Kasprzyk-Hordern, B., Catalytic ozonation for the removal of organic contaminants in water on ZSM-5 zeolites, pg110-122. Copyright (2014), with permission from Elsevier).

On the opposite side, Zao et al. [151] reported a hydroxyl radical formation and reaction when using a cordierite material ($2MgO\text{-}2Al_2O_3\text{-}5SiO_2$) to catalyze the ozonation of nitrobenzene:

$$\text{Catalyst–OH} + O_3 \rightarrow HO^{\bullet} \tag{30}$$

$$HO^{\bullet} + NB \rightarrow \text{Products} \tag{31}$$

The authors confirmed the presence of hydroxyl radicals with the Spin trapping/ EPR technique.

Many examples are also given in literature on the use of carbonaceous materials such as activated carbon, multi-walled carbon nanotubes, xerogels, and graphene oxide [152–156] where these materials were used as main catalyst or as support of metal oxides. Again, discrepancies are observed in the mechanism proposed. Thus, some authors have reported that some functional groups of the catalyst surface can be modified by ozone, by enhancing the specific surface area, and decreasing the total volume of the catalyst [157]. However, some others showed that, once the modified carbon was ozonated, its specific surface area was slightly diminished, and the total volume of the pores remained unchanged [158]. Another important issue is the in situ regeneration of the catalyst, or the predominance of catalytic properties and adsorption–oxidation processes in the catalytic ozonation process [157]. Other works have indicated that ozonation reduces the catalytic properties of activated carbon due to a decrease of basic groups and an increase in the number of oxygenated surface functional groups such as hydroxyl and carboxylic acid groups and nitro aromatic compounds [159].

Most of the research works have been carried out with the objective of wastewater treatment and with different organic pollutants such as dyes, pharmaceuticals, personal care products or model molecules such as oxalic and phenol. Regarding to DBP precursors or NOM removal, Table 6 summarizes the main results of reports on catalytic ozonation for drinking water highlighting those in which DBPs are analyzed.

7.1.1. Catalysts

The main catalytic materials used for NOM or humic/fulvic acids removal were metal oxides, ceramic materials, and carbonaceous materials and the properties more frequently analyzed were composition (by XRD, XRF, or ICP), textural properties by N_2 adsorption-desorption, pHpzc, and crystallinity by XRD. Some studies have also reported surface composition by XPS or FTIR.

Among metal oxides, titanium dioxide, TiO_2 has been used both in slurry reactors in the form of suspended nanoparticles (mainly the commercial TiO_2 P25) or supported in different materials. The characterization of powder TiO_2 P25 has been widely studied and the main features were 50 $m^2\ g^{-1}$ BET surface area, anatase/rutile ratio 80/20, crystallite size 21 nm and pHpzc = 6.7 [167]. These studies [175] probed that at the conditions used, no improvement was observed in THMFP removal from surface water compared to ozonation, but less brominated species were formed in catalytic ozonation. On the contrary, in another work [173], with higher catalyst loading but synthetic humic acid solution, a positive effect in the THMFP removal of water treated by catalytic ozonation was observed. The catalyst loading and the composition of NOM may play a key role in the process (more information in Sections 7.1.3 and 7.1.4).

TiO_2 has also been used supported onto different materials, mainly Al_2O_3, clays, and zeolites. Many of these supports have also catalytic activity for ozone decomposition [143]. For the first time, Allemane et al. [187] studied TiO_2/Al_2O_3 combined with ozone for the degradation of fulvic acid and, at high O_3 doses, the catalytic activity was demonstrated but no DBPs were analyzed. In the same line, Volk et al. [186] observed higher mineralization of fulvic acid using TiO_2/Al_2O_3 than ozone alone with a decrease in the chlorine demand of the treated water though no DBPs were detected. No characterization studies were reported here. On the contrary, Gracia et al. [184,185] compared the catalytic activity of TiO_2 supported on Al_2O_3, attapulgyte (a clay) and silica gel for the degradation of NOM. The best results in terms of mineralization of precursors were obtained with the TiO_2/Al_2O_3 catalyst whose main characteristics can be observed in Table 6, highlighting again the importance of adsorption mainly related with the role of Al_2O_3, BET surface area, pH_{pzc} of the solid, characteristics of NOM and pH of the solution. Only the works reported by Chen et al. [177] and Chen and Wang [170], using the same experimental set up, checked the efficiency of TiO_2/Al_2O_3 in the DBPs formation potential elimination from surface water, obtaining a significant reduction up to 50% of THMFP and HAAFP in the water treated by catalytic ozonation at the conditions tested (see Table 6).

Table 6. Works on catalytic ozonation of DBPs precursors.

Target Water	Catalyst and Main Properties	Reactor and Experimental Conditions	Main Results	Ref.
Humic acid	α-Al_2O_3 SBET = 45.8 $m^2\ g^{-1}$ Pore volume 0.162 $cm^3\ g^{-1}$ pHpzc = 4.2 Crystallite size 41.7 nm Mn_2O_3 SBET = 15.6 $m^2\ g^{-1}$ Pore volume 0.008 $cm^3\ g^{-1}$ pHpzc = 5.9 Crystallite size 15.9 nm XPS before and after ozonation indicates the change in the oxidation state of Mn and Al in both materials and the hydroxylation of the surface.	Semi-batch column reactor H = 68 cm, d = 8 cm V = 1 L Ozone flow rate = 0.063 $m^3\ h^{-1}$ C_{O3l} = 4–8 mg L^{-1} Catalyst loading 0.1–0.5 g L^{-1} C_{HA} = 50 mg L^{-1} T = 25 °C pH = 5.5	Al_2O_3 nanocatalysts showed better performance than Mn_2O_3 in HA removal using lower ozone dosage due to favorable surface charge. Mn_2O_3 decomposes O_3 faster tan Al_2O_3. Adsorption of HA contributes to its catalytic ozonation. Higher adsorption capacity of Al_2O_3. Some deactivation observed in 4 consecutive runs. No post-chlorination. No DBPs were analyzed.	[160]
Humic acid	Fe coated pumice Prepared from natural pumice with $FeCl_3$ impregnation. BET, TEM, XRD, DLS, FTIR and pHpzc XRD: α-FeOOH Particle size 200–250 nm. SBET = 10.56 $m^2\ g^{-1}$ pHpzc = 7.13	Semi-batch column reactor V = 1 L Ozone dose 0.333 g min^{-1} C_{O3ge} = 6–12 mg L^{-1} Catalyst loading 25–100 mg L^{-1} C_{HA} = 10 mg L^{-1} T = 22 °C pH = 3–10, 6.72	Improved efficiency of catalytic ozonation (80% DOC removal vs. 20% ozonation). Contributions: *Surface adsorption 21.3% *HO• radicals 66.2% *Sole ozonation 12.5% Scavenger experiments with t-BuOH and phosphate confirmed the role of hydroxyl radicals in solution. Iron leaching < 13 µg L^{-1} No post-chlorination. No DBPs were analyzed.	[161]
Humic acid	Ce-Ti composites Sol-gel synthesis Ce/Ti = 0.2–1.0 Calcination at 600 °C XRD showed cubic fluorite CeO_2 structure. Crystallite size decreased with Ce content, increasing SBET (not reported).	Semi-batch column reactor (h = 500 mm; D = 60 mm) V = 1 L Q = 0.5 L min^{-1} 30 min reaction time C_{O3ge} = 16.91 mg L^{-1} Catalyst loading 0.3 g L^{-1} C_{HA} = 30 mg L^{-1} COD = 260 mg L^{-1} pH = 6.91	The highest efficiency was found for Ti-Ce (1/0.8) composition due to its high surface area and low pHpzc. Increased the ozone efficiency by 62%. Apparent first order rate constants for homogeneous (0.054 g L^{-1} min^{-1}) vs. heterogeneous (0.067 g L^{-1} min^{-1}). Less of 50% heterogeneous contribution. The distribution of molecular weights tends to lower values in catalytic ozonation. No post-chlorination. No DBPs were analyzed.	[162]

Table 6. *Cont.*

Target Water	Catalyst and Main Properties	Reactor and Experimental Conditions	Main Results	Ref.
Humic acid	Fe coated Zeolite (ICZ) Zeolite clinoptilolite Synthesis by impregnation and precipitation at basic pH, dried at 105 °C. SEM and XRD analyses showed 2.132% Fe and the ICZ morphology. Granular activated carbon (GAC, from Merck) Dp = 1.5 mm SBET = 971.7 $m^2\ g^{-1}$	Semi-batch stirred reactor V = 2 L 1 h reaction time C_{O3l} = 10 mg L^{-1} (monitored) Q = 2 L min^{-1} Catalyst loading 0.75 g L^{-1} C_{HA} = 30 mg L^{-1} DOC = 8.52 mg L^{-1} pH = 6.50–11	DOC removal by ozone alone was 21.4% and increased up to 62% for ICZ-O_3 and 48.1% for GAC-O_3. The efficiency of the process was also tested with different humic acid fractions (<100 and <50 kDa). Fe leaching was analyzed near 60 µg L^{-1} at the beginning of the reaction. Kinetics was studied and showed the efficiency of both catalysts. Better results with ICZ in general. No post-chlorination. No DBPs were analyzed.	[163]
Humic acid	ZnO (Sigma Aldrich) 99.9% Density 5,61 g cm^{-3} Particle size < 5 µm	Semi-batch stirred reactor V = 2 L Inner diameter 12 cm Ozone dose 0.190 mg min^{-1} 2 h reaction time Catalyst loading = 0.25 g L^{-1} DOC_0 = 5.86 mg L^{-1} pH = 5.33 Ambient temperature	The degradation of HA by catalytic ozonation was much more effective (60% mineralization) than ozonation alone (30%). DFT modelling showed O_3 disproportionation on the ZnO surface to form reactive oxygen species. No post-chlorination. No DBPs were analyzed.	[164]
NOM River water (Harbin Mo Panshan, Harbin, China)	FeOOH-goethite pHpzc = 6.8 SBET = 97 $m^2\ g^{-1}$ Average pore size 23.2 nm CeO_2 pHpzc = 6.7 SBET = 116 $m^2\ g^{-1}$ Average pore size 12.6 nm MgO pHpzc = 11.1 SBET = 105 $m^2\ g^{-1}$ Average pore size 16.7 nm SEM and XRD analyses to confirm structure and morphology.	Ozonation in semi-continuous mode T = 20 °C V = 1 L Catalyst loading 100 mg L^{-1} Qg = 150 mL min^{-1} DOC_0 = 2.68 mg L^{-1} pH = 7.53 tert-BuOH as $HO^{\bullet}$ scavenger	UV254 removal was mainly from direct ozonation. CeO_2 was the best system in UV254 removal (69%). FeOOH (27.24%) and MgO (18.66%) better in mineralization than O_3 alone (10.7%) and O_3/CeO_2 (2.24%). Adsorption plays a negative effect in mineralization. Fractionation of DOC. Catalytic stability was high, with very low ions leaching (up to 11 µg L^{-1}) and SEM, XRD of used catalysts similar than for fresh samples. No post-chlorination. No DBPs analysis.	[165]

Table 6. *Cont.*

Target Water	Catalyst and Main Properties	Reactor and Experimental Conditions	Main Results	Ref.
Oxalic acid (OA) Dimethyl phthalate (DMP) and NOM surface water (Jingmi Cannel, Beijing, China)	RuO_2/ZrO_2-CeO_2 pHpzc = 6.0 Powder (less than 4 um) and pelletized (2 mm) 0.5 wt.% Ru SBET = 170 m^2 g^{-1} By XRD: Cubic CeO_2, RuO_2 detected. RuO_2/Al2O_3 0.5–5 mm 0.1 wt.% Ru SBET = 113–183 m^2 g^{-1} RuO_2/AC Particle size 6–10 mesh 1–1.5 mm 0.5 wt.% Ru AC ash content 7.8 wt.% (mainly SiO_2, Al_2O_3 and Fe_2O_3) SBET = 945.2 m^2 g^{-1}	Cylinder reactor, bubble column D = 5 cm L = 120 cm V = 1 L Semi-batch for catalytic activity: Cat. Dosage = 2 g L^{-1} Ozone dosage 116 mg h^{-1} Qg = 400 mL min^{-1} Continuous operation for stability tests: Qw = 20 mL min^{-1} Hydraulic time 60 min Experiment time 48 h Cat. Dosage = 40 g TOC_0 = 2.25–3.20 mg L^{-1} pH = 7.2–8.5 T = 15 °C Chlorination: pH = 7 Excess of chlorine T = 25 °C, 7 days	High efficiency in catalytic ozonation for OA and DMP, adsorption of intermediates. NOM mineralization but high adsorption capacity is observed. RuO_2/ZrO_2-CeO_2 had higher stability than Ru/AC and Ru/Al_2O_3. Post-chlorination and DBPs. HAAFPs removal improved from 38–57%. THMFPs from 50–64% but O_3 alone reacts fast with THMs precursors.	[166]
Humic acid and NOM Porsuk River (Turkey)	Degussa P-25 TiO_2 80/20 anatase/rutile SBET 50 m^2 g^{-1} Crystal size 21 nm.	Semi-batch stirred reactor V = 2 L Inner diameter 12 cm 2 h reaction time DOC (HA) = 5.8 mg L^{-1} DOC (NOM) = 3.76 mg L^{-1} pH = 5.5–6.0 Catalyst dose 0.25–1 g L^{-1} Ambient temperature O_3 dose 0.190 mg min^{-1}	DOC removal of humic acid was highly improved by catalytic ozonation (from 30% to 70%). The process was less effective with NOM. Adsorption plays an important role. By DFT modelling, $HO^\bullet$ are supposed to be formed from adsorbed O_3 and its decomposition onto TiO_2 surface in the presence of H_2O. No post-chlorination. No DBPs were analyzed.	[167]

Table 6. *Cont.*

Target Water	Catalyst and Main Properties	Reactor and Experimental Conditions	Main Results	Ref.
Humic acid	CuO (nanopowder, Sigma-Aldrich) SBET = 0.64 m^2 g^{-1} Pore volume = 0.002 cm^3 g^{-1} Monoclinic structure.	Semi-batch stirred reactor V = 2 L Inner diameter 12 cm Ozone dose 0.190 mg min^{-1} 60 min reaction time DOC = 5.86 mg L^{-1} pH = 5.5–6 T = 23 °C Catalyst dose 0.25 g L^{-1}	CuO presented catalytic activity in the ozonation of humic acid, with 80% of DOC removal in a short time. It was fond that adsorption, desorption, oxidation and chelating reactions took place in solution. The DFT approach proposed that HO• radicals formed in the catalyst surface initiate the heterogeneous catalytic ozonation reaction. No post-chlorination. No DBPs were analyzed.	[168]
Bromide in UP water Humic acid	TiO_2 P25 Anatase/rutile 80/20 Nano-SnO_2 Tetragonal pase Dcrystal) 9.7 nm	Cylindrical bubble column (d = 8 cm, h = 40 cm) V = 2 L Vr = 1.5 L Br- = 0.40 mg L^{-1} C_{O3} liquid = 3.38 mg L^{-1} pH = 6 Ccat = 100 mg L^{-1} T = 26 °C CHumic acid = 0–3.0 mg L^{-1}	Analysis of the BrO_{3-} formed by O_3 and catalytic O_3 The presence of both catalysts reduces the formation of BrO_{3-} during ozonation. Inhibition of BrO_{3-} formation with increasing humic acid concentration. No other DBPs analyzed.	[169]
NOM surface water (Wu-Lo River) pCBA	Fe-Mn oxide 23.0 % Fe 8.17% Mn 68.77% O SBET = 262.0 m^2 g^{-1} pHpzc = 5.9 Acidic groups = 181 ueq g^{-1} Basic groups = 636 TiO_2-αAl_2O_3 7.69% Ti 29.29% Al 63.02% O SBET = 14.5 m^2 g^{-1} pHpzc = 8.3 Acidic groups = 2074 ueq g^{-1} Basic groups = 0.71 ueq g^{-1}	Fluidized bed reactor Semi-continuous operation D = 6 cm L = 38 cm Gas flow rate 100 mL min^{-1} Vr = 1.04 L Reaction time 60 min O_3 concentration 2.5 mg L^{-1} Catalyst loading 1.25 g L^{-1} DOC_0 = 8.22–11.26 mg L^{-1} pH = 8.15–8.57 THMFP = 216–266 μg L^{-1} HAAFP = 509–553 μg L^{-1} T = 20 °C Chlorination: According to SM 2350 Residual chlorine 0.2–1 mg L^{-1}, pH 7, 48 h contact time.	Kinetics of O_3 decomposition, Rct Fe-Mn catalyst presented high hydroxyl radical exposure. Fe-Mn catalyst showed a best performance in O_3 decomposition, DOC (30%) and UV254 (85%) removal. Post-chlorination and DBPs analyses demonstrated that catalytic ozonation reduced THMs (70%) and HAA9 (75%) formation.	[170]

Table 6. *Cont.*

Target Water	Catalyst and Main Properties	Reactor and Experimental Conditions	Main Results	Ref.
NOM-surface water polluted with DOC from domestic and agricultural effluents (Wu-Lo River, Pintung, Taiwan)	α-FeOOH goethite Precipitation synthesis SBET = 61.9 m^2 g^{-1} SEM morphology micro needles.	Fluidized bed reactor Semi-continuous operation D = 6 cm L = 38 cm Gas flow rate 50 mL min^{-1} Vr = 1.04 L Reaction time 60 min O_3 concentration 2.5 mg L^{-1} Catalyst loading 0.5–1.5 g L^{-1} DOC_0 = 9.21 mg L-1 pH = 7.6 T = 20 °C Chlorination: According to SM 2350 Residual chlorine 0.2–1 mg L^{-1}, pH 7, 48 h contact time.	Higher efficiency in DOC removal by catalytic ozonation. Post-chlorination and DBPs No BrO_{3^-} detected, THMs and 9HAAs highly reduced by catalytic ozonation + biofiltration. The bromine incorporation factor of THMs and HAAs increases with catalyst dosage.	[171]
Humic acid Bromide	Ferrate (VI)	Homogeneous ozonation Flasks, Vr = 0.5 L Ferrate dose 0.1–5 mg L^{-1} C_{Br^-} = 100–1500 μg L^{-1} DOC_0 = 0.1–10 mg L^{-1} C_{O3} = 1.5–4 mg L^{-1} T = 5–40 °C pH = 3–11	Ferrate (VI) reduced BrO_{3^-} during catalytic ozonation. Humic acid content decreased bromate formation. No post-chlorination. No other DBPs were analyzed.	[172]
Humic substances in phosphate buffer.	Degussa P-25 TiO_2 80/20 anatase/rutile SBET 50 m^2 g^{-1}	Catalytic ozonation/Ultrafiltration Reactor membrane module and mixing tank. Injection of O_3 in a Y-type mixer. Gas flow rate 50–150 mL min-1 CO_3 gas = 2.5–10 mg L^{-1} TiO_2 loading 0–5 g L^{-1} DOC_0 = 8 mg L^{-1} pH = 2 (O_3 decomposition) Chlorination: SM 2350 (residual chlorine 0.5–2 mg L^{-1} after 48 h, pH 7).	Reduction of THMs formed during post-chlorination after catalytic ozonation. Positive effect of the TiO_2 loading. Membrane permeate recovered 93% by catalytic ozonation vs. 78% in ozonation alone.	[173]

Table 6. *Cont.*

Target Water	Catalyst and Main Properties	Reactor and Experimental Conditions	Main Results	Ref.
Bromide	Ferrate (VI)	Homogeneous ozonation Flasks, Vr = 0.5 L Ferrate dose 1–5 mg L^{-1} C_{Br^-} = 100–1500 μg L^{-1} C_{O3} = 0–5 mg L^{-1} T = 5–40 °C pH = 3–11	Ferrate (VI) reduced BrO_{3^-} during catalytic ozonation No post-chlorination No other DBPs were analyzed.	[174]
NOM from groundwater (Banat, Serbia)	Aeroxide P25 TiO_2 80/20 anatase/rutile SBET = 50 m^2 g^{-1}	Ozonation: Reactor bubble glass column D = 85 mm V = 2 L Vr = 1.5 L Gas flow rate 8 L h^{-1} Contact time 3–25 min O_3 dosage 0.4–3.0 mg O_3/mg DOC Catalyst dose 1 mg L^{-1} DOC_0 = 9.85 mg L^{-1} pH = 7.46 Chlorination: SM 2350 (residual chlorine 0.5–2 mg L^{-1} after 48 h, pH 7).	Removal THMFP 48% (no improvement observed compared to ozonation) Removal HAAFP 44% Less brominated species and haloacetonitrile after catalytic ozonation.	[175]
Humic substances	Bone charcoal (BC) Prepared at 600 °C for 4 h (bone from cattle and sheep). XRD revealed calcium phosphate (hydroxyapatite form) and amorphous carbon. XRF: CaO 92.9%; P2O5 3%; MgO 0.6%; SiO2 0.095%; MnO 0.008%. Boehm titration: Surface acidic groups 0.71 meq g^{-1} Surface basic groups 0.33 meq g^{-1} SBET = 121 m^2 g^{-1} pHpzc = 8.5	Semi-batch operation V = 1 L Catalyst 2 g C_{HS} = 15–100 mg L^{-1} Ozone 0.5 mg L^{-1} min^{-1} T = 15–40 °C pH = 2–12	The catalyst improves the reaction rate of HS degradation by 1.43- and 1.56- fold compared to ozonation. Kinetics and mechanism analysis. Kinetic studies showed that the rate greatly increased in the presence of BC. The reaction rate is related to BC dosage, HS concentration, pH, temperature. The presence of t-BuOH confirmed the main degradation route by hydroxyl radicals. Ea = 10 kJ mol^{-1} Diffusion controlled reaction. No chlorination. No DBPs were analyzed.	[176]

Table 6. *Cont.*

Target Water	Catalyst and Main Properties	Reactor and Experimental Conditions	Main Results	Ref.
NOM from surface water (Dong-Gang River, Pingtung, Taiwan)	$TiO_2/\alpha Al_2O_3$ No characterization studies reported	Fluidized bed reactor Continuous operation Gas flow rate 100 mL min^{-1} Water flow rate 50 mL min^{-1} O_3 concentration 2.5 mg L^{-1} Catalyst loading 0–25 g DOC_0 = 2.8–4.7 mg L^{-1} pH = 7.82–8.15 T = 20 °C Chlorination: According to SM 2350 Residual chlorine 0.2–1 mg L^{-1}, pH 7, 48 h contact time.	The combined catalytic ozonation significant reduced DOC up to 51.4% and DBPFP up to c.a. 50% (THMs and HAA6 analyses). The amount of catalyst had a positive effect in the process.	[177]
Dimethyl phthalate (DMP) and NOM in surface water (Jingmi Cannel, China)	Ru/AC Particle size 6–10 mesh 1–1.5 mm 0.5 wt.% Ru AC ash content 7.8 wt.% (mainly SiO_2, Al_2O_3 and Fe_2O_3) SBET = 945.2 m^2 g^{-1}	Cylinder reactor, bubble column D = 5 cm L = 120 cm V = 1 L Semi-batch for catalytic activity: Cat. Dosage = 2 g L^{-1} Ozone dosage 118 mg h^{-1} Q 300 mL min^{-1} Continuous operation for stability tests: Qw = 20 mL min^{-1} Cat. Dosage = 40 g COD_0 = 2.25–3.29 mg L^{-1} pH = 7.2–8.5 T = 25 °C Chlorination: pH = 7 Excess of chlorine T = 25 °C, 7 days	Ru/AC active in catalytic ozonation improved DMP mineralization. Stable for 42 h with no Ru leaching and 75% DOC removal. In natural water, Ru/AC-O_3 was better than O_3 alone for DOC removal (40% vs. 10%) and DBPFP reduction (60% vs. 40%); THMFP and HAAFP analyzed.	[178]

Table 6. *Cont.*

Target Water	Catalyst and Main Properties	Reactor and Experimental Conditions	Main Results	Ref.
NOM (Songhua River, Harbin, China)	FeOOH-goethite pHpzc = 7.0 SBET = 68.4 m^2 g^{-1} Average pore size 24 nm Surface OH density = 0.5 mmol g^{-1} CeO_2 pHpzc = 6.6 SBET = 117 m^2 g^{-1} Average pore size 7.9 nm Surface OH density = 0.4 mmol g^{-1} D = 0.075–0.3 mm	Ozonized water (20 mL; 10 mg O_3 L^{-1}) VT = 40 mL DOC = 10 mg L^{-1} O_3/DOC = 1 10 min reaction time pH = 7 T = 18 °C Fractionation of NOM with XAD-8 and XAD-4 amberlite resins: HOA: Hydrophobic acid HON: Hydrophobic neutral HIA: Hydrophilic acid HIB: Hydrophilic base Fluorescence analyses.	Ozonation decrease the aromaticity of humic-like structures and increases the generation of carboxylic groups. Catalytic ozonation improve the destruction of humic-like structures. Ozonation of HOA and HIA yields by-products with low aromaticity and low molecular weight. Catalytic ozonation enhances the formation of these by-products from HIA and improves the destruction of highly polycyclic aromatic structures. No post-chlorination. No DBPs were analyzed.	[179]
Humic and fulvic acid in distilled water	MnO_2 No characterization studies reported.	Cylinder reactor (bubble column, semi-batch operation, total recirculation of water) D = 7 cm L = 1.5 m V = 5 L O_3 dose = 47 mg L^{-1} in 30 min TOC_0 = 28–38 mg L^{-1} pH = 8.9 T = 19 °C	TOC removal 79% in catalytic ozonation vs. 67% in single ozonation. No post-chlorination. No DBPs were analyzed.	[180]
NOM-surface water (Lake Lansing, Michigan USA)	Fe_2O_3 over ceramic membrane (Al_2O_3/ZrO_2/TiO_2) Average Fe_2O_3 particles 4–6 nm (TEM)	Conditions of ozone-membrane system: Water recirculation rate = 2.75 L min^{-1} T = 20 °C O_3 flow rate: 100 mL min^{-1} Transmembrane pressure = 0.5 bar O_3 gas concentration = 2.5 mg L^{-1} Ozone injection before membrane module DOC_0 = 8.6–11.6 mg L^{-1} pH = 7.7–8.6 Chlorination SM 2350 (residual chlorine 0.5 mg L^{-1} after 48 h, pH 7).	Decomposition of O_3 in the membrane surface. DOC reduced > 85% Decrease of THMFP (90%), HAAFP (85%) and also the concentration of aldehydes, ketones and ketoacids.	[181] [182]

Table 6. *Cont.*

Target Water	Catalyst and Main Properties	Reactor and Experimental Conditions	Main Results	Ref.
DOC in river water (Japan)	Powdered AC (membrane as separator)	Ozonation: Reactor characteristics not specified. Column before membrane module O_3 dose = 1–2 mg L^{-1} AC dose 10 mg L^{-1} DOC_0 = 2.4 mg L^{-1} Q water = 5.5 m^3 day^{-1} Chlorination conditions not specified.	Combination of PAC and O_3 improved the performance of the subsequent membrane system. DOC reduced 59% T-THMFP reduced 75%	[183]
Humic acid solution NOM from Ebro River (Zaragoza, Spain)	TiO_2 (anatase Probus, SBET = 8.5 m^2 g^{-1}) 2.5 wt.% supported by adsorption on: Attapulgyte (clay, SBET = 104 m^2 g^{-1}) α-Al_2O_3 (SBET = 164 m^2 g^{-1}) Silica gel (SBET = 362 m^2 g^{-1}) Calcination T = 350–600 °C Granular form, d = 2–4 and 4–6 mm. TiO_2/Al_2O_3 catalyst characterization: SBET = 132 m^2 g^{-1} Pore volume = 0.33 cm^2 g^{-1} Pore size = 4.4 nm (12%); 20.56 um (88%). Anatase phase by XRD	Glass fixed bed reactor Semi-batch mode Humic acid experiments: V = 650 mL O_3 = 405 mg O_3 h^{-1} 30 min contact time C HA = 5.34 mg TOC L^{-1} pH = 7.2 Catalyst loading: 2.5–10 g L^{-1} NOM experiments: V = 2 L O_3 = 396.4 mg O_3 h^{-1} Contact time 2–4 min. Catalyst loading: 2.5 g L^{-1} TOC_0 = 4.46 mg L^{-1} pH = 7.94 Stability tests (4 consecutive runs): C Humic acid = 6.65 mg L^{-1} TOC pH = 7.62 Catalyst loading = 2.5 g L^{-1} V = 650 mL O_3 405 mg O_3 h^{-1} 30 min contact time	TOC and UV254 removal were slightly greater in the presence of a catalyst (60% vs. 32% in O_3 for TOC). The best results were obtained with TiO_2/Al_2O_3 calcined at 500 °C with a concentration of 2.5 g L^{-1} of catalyst and particle diameter 2–4 mm. The importance of adsorption is highlighted. Higher doses of O_3 have a positive effect in catalytic ozonation. Stability tests with the best catalyst showed a good stability with similar TOC, UV254 and pH results after the 4 experiments. No post-chlorination. No DBPs were analyzed.	[184] [185]

Table 6. *Cont.*

Target Water	Catalyst and Main Properties	Reactor and Experimental Conditions	Main Results	Ref.
Fulvic acid extracted from surface water (Cebron, France)	Commercial TiO_2 over Al_2O_3 beads Dbeads = 1.5–2.5 mm No characterization provided	Ozonation: Reactor: Conical glass flask 1.3 L V reaction = 1 L Contact time = 10 min O_3 dose = 2–6.5 mg L^{-1} (homogeneous experiments) C_{CAT} = 10 g L^{-1} DOC_0 = 2.84 mg L^{-1} H_2O_2/O_3 ratio = 0.35 mg H_2O_2/mg O_3 pH = 7 Chlorination: Cl_2 dose = 2 mg L^{-1} Contact time 1 h pH = 7	Catalytic ozonation increased ozone consumption. Higher mineralization (24%) compared with O_3 and H_2O_2/O_3 systems. Chlorine demand was minimized in catalytic ozonation. No DBPs were analyzed.	[186]
Model compounds: Fulvic acid Protein disaccharide	Granular catalyst TiO_2/Al_2O_3 GAT 925 SI D = 4 mm For comparison: BST: TiO_2 stick compacted APF: Granular TiO_2/Clay	Tested TiO_2/O_3 and $TiO_2/O_3/H_2O_2$ Bubble column Semi-continuous reactor with recirculation of the aqueous solution Total height 2 m D = 40 mm Porous glass to support the catalyst in the middle Q = 2.4–2.8 L h^{-1} C_{O3g} = 60–90 mg L^{-1} pH = 8 TOC_0 = 12 mg L^{-1} H_2O_2/O_3 = 0.4 g/g Catalyst loading 30 g L^{-1} Room temperature	Cumulative effect of adsorption and ozonation. At high ozone dosage TOC abatement by catalytic ozonation is observed (80% vs. 50% in O_3). APF was not stable under the conditions tested due to mechanic friction. No chlorination. No DBPs were analyzed.	[187]

Different iron oxides/hydroxides have been studied for catalytic ozonation mainly bare FeOOH, Fe_2O_3 and Fe_3O_4 or supported in different materials [143]. α-FeOOH-goethite has been used in other studies that reported SBET 60–97 $m^2\ g^{-1}$ and pH_{pzc} near 7. The properties of this material for catalytic ozonation have also been deeply discussed in Bai et al. [188]. In general, a higher efficiency in NOM mineralization was observed by catalytic ozonation compared to ozone alone. Only Wang et al. [171] analyzed post-chlorination DBPs in treated surface water and observed that THMs and HAAs were highly reduced by catalytic ozonation (combined with biofiltration).

Regarding the supported Fe-catalysts, Karnik et al. [181,182] used Fe_2O_3 over a ceramic membrane composed by $Al_2O_3/ZrO_2/TiO_2$ (no characterization reported) in the hybrid catalytic ozonation/ultrafiltration system and studied the decomposition of ozone in the membrane surface, obtaining a high removal of THMFP and HAAFP (higher than 85%) from surface water at the conditions shown in Table 6. Other authors supported FeOOH or Fe_2O_3 species over zeolite [163] or natural pumice [161] and observed an important catalytic effect of these materials in the degradation of synthetic humic acid solutions but no DBPs were analyzed in their works.

Other metal oxides and composite materials have been tested as catalysts. Turkay et al. [164,168] checked the effectiveness of CuO and ZnO nanoparticles with high crystallinity in the degradation of humic acid and NOM from surface water studying the possible mechanism by DFT modelling but no DBPs analyses. Wang et al. [166,178] studied the performance of RuO_2 supported onto activated carbon, Al_2O_3 and ZrO_2-CeO_2 using oxalic acid, dimethyl phthalate as target compounds and also NOM from surface water. They fully characterized the catalysts (see Table 6) with main differences between S_{BET} in AC supported catalyst. All the systems showed a high catalytic activity and improved ozonation results. THM precursors reacted fast with ozone alone but the THMFP and HAAFP were highly reduced in catalytic ozonation. Some aspects about the activity and stability of these catalytic systems are commented in the next sections.

Different manganese materials also showed catalytic activity for ozonation of NOM or humic and fulvic acids. Alsheyab and Muñoz [180] studied MnO_2 and at the conditions tested TOC removal improved from 67% in single ozonation to 79% in catalytic ozonation (pH = 8.9). No characterization and no DBPs studies were reported. Recently, Salla et al. [160] tested Mn_2O_3 as catalyst for humic acid ozonation that resulted in rapid ozone decomposition, neither DBPs were studied. Only Chen and Wang [170] prepared a mixed Fe-Mn oxide (8.17% Mn) with high surface area (262 $m^2\ g^{-1}$) for NOM removal and reduced THMs and nine HAA up to 70 and 75% respectively.

Cerium oxide have been used as support but also presented catalytic activity for the process. Zhang et al. [179] synthetized CeO_2 with pH_{pzc} = 6.6 and high S_{BET} = 117 $m^2\ g^{-1}$. The elimination of DOC from surface water was between 30–50% of different NOM fractions compared to 3–6% obtained in ozonation alone. On the contrary, a similar CeO_2 material was tested by Wang et al. [165] that found that their adsorption capacity exerted a negative effect in mineralization (lower than ozonation alone). Later, Zhang and Wang [162] synthetized different CeO_2-TiO_2 composites and studied the Ce/Ti ratio effect. They found that CeO_2 main phase was cubic fluorite and the positive effect of Ce content by increasing BET surface area because of lower crystallite sizes. An optimum Ce/Ti ratio of 0.8 was found due to the high surface area (expected) and low pH_{pzc} that improved adsorption properties and lead to lower molecular weight organics from humic acid degradation. Unfortunately, no post-chlorination DBPs were analyzed.

Regarding the use of carbonaceous materials as catalysts, though they have been widely used in catalytic ozonation of different organic pollutants, a few studies have been found for humic acid and NOM removal with post-chlorination and DBPs analyses. Shioyama et al. [183] combined powder activated carbon with membrane filtration for surface water treatment, which improved the performance of the subsequent membrane system reducing DOC up to 59% and T-THMFP up to 75%. Gümüs and Akbal [163] also studied the degradation of humic acid with granular activated carbon reaching 48% compared to 21% in ozonation alone though no DBPs were analyzed.

7.1.2. Catalytic Activity, Stability, and Reusability

No straightforward conclusions can be reached about the catalytic activity of the different catalysts checked at different operating conditions. However, some of the previous reported studies in Table 6 compare different materials that can differ in the active phase or the support.

Allemane et al. [187] and Gracia et al. [184,185] compared different TiO_2 catalytic systems. In their work, the relevance of the support was pointed out, being Al_2O_3 the best option that increased adsorption of NOM. The latter works also studied the stability of the TiO_2/Al_2O_3 catalysts in four reaction cycles of NOM catalytic ozonation at the operating conditions in Table 6. A good stability was maintained with similar TOC, UV254, and pH results after the four experiments.

Chen and Wang [170] compared the behavior of TiO_2/Al_2O_3 and a Fe-Mn oxide in the degradation of NOM and DBPFP of surface water. Fe-Mn presented the best performance in O_3 decomposition, DOC, UV254, and DBPFP removals. This was attributed to the catalytic activity of Fe and Mn species and to the highest specific surface area of this catalyst which improved the adsorption of NOM.

Also, the series of Wang et al. [166] demonstrated the importance of the support in RuO_2 supported catalysts. RuO_2/ZrO_2-CeO_2 resulted highly efficient compared to Al_2O_3 or AC as support, although a high adsorption capacity can also mask mineralization results. HAAFPs improved from 38–57% and THMFPs from 50–64% when comparing catalytic ozonation and ozonation, but O_3 alone reacts fast with THMs precursors. The best performance is ascribed to the efficiency of the redox Ce^{4+}/Ce^{3+} couple in the material which enhances the catalytic activity of very well dispersed Ru species. This catalyst was also more stable in continuous operation (only 48 h were checked with no Ru leaching).

The three works by Turkay et al. [164,167] have been performed using the same experimental set up and with quasi-similar operating conditions but with different catalysts. They studied the catalytic behavior of CuO, TiO_2 and ZnO nanopowders, respectively. All the materials presented an improved DOC removal (from humic acid with the same concentration and pH) which increased from 30% for ozone alone to 80% for CuO, 70% TiO2 and 60% ZnO. Thus, CuO seems the most active catalytic system but, in terms of stability, although not studied, probably TiO_2 is a better candidate for future works since Zn and Cu ions are easily leachable.

On the other hand, Wang et al. [165] compared the catalytic activity of FeOOH, CeO_2, and MgO. Whereas CeO_2 presented a better UV254 removal, FeOOH and MgO showed the best DOC mineralization results. In this work, adsorption onto CeO_2 seems to play a negative effect in its catalytic performance. The stability of all the materials was checked by ion leaching and SEM and XRD of the used catalysts with no significant changes in morphology or structural properties.

Recently, Salla et al. [160] compared classical Al_2O_3 with Mn_2O_3 catalysts in humic acid degradation. The favorable surface charge of Al_2O_3 led to a better performance in humic acid removal, but Mn_2O_3 promoted a faster decomposition of O_3. This last catalyst led to higher ozone consumption. Catalytic stability was studied in four consecutive runs. Some catalyst deactivation is observed with changes in the oxidation state of Mn and Al and the surface hydroxylation of both catalysts was observed by XPS.

Therefore, some controversial effects are still unknown such as the role of the adsorption of some NOM species (NOM fractionation) onto the catalysts surface or the true catalytic activity of the active species in comparable conditions. In terms of stability, long term and deep characterization would be desirable to check the real applicability of these materials.

7.1.3. Reactors and Variables Studied

Bubble columns or vessels were used in semi-batch mode operation, batch for liquid phase, continuous for gas phase have been usually selected for catalytic ozonation processes. The catalyst, depending on the particle size, has been suspended in slurry or fluidized bed reactors, or packed in fixed bed reactors. In any case, they are three-phase reactors with gas–liquid–solid phases in which matter transfer will play a key role. Figure 3 shows a scheme of the typical ozonation experimental set up.

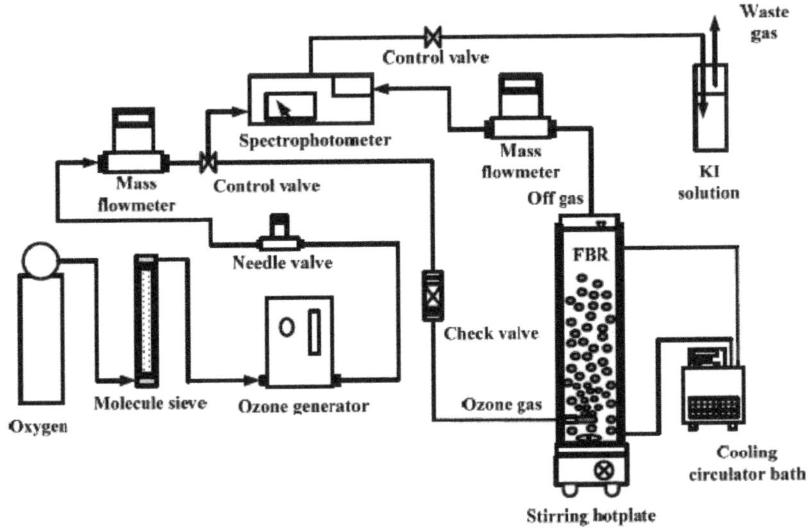

Figure 3. Experimental set up for catalytic ozonation experiments at lab scale [171] (From Wang, Y.H.; Chen, K.C. Removal of disinfection by-products from contaminated water using a synthetic goethite catalyst via catalytic ozonation and a biofiltration system. *Int. J. Environ. Res. Public Health* **2014**, *11*, 9325–9344. doi.org/10.3390/ijerph110909325).

In general, the installation is composed by an air or oxygen cylinder, an ozone generator which sends the ozone gas into the reactor, in line ozone analyzers (at gas inlet and/or gas outlet), mass flow controllers, and ozone destruction systems (KI solution, AC filter or UV analyzers with catalytic destruction). The reactor usually comprises a glass or acrylic column or vessel with variable volume (mainly 1–2 L) provided with a bottom diffuser for ozone inlet which favors mass transfer, magnetic or mechanical stirring, gas inlet, and gas outlet and sampling port. Some reactors are provided with thermostatic baths for temperature control, and temperature or analytic probes (ozone, pH) can be included in the reactor. Usually, the reactor is filled with the water to treat having some head space for gas outlet. The catalyst is usually added, and a pre-adsorption or homogenization time is spent before ozonation. Then the containing ozone stream is bubbled into the reactor and the catalytic ozonation experiment begins.

Among the typical operating conditions studied are DOC concentration, pH, catalyst loading, and ozone dose (Table 6). Ambient temperature was used in most of the works. Regardless of natural surface water or synthetic humic or fulvic acid solutions, DOC values are usually in the range 2–10 mg L^{-1}. Mortazavi et al. [176] detected a positive effect of DOC increasing the apparent reaction rate of catalytic ozonation as DOC increased.

It is well-known the role of pH on ozone reactions that has been usually described and commented in the introduction section. The application of the process for drinking water makes sense at the pH of natural waters (pH 6.0–8.5) and most of the studies are performed at natural pH. However, the relevance of pH in the NOM and catalyst charge has been evaluated. In this line, Salla et al. [160] assigned the best catalytic performance of Al_2O_3 with higher humic acid adsorption and lower ozone dosage due to favorable negative surface charge at the conditions studied ($pH = 5.5 > pH_{pzc} = 4.2$). This also affects the adsorption capacity of the catalysts for different NOM fractions (acidic, neutral, hydrophilic or hydrophobic). On the other hand, Mortazavi et al. [176] in their study evaluated the influence of pH on the catalytic ozonation performance of bone charcoal to remove humic substances in the range of pH 2–12. They found that heterogeneous catalytic ozonation contribution was much higher at low pH and ozonation alone contribution increases with alkalinity due to the relative importance of indirect ozone reactions. In addition, at $pH = 8 < pH_{pzc} = 8.5$ of the catalysts, a positive catalytic effect is still observed due to hydroxyl groups in the catalyst surface that act as Lewis acid sites for ozone decomposition. Therefore, depending on the catalyst nature, the catalytic effect could be improved also in alkaline solutions.

Catalyst loading is also an important parameter to optimize. The optimum dose is highly dependent on the catalyst nature and on the particle size used. The range used in the works reported in Table 6 for semi-batch experiments are as wide as 1 mg L^{-1} for TiO_2 in slurry, 10 mg L^{-1} for activated carbon or 30 g L^{-1} of TiO_2/Al_2O_3. In continuous operation, the space time (mass of catalyst/flow of water) ranged from 500–2000 g min L^{-1} [177]. In general, at sufficient ozone dose, the catalyst loading exerted a positive effect on the degradation rate of NOM. However, experimental check is needed in any different case.

In the same line, the optimization of ozone dose will mark the economic feasibility of the process in terms of ozone consumption/power energy required for ozone production. In general, research works reported in Table 6 applied high ozone dose and high ozone concentration in the gas phase with common values ca 10 mg L^{-1}. However, ozone concentrations used were as varied as 2–90 mg L^{-1}. Molnar et al. [175] observed that DOC removal did not improve at high ozone doses (0.1–3 mg O_3/mg DOC) using TiO_2/O_3 catalytic ozonation compared to ozonation alone, but attributed this effect to the pH of the natural water and the scavenging effect of carbonate/bicarbonate. On the contrary, Wang et al. [173], found a positive effect in the ozone dosage in the permeated flux of their hybrid system. However, no relevant changes were observed in TOC or UV254 removal or THMFP.

7.1.4. Mechanisms and Kinetics for the Removal of DBPs Precursors

General mechanisms such as presented in reactions (21) to (31) have been accepted for NOM or humic/fulvic acids degradation by catalytic ozonation depending on the catalyst used.

In general, humic/fulvic substances are easily degraded by ozone direct reactions due to the presence of unsaturated bonds and aromatic rings to produce other compounds less reactive towards ozone but also with different reactivity towards subsequent chlorine treatment. These other compounds, carboxylic acids, or aliphatic molecules, can be degraded by reactive oxygen species formed by catalytic decomposition of ozone (hydroxyl radical, ozonide radical, etc.) both in the liquid phase or near the catalyst surface/adsorbed. They may react with organic molecules with low selectivity, thus leading to a high mineralization of NOM.

Considering the transformation of NOM in different species prior to mineralization, it is important to know the chlorine reactivity of the different fractions formed during catalytic ozonation. Thus, Zhang et al. [179] observed that ozonation alone decreases the aromaticity of humic-like structures and increases the generation of carboxylic groups. Catalytic ozonation with FeOOH and CeO_2 catalysts also improved the destruction of humic-like structures and enhanced the destruction of the hydrophilic acid fraction destroying polycyclic aromatic structures. Unfortunately, no DBPFP was studied in this work.

Allemane et al. [187] proposed the adsorption of O_3 and NOM onto TiO_2/Al_2O_3 surface and the generation of oxidizing species in the surface. This trend has been observed in many of the studies dealing with TiO_2/Al_2O_3 [184,185]. With respect to the contribution of adsorption, direct reactions, indirect reactions during catalytic ozonation, Alver and Kilic [161] studied the mechanism of humic acid removal by Fe coated pumice catalytic ozonation in the presence of different scavengers (t-BuOH and phosphate). They observed the high contribution of hydroxyl radicals in solution to the overall process. The studies of Turkay et al. [164,167] for the degradation of humic acid are based in DFT modelling calculations to stablish the interaction of reactants with the catalyst surface. They concluded that $HO^{\bullet}$ radicals formed in the catalyst surface initiate the heterogeneous reaction in CuO with important contributions of adsorption-desorption and chelating [168]. For TiO_2, $HO^{\bullet}$ radicals are supposed to be formed from adsorbed O_3 and its decomposition onto TiO_2 surface in the presence of water molecules [167]. Finally, they proposed the O_3 disproportionation over ZnO surface to form reactive oxygen species [164].

In the same line, Gümüs and Akbal [163] studied the degradation of humic acid using GAC and Fe-coated zeolite as catalysts. For GAC they proposed a general mechanism reported by Beltrán et al. for the degradation of diclofenac [189,190]. In this mechanism adsorbed ozone reacts with adsorbed organic compounds to yield hydrogen peroxide which eventually might react with adsorbed ozone to form hydroxyl radicals:

$$DOC_1 + GAC \rightleftarrows DOC_1\text{–}GAC \tag{32}$$

$$O_3 + GAC \rightleftarrows O_3\text{–}GAC \tag{33}$$

$$DOC_1\text{–}GAC + O_3\text{–}GAC \rightarrow DOC_2 + H_2O_2 + GAC \tag{34}$$

$$H_2O_2 + GAC \rightleftarrows H_2O_2\text{–}GAC \tag{35}$$

$$H_2O_2\text{–}GAC + O_3\text{–}GAC \rightarrow HO^{\bullet} + GAC \tag{36}$$

$$DOC_1 + HO^{\bullet} \rightarrow DOC_2 + HO_2^{\bullet} \tag{37}$$

For Fe-coated zeolite (ICZ), they proposed the following mechanism:

$$ICZ\text{–}OH + O_3 \rightleftarrows ICZ(O_3)_S \rightleftarrows ICZ\text{–}O^{\bullet} + HO_3^{\bullet} \tag{38}$$

$$ICZ\text{–}O^{\bullet} + H_2O \rightarrow ICZ\text{–}OH + HO^{\bullet} \tag{39}$$

$$HO_3^{\bullet} \rightleftarrows H^+ + O_3^{\bullet -} \tag{40}$$

$$HO_3^{\bullet} \rightarrow HO^{\bullet} + O_2 \tag{41}$$

In this mechanism ozone is adsorbed in terminal –OH groups of the catalyst and then reactive oxygen species are produced that eventually would react with organic compounds (adsorbed or in the bulk liquid phase).

Then, despite the variety of catalysts and the few studies directly dealing with the determination of the mechanisms involved in the catalytic ozonation of humic acid or NOM, some general steps can be hypothesized:

$$Cat + DOC_1 \rightleftarrows Cat\text{–}DOC_1 \tag{42}$$

$$Cat + O_3 \rightleftarrows Cat\text{–}O_3 \rightarrow Cat\text{–}ROS \rightarrow Cat + HO^{\bullet} \tag{43}$$

$$Cat + DOC_1 + Cat\text{–}O_3 \rightarrow DOC_2 + Cat + [H_2O_2] \tag{44}$$

$$Cat\text{–}ROS + Cat\text{–}DOC_1 \rightarrow DOC_2 + Cat \tag{45}$$

$$DOC_1 + O_3 \rightarrow DOC_2 + [H_2O_2] \tag{46}$$

$$DOC_1 + HO^{\bullet} \rightarrow DOC_2 + HO_2^{\bullet} \tag{47}$$

$$DOC_2{+}HO^{\bullet} \rightarrow CO_2{+}H_2O \tag{48}$$

where Equations (42)–(45) represent heterogeneous reactions and Equations (46)–(48) are some of the well-known homogeneous ozone/hydroxyl radical reactions. This sequence will be completed by the general ozonation and $HO^{\bullet}$ mechanisms in water [26,191,192]. DOC_1 represents initial humic/fulvic substances that readily react with ozone or hydroxyl radicals to form organic compounds with lower molecular weight (DOC_2 accounting for carboxylic, aldehydes, aliphatic compounds, etc.) usually refractory to direct ozone attack that can be further mineralized by hydroxyl radicals. The reactivity of different DOC fractions towards chlorine and their subsequent DBPFP is crucial to stablish the need of achieving an advanced oxidation degree or a high mineralization. Therefore, the importance of each reaction should be the objective of future works to tailor the catalytic properties and to modify operating conditions for the degradation of NOM of specific characteristics to reduce DBPFP. It is necessary to point out that the analysis of the latter will be mandatory regardless of the mineralization achieved.

Regarding the kinetics of the process, only a few works dealing with kinetics beyond calculating apparent rate constants have been carried out. Mortazavi et al. [176] proposed the following kinetics for the degradation of humic acid using bone charcoal as catalysts. They worked reaching ozone saturation before humic acid injection in the reaction medium and, therefore, the system is considered homogeneous from the point of view of ozone. The contribution of both heterogeneous reactions taking place on the catalyst surface and homogeneous reactions by ozone and radicals in the liquid phase are represented:

$$-\frac{dC_{HA}}{dt} = (k^{HA}_{homo} + k^{HA}_{hetero} C_{S-OH}) C_{HA} \tag{49}$$

$$k^{HA}_{homo} = k^{HA}_1 C_{O3} + k^{HA}_2 C_{HO^{\bullet}} \tag{50}$$

$$k^{HA}_{hetero} = k^{HA}_3 C_{O3} + k^{HA}_4 C_{HO^{\bullet}} \tag{51}$$

$$\ln \frac{C_{HA}}{C_{HA0}} = -(k^{HA}_{homo} + k^{HA}_{hetero} C_{S-OH}) t = -kt \tag{52}$$

$$k = k^{HA}_{homo} + k^{HA}_{hetero} C_{S-OH} \tag{53}$$

where k represents the overall reaction rate constant and $k_1{}^{HA}$, $k_2{}^{HA}$, $k_3{}^{HA}$, $k_4{}^{HA}$, $k_{homo}{}^{HA}$, and $k_{hetero}{}^{HA}$ represent the HA degradation rate constants corresponding to a homogeneous reaction with ozone, a homogeneous reaction with hydroxyl radical, a heterogeneous reaction with ozone, a heterogeneous reaction with hydroxyl radicals, and global homogeneous and heterogeneous reactions, respectively; C_{HA}, and C_{HA0} are the concentrations of HA at any time and time zero; C_{HO} and $C_{S\text{-}HO}$ represent the hydroxyl radical concentration in liquid phase or in the catalyst surface and C_{O3} is de dissolved ozone concentration. This study is based on UV254 nm measurements and no mineralization is considered. As can be deduced from Equation (53), the rate of humic acid degradation resulted in apparent first order kinetics whose apparent rate constant k was determined at different operating conditions. With experiments at different temperature an activation energy value (E_a) of 10 kJ mol^{-1} was calculated.

Chen and Wang [170] studied the catalytic decomposition of ozone over Fe-Mn oxide and TiO_2/Al_2O_3 catalysts through simple first order kinetics approximation:

$$-\frac{dC_{O3}}{dt} = k_d C_{CATALYST} C_{O3} \tag{54}$$

$$\frac{C_{O3,t}}{C_{O3,0}} = \exp(-k_d t) \tag{55}$$

where k_d is the observed pseudo first-order reaction rate constant of O_3 decomposition of the catalytic ozonation system, C_{O3} is the dissolved ozone concentration and t is the reaction time. They found values of 2.8×10^{-3}, 9.0×10^{-3} and 4.1×10^{-3} s^{-1} for ozonation and Fe-Mn oxide, TiO_2/Al_2O_3 catalytic ozonation, respectively. They ascribed the low decomposition rate found for TiO_2/Al_2O_3 to the presence of phosphate that may be adsorbed in the catalyst surface.

To study the ozone exposure, hydroxyl radical exposure and the ratio between $HO^{\bullet}$ and O_3 concentration, Chen and Wang [170] also applied the R_{ct} concept from Elovitz and von Gunten [193] using data from experiments in the presence of p-chlorobenzoic acid (pCBA) at low concentration through the following equations:

$$\frac{dC_{pCBA}}{dt} = -k_{HO-pCBA}C_{pCBA}C_{HO^{\bullet}} \tag{56}$$

$$\ln\frac{C_{pCBAt}}{C_{pCBA0}} = -k_{HO-pCBA}\int C_{HO^{\bullet}}dt \tag{57}$$

$$R_{CT} = \frac{\int C_{HO^{\bullet}}dt}{\int C_{O3}dt} \tag{58}$$

$$\ln\frac{C_{pCBAt}}{C_{pCBA0}} = -k_{HO-pCBA}R_{CT}\int C_{O3}dt. \tag{59}$$

In this scheme, pCBA is considered to react only with hydroxyl radicals due to the low rate constant of its reaction with ozone. C_{pCBAt} and C_{pCBA0} are the concentrations of pCBA at any time and time zero, respectively, $k_{OH\text{-}pCBA}$ is the rate constant of HO-pCBA reaction, $C_{HO^{\bullet}}$ is the concentration of hydroxyl radical and C_{O3} is the dissolved ozone concentration. They found values for R_{ct} from 3.6×10^{-8} mol $HO^{\bullet}$/mol O_3 for single ozonation up to 14×10^{-8} and 9.8^{-8} mol $HO^{\bullet}$/mol O_3 for Fe-Mn oxide and TiO_2/Al_2O_3 catalytic ozonation, respectively. This is indicative of the improved capacity of Fe-Mn oxide catalyst to generate hydroxyl radicals during the process.

Gümüs and Akbal [163] used a second-order kinetic model for humic acid catalytic ozonation (kinetic equations not shown) using Fe coated zeolite (ICZ) and GAC as catalysts. They reported different rate constants for ozonation at different dissolved O_3 concentrations, catalyst doses, pH and molecular weight of initial DOC (<100 and <50 kDa). They observed the high catalytic activity of ICZ vs. GAC and also demonstrated the highest reactivity of the humic acid fraction <100 kDa with ozone, being catalytic processes much more effective for the fraction <50 kDa.

In general, no complex mechanistic kinetics has been developed for catalytic ozonation of NOM, being an important gap for catalytic systems optimization. In addition, taking into account every topic of these section, NOM or humic/fulvic acids degradation has been evaluated but there is a lack of studies covering the formation of different DBPs also with global parameters determination such as AOX that are important to check the real applicability of this process for drinking water treatment.

7.2. Photocatalytic Ozonation

As a difference from catalytic ozonation, so far literature only reports just one work dealing with the combination of ozone, catalyst and light, that is, photocatalytic ozonation (PhCatOz) to remove natural organic matter, NOM, or humics from water [194]. Photocatalytic ozonation is an emerging AOP where hydroxyl radicals may generate from different mechanisms including adsorption and oxidation of water or hydroxyl groups on the valence band of the catalyst, ozone decomposition, reactions of ozone and oxygen with electrons of the conduction band of the catalyst or even reaction of ozone with possible hydrogen peroxide formed from superoxide ion recombination or from direct ozone-organics reactions [27]. The basic mechanism NOM PhCatOz is as follows:

Direct ozone reactions with NOM:

$$O_3 + NOM \rightarrow O_3\text{–}DBP + H_2O_2 \tag{60}$$

Direct photolysis of NOM (usually at $\lambda < 300$ nm):

$$NOM \overset{h\gamma}{\rightarrow} UV\text{–}DBP \quad (61)$$

In these photolytic reactions, ozone and UV radiation disinfection byproducts are formed (see Section 3). These DBPs eventually, after subsequent ozone and photolysis reactions, are transformed in biodegradable compounds but some may have certain toxicity (i.e., aldehydes, ketones). Bromate can be formed if bromide was initially present.

In addition to reactions (60) and (61), given the presence of aromatic ring structures with hydroxyl substituents groups in NOM, and, hence, the presence of nucleophilic points, ring breakings give rise to the appearance of hydrogen peroxide as has been reported previously for the ozonation of phenol compounds.

Hydroxyl radical formation reactions via direct ozone decomposition, mainly from:

$$O_3+HO_2^- \rightarrow O_3^{\bullet -}+HO_2^{\bullet} \rightarrow O_3^{\bullet -}+O_2^{\bullet -} \ldots \overset{O_3}{\longrightarrow} \ldots \rightarrow 2HO^{\bullet} \quad (62)$$

Catalyst activation via light incidence:

$$\text{Catalyst} \overset{h\nu}{\longrightarrow} h^+ + e^- \quad (63)$$

Hydroxyl radical formation reactions via hole (h+) reduction from the valence band and/or superoxide and ozonide ion radical formation by capturing electrons from the conduction band:

$$h^+ + H_2O, OH^- (\text{adsorbed}) \rightarrow HO^{\bullet} \quad (64)$$

$$O_3+e^- \rightarrow O_3^{\bullet -} \rightarrow \ldots \rightarrow HO^{\bullet} \quad (65)$$

$$O_2+e^- \rightarrow O_2^{\bullet -} \overset{O_3}{\rightarrow} O_3^{\bullet -}+O_2 \rightarrow \ldots \rightarrow HO^{\bullet} \quad (66)$$

$$2HO_2^{\bullet} \rightarrow H_2O_2 \quad (67)$$

$$H_2O_2+e^- \rightarrow HO^{\bullet}+OH^-. \quad (68)$$

Hydroxyl radical formation from ozone and hydrogen peroxide photolysis (at $\lambda < 320$ nm):

$$O_3 \xrightarrow{h\nu, H_2O} H_2O_2+O_2 \xrightarrow{h\nu} 2HO^{\bullet} \quad (69)$$

and reactions of NOM with hydroxyl radicals:

$$NOM + HO^{\bullet} \rightarrow \text{Intermediates} \xrightarrow{+HO^{\bullet}} \ldots \rightarrow CO_2+H_2O + \text{Inorganic Anions} \quad (70)$$

In spite of being an incipient process compared to other AOP such as O_3/H_2O_2, O_3/UVC-UVB, photocatalytic oxidation or Fenton process, PhCatOz has already been the subject of reviews since 2005 [195]. Many works of PhCatOz deal with the removal of model compounds such as pharmaceuticals and pesticides [196] or wastewater [197], with the use of visible or solar light and catalysts such as TiO_2, and others (ZnO, WO_3, C_3N_4, etc.), or TiO_2 composites that allow TiO_2 activation with visible light [198]. For instance, among catalysts checked, MOFs (already quoted in Section 7.1) [199] and graphitic carbon nitride, gC_3N_4 [200] has emerged to improve photocatalytic ozonation rate of pollutants. The use of gC_3N_4 is particularly interesting since this material presents a very high conduction band potential (-1.3 V_{NHE}) that greatly improves the ozone photoelectron capture to diminish electron-hole recombination and improve the formation of hydroxyl radicals [201]. According to WOS data base there are 160 works on photocatalytic ozonation so far published. As said before, however, only the work of Yuan et al. [194] treats photocatalytic ozonation for the removal of humics to reduce THM and TOX formation potential. Nonetheless, treatment of humics or NOM with photocatalysis and ozone as a sequence of processes has some antecedents due to Kerc et al. [202], Bekbolet et al. [203], and Uyguner et al. [204]. In their works, removal of humic acids is accomplished by ozonation, photocatalytic oxidation and the sequential use of the two latter, that is, they used ozonation followed by photocatalytic oxidation in separate processes, and they do not use PhCatOz as the simultaneous application of both processes. Degradation of humics is followed by measuring the absorbance of remaining water at 254 and 436 nm, corresponding to aromatics and colored compounds, respectively. In some cases, the treated water later undergoes an adsorption or coagulation process to improve organic matter removal. In these works, it is observed the better efficiency of the sequential process to remove the humic matter from water. Nonetheless, only the photocatalytic oxidation is an AOP since ozonation, in most cases, is finished when dissolved ozone was not yet present in water. This means that during the ozonation step humic acids were mainly removed by direct ozone reactions [27]. The work of Bekbolet et al. [203] also gives results of formation potential of THM and some other halogen organic compounds (of chlorine and bromine) such as haloacetic acids, halonitriles and others. The results show the better performance of the sequential oxidation process (ozonation followed by photocatalytic oxidation) to decrease TOXFP. Coming back to the work of Yuan et al. [194], PhCatOz is applied, in this case, to remove a humic acid (its origin is not reported) with ozone, UVC radiation (250–260 nm with 254 nm as main wavelength) and different TiO_2 catalysts in powder or supported form. Supported materials were ceramsite, zeolite, alumina, and activated carbon (AC), while the composite prepared was Fe on TiO_2 nanotubes (Fe-TNT). The catalyst was characterized through XRD, SEM, TEM, EDS, XPS, and N_2 adsorption and UVVis-DRS. With the latter, they observed band gaps of 3.23 and 3.03 eV for their synthesized TNT and Fe-TNT with no influence of the supported material, a logical consequence of its lack of action on optical properties. The authors observed a significant increase of humic acid removal with PhCatOz when using AC as the supported material likely due to the BET surface area, the highest among the synthesized catalysts. They also observed after five consecutive runs with the same catalyst that removal efficiency of humic acid decreased from 94 to 78%. They concluded with the existence of a synergism between AC adsorption and photocatalytic ozonation. However, no data is given about TOXFP or THMFP. However, the results, suggest the importance of PhCatOz to improve the removal of TOX precursors.

8. Elimination of DBPs by Catalytic/Photocatalytic Ozonation Processes

As it can be seen from Table 7, as far as these authors know, there are only six works dealing with catalytic/photocatalytic ozonation processes of DBPs of drinking water. As it occurs with Section 6.2, chloroacetic acids (CAAs) are the most studied.

Table 7. Works on DBPs removal by catalytic/photocatalytic ozonation processes.

Target DBPs	Catalyst and Main Properties	Reactor and Experimental Conditions	Main Results	Ref.
MCA DCA Deionized water	TiO_2 photocatalyst coated on the commercial product "Pilkington ActiveTM glass" (PAGs).	Planar falling film reactor. Seven UVA lamps (15 W, 360 nm) fixed inside the reactor. Intensity UV light: 1 mW cm^{-2}; water flow rate: 1 L min^{-1}, volume 0.5 L; [CAA] = 1 mM; pH 3, pure gaseous oxygen rate of 10 L h^{-1}, ozone gas with 130 ± 5 mg L^{-1} ozone at a power of 30 W.	Single ozonation was ineffective method for the removal of chloroacetic acids. However, a significant enhancement in the degradation and mineralization efficiencies was observed by the combination of ozone with photocatalysis.	[106]
DCA	FMOG: Flower-like nanocomposite (FMOG) consisted of pure β -MnO_2 (PMO) and reduced graphene oxide nanosheets (RGO). Atomic ratio between Mn and O element is ca. 1:2.1.	500 mL glass conical flask. 110 min; flow rate of O_3: 4.47 mmoL min^{-1}, T = 20 °C, concentration of catalyst: 50 mg L^{-1}, and pH = 4.4. $[DCAA]_0$ = 100 mg L^{-1}.	FMOG displayed higher catalytic performance compared with ozonation, and PMO catalytic ozonation.	[205]
DCAcN	Titanium dioxide (Degussa, P-25, anatase/rutile = 75/25%).	Bench and outdoor photocatalytic ozonation systems consisted of the combination of CPC and ozone reactor. Metal halide lamps of different powers (100, 250, and 400 W); UV-solar, 300–800 nm. pH 6.5, 20 ± 1 C, TiO_2 dose 0.4 g L^{-1}, ozone dose 1 g L^{-1} h^{-1}, UV intensity 33.8 W m^{-2}. pH: 3, 6.5, and 10.	Compared to the single process, the UV-solar/TiO_2/O_3 process had the highest DCAN-removal rate. Optimal conditions: pH neutral; 1 g L^{-1} TiO_2 and 1.13 g L^{-1} h^{-1} ozone; T = 20 °C; UV-solar intensity: 4.6–25 W m^{-2}.	[206]
DCA	Immobilized TiO_2 in the form of a well-known commercial product called "Pilkington ActiveTM glass" (PAG) sheet with a contact area of 30 × 5 cm^2.	A planar reactor: Polymethylmethacrylate box (160 cm^3 vol) covered by an optical window made from the same material. Reactor connected to an ozonation chamber (500 cm^3 vol). 400 mL of dichloroacetic acid solution. 30 W lamp (300–420 nm and a maximum at about 360 nm). T = 25 °C, pH = 3, [DCA]o = 1 Mm. O_3 input concentration: 135 ± 5 mg L^{-1}.	(PAG/O_3/UVA) showed highly modified oxidation properties in the decontamination of DCA in aqueous solutions. Kinetics of first order reactions with respect to dichloroacetic acid were found. Higher initial concentrations of DCA and higher temperatures increased the initial degradation rate. More than 90% of DCA was removed.	[207]

Table 7. *Cont.*

Target DBPs	Catalyst and Main Properties	Reactor and Experimental Conditions	Main Results	Ref.
DCA in simulated water: deionized water with different amounts of humic acid (HA).	Natural bentonite, composed primarily of Ca^{2+}-montmorillonite.	Cylindrical reactors with an inner diameter of 60 mm and a length of 500 mm. Gas flow rate 20 mL min^{-1}; ozone gaseous concentrations of 0.90 mg L^{-1}; 298 K; 2 L simulated water: Deionized water containing 1 mg L^{-1} DCA or different amounts of HA, pH adjusted to 6.0.; Fe^{3+} dosage: 5 mg L^{-1}.	Under the combined effects of adsorption, ozonation and catalytic oxidation, high DCA removal is obtained.	[208]
DCA	Nanometer ZnO powder: Size 90 nm, and surface area (11 m^2 g^{-1}).	Batch experiments. A glass flat-bottomed flask with the inside diameter of 45 cm and the volume of 1.2 L. T = 20 ± 1 °C; initial pH 6.88; [DCA]o = 100 μg L^{-1}; initial ozone concentration 1.96 mg L^{-1}; catalyst dosage 100 mg L^{-1}.	ZnO as catalyst in water significantly improved the ozonation removal of DCA. The degradation efficiency of DCA increased with the increasing pH of solution, catalyst dosage and ozone dosage. The presence of t-BuOH had a negative effect on catalytic ozonation of DCA.	[209]

8.1. Catalytic Ozonation

Due to the selective oxidation property of the ozone molecule, highly structured organic substances cannot be thoroughly mineralized in ozonation alone and low-mass compounds like chloroacetic acids tend to be discharged. In addition, the efficiency of the oxidation can be strongly affected by the presence of natural organic matter which is ubiquitous in drinking or fresh water [208]. Hence, catalytic ozonation, an efficient AOP that introduces homogeneous or heterogeneous catalyst to single ozonation, can increase ozonation efficacy and ozone utilization degree. Especially, heterogeneous catalytic ozonation processes have received increasing attention recently due to the potentially higher effectiveness in the degradation and mineralization of refractory organic pollutants and easier recovery processing [209].

8.1.1. Catalysts

Li et al. [205] studied the potential use of manganese oxide-based nanocomposite in catalytic ozonation of DCA. Manganese oxide (MnOx) is an eco-friendly and inexpensively used catalyst in water treatment. Among various crystal phases of MnOx, β-MnO_2, not only has the highest stability and the lowest water-solubility but can also be easily fabricated in various morphologies such as nanowires, nanorods, nanotubes, and nanoflowers. However, the catalytic performance of single-component β -MnO_2 so far is unsatisfactory. In order to enhance the catalytic property of β -MnO_2, some promising methods such as modifying β -MnO_2 with noble metals or combining β -MnO_2 with other oxides have been developed. In this sense, graphene and its partial oxide counterpart-reduced graphene oxide (RGO) have emerged as promising candidates for fabricating new materials due to their high specific surface area, chemical stability, as well as biocompatibility. It has been proved that composite fabricated by coupling graphene or RGO with semiconductors can achieve a higher catalytic activity. Herein, Li et al. [205] reported a unique three-dimensional (3D) flower-like nanocomposite (FMOG) consisted of pure β -MnO_2 (PMO) and reduced graphene oxide (RGO). Interestingly, FMOG displayed higher catalytic performance compared with ozonation, and PMO catalytic ozonation. Gu and co-workers [208] used bentonite and Fe^{3+} for dichloroacetic acid (DCA) removal from drinking water. Bentonite is a 2:1 type clay mineral. Its unit layer structure consists of one Al^{3+} octahedral sheet placed between two Si^{4+} tetrahedral sheets. It should also be noted that the addition of hydrolysable metal ion species (iron or aluminum) can be rapidly hydrolyzed around the surface of bentonite particles, resulting in a hydroxyl surface. In addition, the hydrozed iron (or iron ion) possesses the ability to catalyze the transformation of O_3 to hydroxyl radicals, and hence further promote the removal of dissolved target organic compounds. Zhai et al. in their study [209] used nanometer ZnO powder as a heterogeneous catalyst for catalytic ozonation of dichloroacetic acid (DCA) in aqueous solution, which is non-toxic, insoluble, and a cheaper transition metal oxide widely used in various processes.

8.1.2. Catalytic Activity, Stability, and Reusability

To demonstrate the catalytic activity of PMO and FMOG, Li et al. [205] compared these catalysts in catalytic ozonation of DCA from drinking water sources. Ozonation alone of DCA achieved 32.7% elimination in 60 min. When the PMO or FMOG were added into the system, the elimination ratio of DCA increased to 39.2% and 46.8% in 60 min, respectively. In this research, DCA was degraded completely in FMOG catalytic reaction after 110 min, while in the same reaction conditions the residual concentration ratio of DCA was 30% and 13.3% for ozonation alone and PMO catalytic reaction. These results indicate an evident catalytic ozonation effect of FMOG for DCA degradation. Also, reusing experiments were carried out to estimate the stability of PMO and FMOG. The used catalyst was separated from the solution by centrifugation. The catalytic activity of FMOG decreases gradually in the first four runs. However, the catalytic activity of PMO decreases more obviously compared with the one of FMOG. In the fifth run, the DCA elimination ratio of PMO dropped to 32.5%, demonstrating a continuous downtrend; while the DCA elimination ratio of FMOG in the fifth run was 37.7%, which was almost the same as that of the fourth run. SEM of FMOG after fifth catalytic ozonation showed that the flower-like nanostructure was still distinct. These facts indicate that FMOG is a relatively robust catalyst for catalytic ozonation of DCA in practical applications. Gu et al. [208] compared ozonation alone, ozone/bentonite and combined addition of ozone/bentonite/Fe^{3+} (adsorptive ozonation) for removing DCA from water in the presence and absence of HA. In this work, the combination of ozone/bentonite and Fe^{3+}, significantly promotes DCA removal achieving a percentage of about 73% after 40 min reaction time. This improvement was due to the ability of Fe^{+3} in catalyzing ozone decomposition to hydroxyl radical. Hence, in the process of ozonation/bentonite/Fe^{+3}, more $HO^{\bullet}$ radicals are generated. ZnO as catalyst in water significantly improved the ozonation removal of DCA compared with ozonation alone [209]. The addition of ZnO catalyst improved the degradation efficiency of DCA during ozonation, which caused an increase of 22.8% for DCA decomposition compared to the case of ozonation alone after 25 min. Under the same experimental conditions, DCA decomposition was enhanced by increasing catalyst dosage from 100 to 500 mg L^{-1} and ozone dosage from 0.83 to 3.2 mg L^{-1}. The catalytic ozonation process is more pronounced than the ozonation process alone at pH 3.93, 6.88, and 10. With increasing the concentration of t-BuOH from 10 to 200 mg L^{-1}, the degradation of DCA was significantly inhibited in the process of catalytic ozonation, indicating that the degradation of DCA by adding nanometer ZnO powder follows a radical-type mechanism.

8.1.3. Reactors and Variables Studied

In the work of Li et al. [205] catalytic ozonation of DCA was carried out into a 500 mL glass conical flask containing 200 mL of DCA aqueous solution (100 mg L^{-1}) with 10 mg of the as-prepared catalyst at room temperature under magnetic stirring. The effect of pH, the reaction temperature and catalyst dose on the catalytic DCA elimination by using FMOG was studied. Cylindrical reactors were used by Gu et al. [208], with an inner diameter of 60 mm and a length of 500 mm and 2 L working volume. In this work, authors study the efficiency of ozonation alone and catalytic ozonation with bentonite on DCA removal in the presence of different humic acid (HA) concentrations. HA was applied to test the effect of NOM concentrations on catalytic ozonation of DCA. Zhai and co-workers [209] performed the experiments in a laboratory batch reactor, which consisted of a glass flat-bottomed flask with the inside diameter of 45 cm and the volume of 1.2 L. In this study, authors checked the influencing factors on the degradation efficiency of DCA in aqueous solution by ZnO catalytic ozonation, including catalyst dosage, ozone concentration, initial solution pH, and tert-butyl alcohol.

8.1.4. Kinetics and Mechanisms

In all experiments carried out by Li et al. [205], ozone was supplied into the reaction system superfluously and continuously. Therefore, the pseudo first order reaction was employed by authors to investigate kinetics of single and catalytic ozonation, respectively, of DCA. For the DCA ozonation process, the apparent rate constant was 0.0062 min^{-1}. For the catalytic ozonation of DCA process in the presence of PMO, the apparent rate constant augmented to 0.0081 min^{-1}. However, when using FMOG as catalyst, the apparent rate constant increased to 0.0104 min^{-1}. Compared with the PMO catalytic process and the ozonation process, the apparent rate constant of FMOG catalytic process increases 28% and 68% respectively. The remarkable catalytic capacity of FMOG revealed that it can be used as an effective catalyst for the ozonation degradation of DCA. Based on literature, RGO has π electron donating character of the sp^2-bonded carbon structure, which can make FMOG more effective than PMO in transferring electron to ozone, and the polylaminate structure of RGO improved the surface area of the catalyst, which provides more ozonation reaction centers. The synergistic effect of β -MnO_2 and RGO in catalytic ozonation may contribute to the outstanding performance of FMOG. Gu et al. [208] proposed a pseudo first order degradation model. Results showed that the rate of DCA removal with ozonation is close to that of the ozone/bentonite, and the rate of the adsorptive ozonation is much higher than that of the other two processes. The presence of HA in aqueous solution lowers the rate of the DCA degradation. In ozonation alone, when 4 mg L^{-1} HA was added, the rate is lowered to about 0.006 min^{-1}, which is only 28.6% of the rate when HA is absent. However, in the process of adsorptive ozonation, the negative effect of HA on DCA removal is weakened. When 4 mg L^{-1} HA was added, the rate constant is lowered to 0.052 min^{-1}, about 53% of that in the absence of HA. Zhai et al. [209] supposed that hydroxyl radicals in solution promote the oxidation degradation of DCA. Here, fundamental study was performed using DFT to explore the mechanism of generating hydroxyl radical on the ZnO surface. The DFT calculation results further verified the decomposition of the adsorbed ozone on the catalyst surface and the enhancement of generation of OH responsible for high ZnO catalytic activity, leading to the increase of degradation efficiency of the model pollutant DCA.

8.2. Photocatalytic Ozonation

So far only three works on photocatalytic ozonation of DBPs already formed have been published (see Table 7). As happens in catalytic ozonation works, chloroacetic acids are the chlorine disinfection by-products most studied.

8.2.1. Catalysts and Radiation Use

Degradation of DCA in aqueous solutions using photocatalytic ozonation has been investigated by Hama Aziz [106] and Mehrjouei et al. [207] (see Table 7). The photocatalyst used in both works was titanium dioxide immobilized in the commercial product "Pilkington ActiveTM glass" (PAGs). The great superhydrophilicity of TiO_2 photocatalyst provided a homogeneous and stable falling liquid film along the glass sheets. In both works, the application of irradiation with wavelengths in the region of UVA instead of shorter wavelengths promotes the idea of moving towards the use of natural solar light. Hence, radiation use for the photocatalytic experiments was in a range of wavelengths between 300 nm and 420 nm and a maximum at about 360 nm. On the other hand, Shin et al. [206] studied the removal of dichloroacetonitrile (DCAcN) with solar PhCatOz and P25 TiO_2 as catalyst (slurry mode). The authors use both UVA-visible lamps of different intensity and solar light as radiation source as an energy-saving and environmental-friendly process.

8.2.2. Catalytic Activity, Stability, and Reusability

Hama Aziz [106] found that the combination of ozonation with photocatalysis (TiO_2 on the surface of PAGs irradiated by the UVA) had shown a significant synergistic effect on DCA degradation due to the production of highly reactive hydroxyl radicals either by the direct photolysis of ozone with UVA or by interaction of ozone with the conduction band electrons on the illuminated TiO_2. At the same time, a dramatic increase in the mineralization level was achieved. Thus, comparing the degradation efficiencies of DCA by the photocatalytic ozonation ($UVA/TiO_2/O_3$) and photocatalytic decomposition of H_2O_2 ($UVA/TiO_2/H_2O_2$) processes in terms of the observed rate constants (the degradation of chloroacetic acids was described by pseudo first-order kinetics) indicates that the photocatalytic ozonation provides remarkably higher degradation efficiency than the photocatalytic-H_2O_2 [106]. Similar results were found by Mehrjouei et al. [207] where photocatalytic ozonation treatment by means of Pilkington ActiveTM glass as a commercial product irradiated by UVA light and combined with ozone ($PAG/O_3/UVA$) showed highly modified oxidation properties in the decontamination of DCA in aqueous solutions compared to both photocatalytic oxidation and ozonation separately. Also, photocatalytic ozonation exhibited high potential in the mineralization of DCA. It was observed that more than 90% of DCA decomposed during the oxidation period was mineralized to carbon dioxide molecules and chloride anions. In Mehrjouei study [207], the degradation of DCA by photocatalytic ozonation in heterogeneous system showed good agreement with the kinetics of first order reactions. However, at initial concentrations of DCA both higher and lower than 1 mM, the initial degradation rates were found to be slightly different. In the case of DCAN degradation [206], compared to the single process, the UV-solar/TiO_2/O_3 process had the highest DCAN-removal rate. This was due to additional $HO^{\bullet}$ production by the reaction of ozone with the electrons of the conduction band in the catalyst. The UV-solar/TiO_2/O_3 system showed an enhanced DCAcN-removal rate, and the synergetic index calculated according to equation (71) for a single process was 3.8:

$$SI = \frac{k_{Sun/O3/Cat}}{k_{Sun/Cat} + k_{O3}} \tag{71}$$

where k values are the apparent pseudo first order rate constants of PhCatOz, solar photocatalysis and ozone processes, respectively.

Works quoted in Table 7 have also studied the photocatalyst stability and activity by conducting cycles of photocatalytic ozonation runs of fresh compound aqueous solutions with the same photocatalyst.

8.2.3. Reactors, Radiation Source, and Variables Studied

A planar falling film reactor was used by Hama Haziz [106] for the photocatalytic experiments. The reactor consists of two photoactive self-cleaning Pilkington Active^{TM} glass (PAGs) sheets, each with a 68 × 29 cm surface area, connected by a frame of PVC. The falling liquid film was established on the inner side of the PAGs. Seven UVA lamps (each with 15W energy consumption and the maximum wavelength at 360 nm) were fixed inside the reactor and used for UVA irradiation. The incident light intensity of the UV light was 1 Mw/cm^2, and no significant aging effect was found during the experiments. The optimization of water flow rate to generate a stable and homogeneous falling film along the PAG sheets was tested being 1 Lmin^{-1}. The planar reactor used by Mehrjouei et al. [207] was a polymethylmethacrylate box with an internal volume of 160 cm^3 covered by an optical window made from the same material in order to let the irradiation of UVA light pass through and reach the photocatalytic surface of a Pilkington ActiveTM glass sheet with a contact area of 30.5 cm^2, which was embedded and fixed inside the reactor. For all runs, a volume of 400 mL of DCAA solution was injected through the bottom inlet of the reactor to form a 3 mm liquid layer over the semiconductor surface, and it left the reactor through the top outlet point. The UVA-light source employed in this study was a 30 W lamp with a range of wavelengths between 300 nm and 420 nm and a maximum at about 360 nm. The incident light intensity of this source, as in the case of Hama Haziz [106] was ca. 1 MW/cm^2. In [207] influence of initial concentration and temperature on the degradation rate of DCAA and the ozone consumption level during the oxidation process were investigated. Higher initial concentrations of DCAA and higher temperatures increased the initial degradation rate and, the level of ozone consumption during the photocatalytic ozonation treatment.

Shin et al. [206] for the photocatalytic experiments (bench and outdoor) used the combination of CPC and ozone reactor. CPC contained three quartz tubes (length: 40 cm, diameter: 3 cm, thickness: 1.5 mm) and three stainless modules with reflectors made of polished aluminum. A cylindrical ozone reactor with a cooling system was made of stainless steel (length: 110 cm, diameter: 10.5 cm, and total volume: 9.5 L). For the bench system, UV radiations were obtained from three metal halide lamps of different powers (100, 250, and 400 W), which were placed 60 cm above the CPC reactor at the top of the chamber as artificial solar light. Figure 4 represents the experimental set up used in this work.

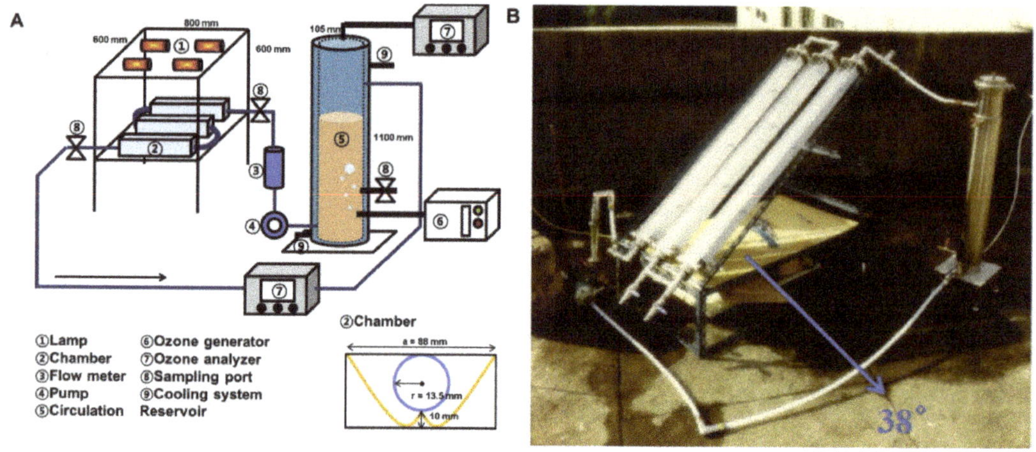

Figure 4. Experimental set up for photocatalytic ozonation experiments at lab and bench scale [206]. (**A**): Flow diagram. (**B**): CPC reactor. (Reprinted with permission from *Chemosphere*, 93, Shin, D.; Jang, M.; Cui, M.; Na, S.; and Khim, J. Enhanced removal of dichloroacetonitrile from drinking water by the combination of solar-photocatalysis and ozonation, 2901–2908. Copyright (2013) With permission from Elsevier).

The UV intensity in the sunlight wavelength range was measured using a spectroradiometer. The irradiation wavelengths of the metal halide lamps (designated as UVsolar, 300–800 nm) and solar light (>300 nm) were similar in terms of the light spectrum. In this research, the major operating parameters such as pH, catalyst (i.e., TiO_2) and ozone doses, temperature, and UV intensity were investigated in the removal of DCAN by photocatalytic ozonation. Shin et al. [206] found that the optimal pH for DCAN removal by UV-solar/TiO_2/O_3 was neutral because of the maximum interaction between TiO_2 and DCAN. According to the results of TiO_2 and ozone-dose tests, 1 g L^{-1} TiO_2 and 1.13 g L^{-1} h^{-1} ozone in photocatalytic ozonation were optimal to afford the highest rate constant. From the temperature variation trials, the highest rate constant (0.033 min^{-1}) was obtained at 20 °C. Temperature simultaneously affected several parameters such as DCAcN adsorption and particle-size change of TiO_2, as well as the decomposition and dissolution rates of ozone. Nevertheless, temperatures higher than 20 °C negatively influenced the DCAcN-removal kinetics. In the assays regarding the UV-solar intensity, the kinetic rate constants increased linearly when the UV-solar intensity increased in the range 4.6–25 Wm^{-2}; however, the increasing trend of rate constants were gradually reduced above 25 Wm^{-2}. The test results of the outdoor system indicated that the solar/TiO_2/O_3 process showed complete removal with rates that are two orders greater than those obtained with solar/TiO_2.

To conclude this section, DBPs elimination by catalytic/photocatalytic ozonation has been demonstrated to proceed by indirect or radical ozonation pathways more than direct ozonation. However, the limited studies carried out mainly with DCA, TCA, and DCAN as model DBPs at high concentrations, many of them in ultrapure water opens some questions about the effect of a real water matrix, the optimization of the catalytic systems to improve the degradation, the use of solar radiation or LEDs, long-term behavior of the catalytic systems, etc. that opens the research possibilities in this field. In addition, the economical assessment of these processes to remove DBPs compared to the elimination efficiency of DBPs precursors needs to be balanced to choose a treatment strategy.

9. The Case of Bromate in Ozonation Processes

An additional survey deserves bromate formation/inhibition during ozonation processes when bromide ions are present in surface water. Bromate ion (BrO_{3}-) is a well-known DBP with a maximum allowed level of 10 μg L^{-1} by many regulations (USEPA, WHO, ECA, etc.) suspected to be a carcinogenic agent. Yang et al. [210] recently reviewed the formation of bromate in chemical oxidation and its control strategies.

Fischbacher et al. [211] reported a multistep mechanism of bromate formation during ozonation in the presence of bromide in which $HO^{\bullet}$ radicals are formed and also contribute to bromate formation according to Scheme 7.

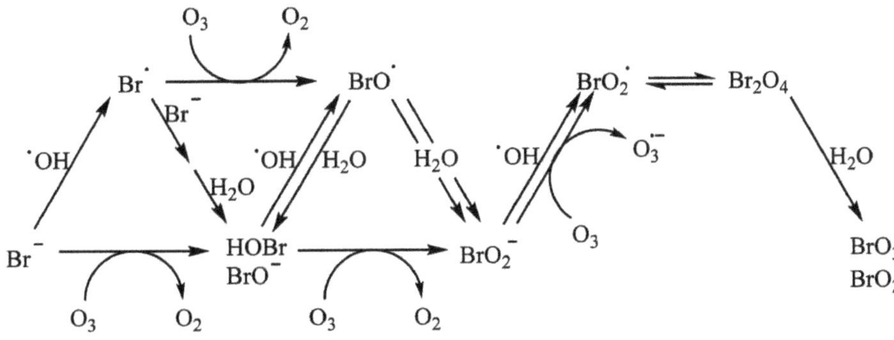

Scheme 7. New reaction pathway of the ozonation of bromide in water [211], (Reprinted with permission from Fischbacher, A.; Löppenberg, K.; von Sonntag, C.; Schmidt, T.C. A new reaction pathway for bromite to bromate in the ozonation of bromide. Environ. Sci. Technol. 2015, 49, 11714–11720, Copyright (2015) American Chemical Society).

Different strategies for the inhibition of bromate formation during ozonation have been extensively studied such as pH depression, ammonia addition, Cl_2-ammonia addition, H_2O_2 addition or optimization of reactor configuration. The main effects of these can be found in Yang et al. [210] and references herein. However, during catalytic/photocatalytic ozonation, the presence of the catalyst can bring new benefits for bromate control. Thus, Han et al. [172,174] in their works of catalytic ozonation using ferrate (VI) as catalyst (see conditions in Table 6) demonstrated that low doses of ferrate up to 5 mg L^{-1} can lead to the reduction or flocculation of BrO_3- or HBrO/BrO- by the produced intermediates Fe(III), Fe(II), H_2O_2, and Fe(III) hydroxydes. Humic acid content decreased bromate formation and also high ambient temperature had a positive effect.

Wu et al. [169] studied the inhibition of nano TiO_2 and SnO_2 on bromate formation during catalytic ozonation of humic acid and observed that the presence of both catalysts reduced the formation of BrO_3- and also a positive effect of increasing humic acid concentration. They hypothesized that the minimization using TiO_2 was due to the decomposition of ozone into $HO^{\bullet}$ which rapidly generated H_2O_2. This hypothesis agrees with the effect of humic acid as $HO^{\bullet}$ and O_3 scavenger. This effect has also been observed in other studies with different metal oxides although in ultrapure water or with other organic pollutants.

Regarding photocatalytic ozonation, Parrino et al. [212] studied the formation or reduction of bromate by photocatalysis, ozonation and photocatalytic ozonation using TiO_2 as catalyst and UVA radiation. They observed that the sole photocatalysis did not produce bromate ions and in the case of its presence, it was able to reduce bromate to bromide ions. On the contrary, in the combined process, bromate ions were not produced when organics still present. They proposed that photo-generated electrons reduced adsorbed bromate to bromide over the catalyst surface.

Thus, the possibility of controlling bromate by tailoring the catalyst has relevance for real application of catalytic/photocatalytic ozonation in drinking water treatment.

10. Concluding Remarks and Future Challenges

Main conclusions of this work are:

DBPs, specifically THMs, were discovered more than 40 years ago but, still, there is much research on their nature, since many recent papers deal with the founding of new DBPs. Already in 1980 about 400 DBPs were identified in chlorinated fulvic acids. At present, only a few DBPs are regulated with a maximum level concentration (MCL): Total THMs, total HAAs, bromate and chlorite. Nonregulated DBPs form different family groups: Halogenated compounds, ketones, aldehydes, and nitrosamines to quote the most representatives with the first ones as the most abundant formed in DWT. The halogenated compounds, (mainly chlorinated but also brominated and iodinated compounds) constituted by far the main family of DBPs from WTP. Today, many of these compounds have been classified as rodent carcinogens, genotoxic, and mutagenic. This reveals the importance in the development of new analytical methods to quantify also nonregulated DBPs or surrogate parameters to fulfil future regulation in order to produce healthy drinking water.

Both DBPs precursors and DBPs have been treated with different oxidation processes where AOPs are the most representative. Because of the double way of ozone reactions in water, ozone has attracted the interest of many researchers. NOM and bromide ion are the main DBPs precursors treated with AOPs. In the last year, about 40 and 20 works were published on this matter with AOPs and ozone processes, as oxidants, respectively. On the whole, precursors react fast with ozone due to the presence of aromatic rings with hydroxyl substituents groups in humic and fulvic acid macromolecules. However, regarding DPBs removal classical ozone AOPs, such as O_3/UVC or O_3/H_2O_2, have been applied in a few cases to mainly remove a few HAAs (dichloroacetic and trichloracetic acids) and some N-nitrosamines. On the contrary, much more work has been done with ozone free processes such as Fenton, UVC/H_2O_2, and photocatalytic oxidation during the last 25 years. In these works, HAAs are the main DBPs studied. In general, these processes due to the formation of hydroxyl radicals reduce DBP concentrations but high oxidant doses are needed.

Most of the works dealing with catalytic and photocatalytic ozonation are lab-scale studies about catalyst properties, catalytic activity and optimization of operating conditions but much less work has been done to go in deep in the stability and reutilization of the catalysts in long term experiments. These studies are crucial to take a step forward to pilot scale studies with the best catalytic systems mainly for DBPFP removal. In addition, the use of natural radiation or environmentally friendly LEDs should be prioritized in photocatalytic ozonation studies towards sustainable processes. However, comprehensive economic and environmental assessments are also required to balance the benefits of precursors or DBPs removal from drinking water in order to draw the best strategy from economic, environmental and health priorities.

Author Contributions: Conceptualization, F.J.B., A.R. and O.G.; investigation, F.J.B., A.R. and O.G.; writing—original draft preparation, F.J.B., A.R. and O.G.; writing—review and editing, F.J.B., A.R. and O.G.; funding acquisition, F.J.B. and A.R. All authors have read and agreed to the published version of the manuscript.

Funding: This research was carried out through projects CTQ2015-73168-JIN/AEI/FEDER/UE funded by Agencia Estatal de Investigación of Spain and co-financed by the European Funds for Regional Development (FEDER, EU) and project PID2019-104429RB-I00/AEI/10.13039/501100011033 funded by Agencia Estatal de Investigación of Spain.

Conflicts of Interest: The authors declare no conflict of interest.

Abbreviations

Nomenclature

AC	Activated carbon
AOPs	Advanced oxidation processes
AORPs	Advanced oxidation/reduction processes
AOX	Adsorbable organic halides
BAC	Biological activated carbon
BC	Before Christ
BDCA	Bromodichloroacetic acid
BET	Brunauer, Emmet, and Teller (referred to isotherm and surface area)
Br-DBPs	Brominated disinfection by-products
C_i	Concentration of i
CAAs	Chloroacetic acids
Cat	Catalyst
CDBA	Chlorodibromoacetic acid
CH	Chloral hydrate
CPC	Compound parabolic collector
CT	Parameter used in disinfection with C concentration of chlorine, T contact time
D	Diameter (particle size or reactor diameter)
DBPFP	Disinfection by-products formation potential
DBPs	Disinfection by-products
DCA	Dichloroacetic acid
DCAN	Dichloroacetonitrile
DFT	Density functional theory
DLS	Dynamic light scattering
DMP	Dimethyl phthalate
DOC	Dissolved organic carbon
DWT	Drinking water treatment
DWTP	Drinking water treatment plant
EAOPs	Electrochemical advanced oxidation processes
ECA	European Chemical Agency
EDS	Dispersive X-Ray spectroscopy
EPA	Environmental Protection Agency
E-peroxone	Electro-peroxone
EPR	Electron paramagnetic resonance
ER	Eley-Rideal mechanism
FMOG	Flower-like nanocomposite
FTIR	Fourier transformed infrared spectroscopy
GAC	Granular activated carbon
HA	Humic acid
HAAs	Haloacetic acids
HAcAm	Haloacetamides
HANs	Haloacetonitriles
HIA	Hydrophilic acid
HIB	Hydrophilic base
HKs	Haloketones
HOA	Hydrophobic acid
HON	Hydrophobic neutral
HS	Humic substances
ICP	Inductively coupled plasma
ICZ	Iron coated zeolite
I-DBPs	Iodinated disinfection by-products
JCR	Journal Citation Report
k	Kinetic constant
L	Length
LED	Light emitting diodes

LHHW	Langmuir–Hinshelwood–Hougen–Watson mechanism
MBA	Bromoacetic acid
MCA	Monochloroacetic acid
MCL	Maximum concentration level
MOFs	Metal organic frameworks
NB	Nitrobencene
N-DBPs	Nitrogen containing disinfection by-products
NDBA	N-nitrosodibuthylamine
NDMA	N-nitrosodimethylamine
NOM	Natural organic matter
NR-DBPs	Non-regulated disinfection by-products
NTU	Nephelometric turbidity unit
OA	Oxalic acid
PACL	Poly-aluminium chloride
PAG	Pilkington ActiveTM glass
pCBA	p-Chlorobenzoic acid
PMO	Pure manganese oxide
PMS	Peroxymonosulfate
PS	Persulfate
PZC	Potential of zero charge
Q	Volumetric flow
Rct	Ratio of the hydroxyl radicals to the ozone exposure during ozone processes
R-DBPs	Regulated disinfection by-products
RGO	Reduced graphene oxide
RO	Reverse osmosis
ROS	Reactive oxidizing species
SCE	Saturated calomel electrode
SEM	Scanning electron microscopy
SHE	Standard hydrogen electrode
TBA	Tribromoacetic acid
TCA	Trichloroacetic acid
TCNM	Trichloronitromethane
THAAs	Total haloacetic acids
THAAFP	Total haloacetic acids formation potential
THMs	Trihalomethanes
THMFP	Trihalomethane formation potential
TNT	Titanate nanotubes
TOC	Total organic carbon
TOX	Total organic halogen
TOXFP	Total organic halogen formation potential
TTHMFP	Total trihalomethane formation potential
WHO	World Health Organization
USEPA	United States Environmental Protection Agency
USA	United States of America
UV	Ultraviolet radiation
UVA	Ultraviolet A radiation
UVC	Ultraviolet C radiation
UV-Vis-DRS	Ultraviolet-visible diffuse reflectance spectroscopy
UV254	Referred to absorbance at 254 nm
V	Volume
WOS	Web of Science
X	Halogen
XPS	X-ray photoelectron spectroscopy
XRD	X-ray diffraction
XRF	X-ray fluorescence

References

1. Danil De Namor, A.F. Water purification: From ancient civilization to the XXI Century. *Water Sci. Technol. Water Supply* **2007**, *7*, 33–39. [CrossRef]
2. Merlet, N. Contribution á l'Etude du Mécanisme de Formation des TrihaloméThanes et des Composés Organohalogénés non Volatils Lors de la Chloration de Molecules Modèles. Ph.D. Thesis, Université de Poitiers, Poitiers, France, 1986.
3. White, G.C. Current chlorination and dechlorination practices in the treatment of potable water, wastewater and cooling water. In *Water Chlorination: Environmental Impacts and Health Effects*; Ann Arbor Science Publishers: Ann Arbor, MI, USA, 1978; Volume 1, pp. 1–18.
4. Symons, J.M.; Bellar, T.A.; Carswell, J.K. National organics reconnaissance survey for halogenated organics. *J. Am. Water Work. Assoc.* **1975**, *67*, 634–647. [CrossRef]
5. Rook, J.J. Formation of Haloforms during Chlorination of Natural Waters. *Water Treat. Exam.* **1974**, *23*, 234–243.
6. Bellar, T.A.; Lichtenberg, J.J.; Kroner, R.C. Occurrence of Organohalides in Chlorinated Drinking Waters. *J. Am. Water Work. Assoc.* **1974**, *66*, 703–706. [CrossRef]
7. US Environmental Protection Agency. *Lower Mississippi River Facility. New Orleans Area Water Supply Study*; US Environmental Protection Agency: Whasington, DC, USA, 1974.
8. Chang, S.L. The safety of water disinfection. *Annu. Rev. Public Health* **1982**, *3*, 393–418. [CrossRef]
9. Glaze, W.H.; Henderson, J.E., IV. Formation of organochlorine compounds from the chlorination of a municipal secondary effluent. *J. Water Pollut. Control Fed.* **1975**, *47*, 2511–2515.
10. Schnitzer, M.; Khan, S.U. *Humic Subtances in the Environment*; Marcel Dekker Inc.: New York, NY, USA, 1972.
11. Steelink, C. Humates and other natural organic substances in the aquatic environment. *J. Chem. Educ.* **1977**, *54*, 599–603. [CrossRef]
12. Croué, J.P. Contribution á l'étude de l'Oxydation par le Chlore et l'Ozone d'Acides Fulviques Naturels Extraits d'Eaux de Surface. Ph.D. Thesis, Université de Poitiers, Poitiers, France, 1987.
13. Corin, N.; Backhand, P.; Kulovaara, M. Degradation products formed during UV-irradiation of humic waters. *Chemosphere* **1996**, *33*, 245–255. [CrossRef]
14. Beckett, R. The Surface Chemistry of Humic Substances in Aquatic Systems. In *Surface and Colloid Chemistry in Natural Waters and Water Treatment*; Beckett, R., Ed.; Plenum: New York, NY, USA, 1990; pp. 3–16.
15. Sirivedhin, T.; Gray, K.A., II. Comparison of the disinfection by-product formation potentials between a wastewater effluent and surface waters. *Water Res.* **2005**, *39*, 1025–1036. [CrossRef]
16. Doré, M.; Goichon, J. Etude d'une methode d'evaluation globale des precurseurs de la reaction haloforme. *Water Res.* **1980**, *14*, 657–663. [CrossRef]
17. Gilca, A.F.; Teodosiu, C.; Fiore, S.; Musteret, C.P. Emerging disinfection byproducts: A review on their occurrence and control in drinking water treatment processes. *Chemosphere* **2020**, *259*, 127476. [CrossRef]
18. Sun, S.; Jiang, T.; Lin, Y.; Song, J.; Zheng, Y.; An, D. Characteristics of organic pollutants in source water and purification evaluations in drinking water treatment plants. *Sci. Total Environ.* **2020**, *733*, 139277. [CrossRef]
19. Gonsioroski, A.; Mourikes, V.E.; Flaws, J.A. Endocrine disruptors in water and their effects on the reproductive system. *Int. J. Mol. Sci.* **2020**, *21*, 1929. [CrossRef] [PubMed]
20. Tak, S.; Vellanki, B.P. Natural organic matter as precursor to disinfection byproducts and its removal using conventional and advanced processes: State of the art review. *J. Water Health* **2018**, *16*, 681–703. [CrossRef]
21. Miller, G.W.; Rice, R.G.; Robson, C.M.; Scullin, R.L.; Kuhn, W.; Wolf, H. *An Assessment of Ozone and Chlorine Dioxide Technologies for Treatment of Municipal Water Supplies*; Environmental Protection Agency, Office of Research and Development, Municipal Environmental Research Laboratory: Cincinnati, OH, USA, 1978; Executive Summary EPA Report; EPA-600/8-78-018 October 1978.
22. Doré, M. Chimie des Oxydants et Traitement des Eaux. *Tech. Doc. Paris* **1989**.
23. Langlais, B.; Reckhow, D.A.; Brink, D.R. *Ozone in Water Treatment: Application and Engineering*; Langlais, B., Reckhow, D.A., Brink, D.R., Eds.; Lewis Publishers: Chelsea, MI, USA, 1991.
24. Glaze, W.H.; Kang, J.W.; Chapin, D.H. The chemistry of water treatment processes involving ozone, hydrogen peroxide and ultraviolet radiation. *Ozone Sci. Eng.* **1987**, *9*, 335–352. [CrossRef]
25. Bailey, P.S. The Reactions Of Ozone With Organic Compounds. *Chem. Rev.* **1958**, *58*, 925–1010. [CrossRef]
26. Staehelin, J.; Hoigne, J. Decomposition of ozone in water in the presence of organic solutes acting as promoters and inhibitors of radical chain reactions. *Environ. Sci. Technol.* **1985**, *19*, 1206–1213. [CrossRef]
27. Beltrán, F.J. *Ozone Reaction Kinetics for Water and Wastewater Systems*; CRC Press: Boca Raton, FL, USA, 2004.
28. Bader, H.; Hoigné, J. Rate Constants of Reactions of Ozone with Organic and Inorganic Compounds in Water -II. Dissociating Organic Compounds. *Water Res.* **1983**, *17*, 185–194.
29. Hoigné, J.; Bader, H. Rate constants of reactions of ozone with organic and inorganic compounds in water-I. Non-dissociating organic compounds. *Water Res.* **1983**, *17*, 173–183. [CrossRef]
30. Kuczkowski, R.L. Ozone and carbonyl oxides. In *1,3, Dipolar Cycloaddition Chemistry*; John Wiley and Sons: New York, NY, USA, 1984; Volume 2A, pp. 197–277.
31. Beltran, F.J.; Gonzalez, M.; GarcÍA-Araya, J.F.; Cabrera, J.L. The use of ozonation to reduce the potential for forming trihalomethane compounds in chlorinating resorcinol, phloroglucinol and 1,3 cyclohexanedione. *Chem. Eng. Commun.* **1990**, *96*, 321–339. [CrossRef]

32. Cavanagh, J.E.; Weinberg, H.S.; Avram, G.; Sangalah, R.; Dean, M.; Glaze, W.H.; Collette, T.W.; Richardson, S.D.; Thruston, A.D. Ozonation Byproducts: Identification of Bromohydrins from the Ozonation of Natural Waters with Enhanced Bromide Levels. *Environ. Sci. Technol.* **1992**, *26*, 1658–1662. [CrossRef]
33. Rav-Acha, C. Review Paper the Reactions of Chlorine Dioxide With Aquatic Organic Materials and. *Water Res.* **1984**, *18*, 1329–1341. [CrossRef]
34. Wajon, J.E.; Rosenblatt, D.H.; Burrows, E.P. Oxidation of Phenol and Hydroquinone by Chlorine Dioxide. *Environ. Sci. Technol.* **1982**, *16*, 396–402. [CrossRef] [PubMed]
35. Rice, R.G.; Gomez-Taylor, M. Occurrence of by-products of strong oxidants reacting with drinking water contaminants—Scope of the problem. *Environ. Health Perspect.* **1986**, *69*, 31–44. [CrossRef]
36. Song, K.; Mohseni, M.; Taghipour, F. Application of ultraviolet light-emitting diodes (UV-LEDs) for water disinfection: A review. *Water Res.* **2016**, *94*, 341–349. [CrossRef]
37. Dalrymple, O.K.; Stefanakos, E.; Trotz, M.A.; Goswami, D.Y. A review of the mechanisms and modeling of photocatalytic disinfection. *Appl. Catal. B Environ.* **2010**, *98*, 27–38. [CrossRef]
38. Fernández, P.; Blanco, J.; Sichel, C.; Malato, S. Water disinfection by solar photocatalysis using compound parabolic collectors. *Catal. Today* **2005**, *101*, 345–352. [CrossRef]
39. Booshehri, A.Y.; Polo-Lopez, M.I.; Castro-Alférez, M.; He, P.; Xu, R.; Rong, W.; Malato, S.; Fernández-Ibáñez, P. Assessment of solar photocatalysis using $Ag/BiVO_4$ at pilot solar Compound Parabolic Collector for inactivation of pathogens in well water and secondary effluents. *Catal. Today* **2017**, *281*, 124–134. [CrossRef]
40. Madaeni, S.S. The application of membrane technology for water disinfection. *Water Res.* **1999**, *33*, 301–308. [CrossRef]
41. Sun, X.; Liu, J.; Ji, L.; Wang, G.; Zhao, S.; Yoon, J.Y.; Chen, S. A review on hydrodynamic cavitation disinfection: The current state of knowledge. *Sci. Total Environ.* **2020**, *737*. [CrossRef] [PubMed]
42. Li, Q.; Mahendra, S.; Lyon, D.Y.; Brunet, L.; Liga, M.V.; Li, D.; Alvarez, P.J.J. Antimicrobial nanomaterials for water disinfection and microbial control: Potential applications and implications. *Water Res.* **2008**, *42*, 4591–4602. [CrossRef] [PubMed]
43. Pina, A.S.; Batalha, Í.L.; Fernandes, C.S.M.; Aoki, M.A.; Roque, A.C.A. Exploring the potential of magnetic antimicrobial agents for water disinfection. *Water Res.* **2014**, *66*, 160–168. [CrossRef] [PubMed]
44. Emile Coleman, W.; Melton, R.G.; Kopfler, F.C.; Barone, K.A.; Aurand, T.A.; Jellison, M.G. Identification of Organic Compounds in a Mutagenic Extract of a Surface Drinking Water by a Computerized Gas Chromatography/Mass Spectrometry System (GC/MS/COM). *Environ. Sci. Technol.* **1980**, *14*, 576–588. [CrossRef]
45. Glaze, W.H. Brogan & Partners Reaction Products of Ozone: A Review. *Environ. Health Perspect.* **1986**, *69*, 151–157.
46. Werdehoff, K.S.; Singer, P.C. Chlorine Dioxide Effects on Thmfp, Toxfp, and the Formation of Inorganic By-Products. *J. Am. Water Work. Assoc.* **1987**, *79*, 107–113. [CrossRef]
47. Richardson, S.D. Disinfection by-products and other emerging contaminants in drinking water. *TrAC Trends Anal. Chem.* **2003**, *22*, 666–684. [CrossRef]
48. Lavonen, E.E.; Gonsior, M.; Tranvik, L.J.; Schmitt-Kopplin, P.; Köhler, S.J. Selective chlorination of natural organic matter: Identification of previously unknown disinfection byproducts. *Environ. Sci. Technol.* **2013**, *47*, 2264–2271. [CrossRef] [PubMed]
49. Bougeard, C.M.M.; Goslan, E.H.; Jefferson, B.; Parsons, S.A. Comparison of the disinfection by-product formation potential of treated waters exposed to chlorine and monochloramine. *Water Res.* **2010**, *44*, 729–740. [CrossRef]
50. Le Roux, J.; Nihemaiti, M.; Croué, J.P. The role of aromatic precursors in the formation of haloacetamides by chloramination of dissolved organic matter. *Water Res.* **2016**, *88*, 371–379. [CrossRef] [PubMed]
51. Padhi, R.K.; Subramanian, S.; Satpathy, K.K. Formation, distribution, and speciation of DBPs (THMs, HAAs, ClO^{2-},andClO^{3-}) during treatment of different source water with chlorine and chlorine dioxide. *Chemosphere* **2019**, *218*, 540–550. [CrossRef]
52. Gan, W.; Huang, S.; Ge, Y.; Bond, T.; Westerhoff, P.; Zhai, J.; Yang, X. Chlorite formation during ClO_2 oxidation of model compounds having various functional groups and humic substances. *Water Res.* **2019**, *159*, 348–357. [CrossRef]
53. Postigo, C.; Zonja, B. Iodinated disinfection byproducts: Formation and concerns. *Curr. Opin. Environ. Sci. Health* **2019**, *7*, 19–25. [CrossRef]
54. Richardson, S.D.; Plewa, M.J.; Wagner, E.D.; Schoeny, R.; DeMarini, D.M. Occurrence, genotoxicity, and carcinogenicity of regulated and emerging disinfection by-products in drinking water: A review and roadmap for research. *Mutat. Res. Rev. Mutat. Res.* **2007**, *636*, 178–242. [CrossRef] [PubMed]
55. Pan, Y.; Wang, Y.; Li, A.; Xu, B.; Xian, Q.; Shuang, C.; Shi, P.; Zhou, Q. Detection, formation and occurrence of 13 new polar phenolic chlorinated and brominated disinfection byproducts in drinking water. *Water Res.* **2017**, *112*, 129–136. [CrossRef]
56. How, Z.T.; Kristiana, I.; Busetti, F.; Linge, K.L.; Joll, C.A. Organic chloramines in chlorine-based disinfected water systems: A critical review. *J. Environ. Sci.* **2017**, *58*, 2–18. [CrossRef]
57. Mian, H.R.; Hu, G.; Hewage, K.; Rodriguez, M.J.; Sadiq, R. Prioritization of unregulated disinfection by-products in drinking water distribution systems for human health risk mitigation: A critical review. *Water Res.* **2018**, *147*, 112–131. [CrossRef]
58. Kimura, S.Y.; Ortega-Hernandez, A. Formation mechanisms of disinfection byproducts: Recent developments. *Curr. Opin. Environ. Sci. Heal.* **2019**, *7*, 61–68. [CrossRef]
59. Tardiff, R.G.; Garlson, G.P.; Simmon, V. Halogenated organics in tap water: A toxicological evaluation in Water Chlorination. Environmental Impact and Health Effects. *Ann Arbor Sci. Ann Arbor Michigan. USA* **1978**, *1*, 195–209.

60. U.S. National Cancer institute. *Report on the Carcinogenesis Bioassay of Chloroform (CAS No. 67-66-3)*; TR-000. NTIS Rpt No PB264018; U.S. National Cancer institute: Bethesda, MD, USA, 1976.
61. King, W.D.; Marrett, L.D. Case-control study of bladder cancer and chlorination by-products in treated water (Ontario, Canada). *Cancer Causes Control* **1996**, *7*, 596–604. [CrossRef]
62. Hildesheim, M.E.; Cantor, K.P.; Lynch, C.F.; Dosemeci, M.; Lubin, J.; Alavanja, M.; Craun, G. Drinking water source and chlorination byproducts II. Risk of colon and rectal cancers. *Epidemiology* **1998**, *9*, 29–35. [CrossRef]
63. Boorman, G.A.; Dellarco, V.; Dunnick, J.K.; Chapin, R.E.; Hauchman, F.; Gardner, H.; Cox, M.; Sills, R.C.; Boorman, G.A.; Dellarco, V.; et al. Brogan & Partners Drinking Water Disinfection Byproducts: Review and Approach to Toxicity Evaluation Source: Environmental Health Perspectives, Vol. 107, Supplement 1: Reviews in Environmental Health, 1999 (Feb., 1999), pp. 207–217 Published by. *Environ. Health Perspect.* **1999**, *107*, 207–217. [PubMed]
64. DeMarini, D.M. A review on the 40th anniversary of the first regulation of drinking water disinfection by-products. *Environ. Mol. Mutagen.* **2020**, *61*, 588–601. [CrossRef] [PubMed]
65. Wagner, E.D.; Plewa, M.J. CHO cell cytotoxicity and genotoxicity analyses of disinfection by-products: An updated review. *J. Environ. Sci.* **2017**, *58*, 64–76. [CrossRef] [PubMed]
66. Demarini, D.M.; Abu-Shakra, A.; Felton, C.F.; Patterson, K.S.; Shelton, M.L. Mutation spectra in salmonella of chlorinated, chloraminated, or ozonated drinking water extracts: Comparison to MX. *Environ. Mol. Mutagen.* **1995**, *26*, 270–285. [CrossRef]
67. Villanueva, C.M.; Cantor, K.P.; Grimalt, J.O.; Malats, N.; Silverman, D.; Tardon, A.; Garcia-Closas, R.; Serra, C.; Carrato, A.; Castaño-Vinyals, G.; et al. Bladder cancer and exposure to water disinfection by-products through ingestion, bathing, showering, and swimming in pools. *Am. J. Epidemiol.* **2007**, *165*, 148–156. [CrossRef] [PubMed]
68. Le Roux, J.; Plewa, M.J.; Wagner, E.D.; Nihemaiti, M.; Dad, A.; Croué, J.-P. Chloramination of wastewater effluent: Toxicity and formation of disinfection byproducts. *J. Environ. Sci.* **2017**, *58*, 135–145. [CrossRef] [PubMed]
69. Bond, T.; Goslan, E.H.; Parsons, S.A.; Jefferson, B. Treatment of disinfection by-product precursors. *Environ. Technol.* **2011**, *32*, 1–25. [CrossRef]
70. Sillanpää, M.; Ncibi, M.C.; Matilainen, A. Advanced oxidation processes for the removal of natural organic matter from drinking water sources: A comprehensive review. *J. Environ. Manag.* **2018**, *208*, 56–76. [CrossRef]
71. Matilainen, A.; Sillanpää, M. Removal of natural organic matter from drinking water by advanced oxidation processes. *Chemosphere* **2010**, *80*, 351–365. [CrossRef]
72. Glaze, W.H.; Peyton, G.R.; Lin, S.; Huang, R.Y.; Burleson, J.L. Destruction of pollutants in water with ozone in combination with ultraviolet radiation. II. Natural trihalomethane precursors. *Environ. Sci. Technol.* **1982**, *16*, 454–458. [CrossRef]
73. Zhang, Y.; Zhao, X.; Zhang, X.; Peng, S. A review of different drinking water treatments for natural organic matter removal. *Water Sci. Technol. Water Supply* **2015**, *15*, 442–455. [CrossRef]
74. Miklos, D.B.; Remy, C.; Jekel, M.; Linden, K.G.; Drewes, J.E.; Hübner, U. Evaluation of advanced oxidation processes for water and wastewater treatment—A critical review. *Water Res.* **2018**, *139*, 118–131. [CrossRef]
75. Hua, G.; Reckhow, D.A. Characterization of Disinfection Byproduct Precursors Based on Hydrophobicity and Molecular Size. *Environ. Sci. Technol.* **2007**, *41*, 3309–3315. [CrossRef] [PubMed]
76. Lamsal, R.; Walsh, M.E.; Gagnon, G.A. Comparison of advanced oxidation processes for the removal of natural organic matter. *Water Res.* **2011**, *45*, 3263–3269. [CrossRef] [PubMed]
77. De Vera, G.A.; Stalter, D.; Gernjak, W.; Weinberg, H.S.; Keller, J.; Farré, M.J. Towards reducing DBP formation potential of drinking water by favouring direct ozone over hydroxyl radical reactions during ozonation. *Water Res.* **2015**, *87*, 49–58. [CrossRef] [PubMed]
78. Petronijević, M.; Agbaba, J.; Ražić, S.; Molnar Jazić, J.; Tubić, A.; Watson, M.; Dalmacija, B. Fate of bromine-containing disinfection by-products precursors during ozone and ultraviolet-based advanced oxidation processes. *Int. J. Environ. Sci. Technol.* **2019**, *16*, 171–180. [CrossRef]
79. Sakai, H.; Autin, O.; Parsons, S. Change in haloacetic acid formation potential during UV and UV/H_2O_2 treatment of model organic compounds. *Chemosphere* **2013**, *92*, 647–651. [CrossRef]
80. Murray, C.A.; Parsons, S.A. Comparison of AOPs for the removal of natural organic matter: Performance and economic assessment. *Water Sci. Technol.* **2004**, *49*, 267–272. [CrossRef]
81. Moncayo-Lasso, A.; Pulgarin, C.; Benítez, N. Degradation of DBPs' precursors in river water before and after slow sand filtration by photo-Fenton process at pH 5 in a solar CPC reactor. *Water Res.* **2008**, *42*, 4125–4132. [CrossRef]
82. Moncayo-Lasso, A.; Rincon, A.G.; Pulgarin, C.; Benitez, N. Significant decrease of THMs generated during chlorination of river water by previous photo-Fenton treatment at near neutral pH. *J. Photochem. Photobiol. A Chem.* **2012**, *229*, 46–52. [CrossRef]
83. Lee, S.; Ohgaki, S. Oxidative degradation of toc and thmfp by fluidized bed photocatalysis reactor. *J. Environ. Sci. Heal. Part A* **1999**, *34*, 1933–1944. [CrossRef]
84. Liu, S.; Lim, M.; Fabris, R.; Chow, C.; Drikas, M.; Amal, R. TiO2 Photocatalysis of Natural Organic Matter in Surface Water: Impact on Trihalomethane and Haloacetic Acid Formation Potential. *Environ. Sci. Technol.* **2008**, *42*, 6218–6223. [CrossRef] [PubMed]
85. Liu, S.; Lim, M.; Fabris, R.; Chow, C.; Chiang, K.; Drikas, M.; Amal, R. Removal of humic acid using TiO_2 photocatalytic process—Fractionation and molecular weight characterisation studies. *Chemosphere* **2008**, *72*, 263–271. [CrossRef]
86. Gerrity, D.; Mayer, B.; Ryu, H.; Crittenden, J.; Abbaszadegan, M. A comparison of pilot-scale photocatalysis and enhanced coagulation for disinfection byproduct mitigation. *Water Res.* **2009**, *43*, 1597–1610. [CrossRef] [PubMed]

87. Rodríguez, E.M.; Gordillo, M.V.; Rey, A.; Beltrán, F.J. Impact of TiO2/UVA photocatalysis on THM formation potential. *Catal. Today* **2018**, *313*, 167–174. [CrossRef]
88. Murray, C.A.; Parsons, S.A. Preliminary laboratory investigation of disinfection by-product precursor removal using an advanced oxidation process. *Water Environ. J.* **2006**, *20*, 123–129. [CrossRef]
89. Murray, C.A.; Goslan, E.H.; Parsons, S.A. TiO2/UV: Single stage drinking water treatment for NOM removal? *J. Environ. Eng. Sci.* **2007**, *6*, 311–317. [CrossRef]
90. Kent, F.C.; Montreuil, K.R.; Brookman, R.M.; Sanderson, R.; Dahn, J.R.; Gagnon, G.A. Photocatalytic oxidation of DBP precursors using UV with suspended and fixed TiO_2. *Water Res.* **2011**, *45*, 6173–6180. [CrossRef]
91. Lu, J.; Dong, W.; Ji, Y.; Kong, D.; Huang, Q. Natural Organic Matter Exposed to Sulfate Radicals Increases Its Potential to Form Halogenated Disinfection Byproducts. *Environ. Sci. Technol.* **2016**, *50*, 5060–5067. [CrossRef] [PubMed]
92. Hua, Z.; Kong, X.; Hou, S.; Zou, S.; Xu, X.; Huang, H.; Fang, J. DBP alteration from NOM and model compounds after UV/persulfate treatment with post chlorination. *Water Res.* **2019**, *158*, 237–245. [CrossRef]
93. Wang, Z.; An, N.; Shao, Y.; Gao, N.; Du, E.; Xu, B. Experimental and simulation investigations of UV/persulfate treatment in presence of bromide: Effects on degradation kinetics, formation of brominated disinfection byproducts and bromate. *Sep. Purif. Technol.* **2020**, *242*. [CrossRef]
94. Wang, L.; Ji, Y.; Lu, J.; Yin, X.; Zhou, Q.; Kong, D. Transformation of iodide and formation of iodinated by-products in heat activated persulfate oxidation process. *Chemosphere* **2017**, *181*, 400–408. [CrossRef]
95. Pisarenko, A.N.; Stanford, B.D.; Snyder, S.A.; Rivera, S.B.; Boal, A.K. Investigation of the use of chlorine based advanced oxidation in surface water: Oxidation of natural organic matter and formation of disinfection byproducts. *J. Adv. Oxid. Technol.* **2013**, *16*, 137–150. [CrossRef]
96. Wang, D.; Bolton, J.R.; Andrews, S.A.; Hofmann, R. Formation of disinfection by-products in the ultraviolet/chlorine advanced oxidation process. *Sci. Total Environ.* **2015**, *518–519*, 49–57. [CrossRef]
97. Liu, Z.; Xu, B.; Zhang, T.Y.; Hu, C.Y.; Tang, Y.L.; Dong, Z.Y.; Cao, T.C.; El-Din, M.G. Formation of disinfection by-products in a UV-activated mixed chlorine/chloramine system. *J. Hazard. Mater.* **2021**, *407*, 124373. [CrossRef] [PubMed]
98. Liao, P.; Al-Ani, Y.; Malik Ismael, Z.; Wu, X. Insights into the role of humic acid on Pd-catalytic electro-fenton transformation of toluene in groundwater. *Sci. Rep.* **2015**, *5*. [CrossRef] [PubMed]
99. Trellu, C.; Péchaud, Y.; Oturan, N.; Mousset, E.; Huguenot, D.; van Hullebusch, E.D.; Esposito, G.; Oturan, M.A. Comparative study on the removal of humic acids from drinking water by anodic oxidation and electro-Fenton processes: Mineralization efficiency and modelling. *Appl. Catal. B Environ.* **2016**, *194*, 32–41. [CrossRef]
100. Mao, Y.; Guo, D.; Yao, W.; Wang, X.; Yang, H.; Xie, Y.F.; Komarneni, S.; Yu, G.; Wang, Y. Effects of conventional ozonation and electro-peroxone pretreatment of surface water on disinfection by-product formation during subsequent chlorination. *Water Res.* **2018**, *130*, 322–332. [CrossRef] [PubMed]
101. Mayer, B.K.; Daugherty, E.; Abbaszadegan, M. Evaluation of the relationship between bulk organic precursors and disinfection byproduct formation for advanced oxidation processes. *Chemosphere* **2015**, *121*, 39–46. [CrossRef]
102. Zhang, Z.; Zhao, Q.; Song, H.; Zhang, J.; Wang, L.; Qi, J.; Liu, Y.; Ma, J. Comparative study about oxidation of trace N-nitrosamines by seven oxidation processes with a sensitivity improved determination method. *Sep. Purif. Technol.* **2020**, *236*, 116009. [CrossRef]
103. Xu, B.; Chen, Z.; Qi, F.; Ma, J.; Wu, F. Inhibiting the regeneration of N-nitrosodimethylamine in drinking water by UV photolysis combined with ozonation. *J. Hazard. Mater.* **2009**, *168*, 108–114. [CrossRef]
104. Lee, C.; Yoon, J.; Von Gunten, U. Oxidative degradation of N-nitrosodimethylamine by conventional ozonation and the advanced oxidation process ozone/hydrogen peroxide. *Water Res.* **2007**, *41*, 581–590. [CrossRef]
105. Lv, J.; Li, Y.; Song, Y. Reinvestigation on the ozonation of N-nitrosodimethylamine: Influencing factors and degradation mechanism. *Water Res.* **2013**, *47*, 4993–5002. [CrossRef]
106. Hama Aziz, K.H. Application of different advanced oxidation processes for the removal of chloroacetic acids using a planar falling film reactor. *Chemosphere* **2019**, *228*, 377–383. [CrossRef]
107. Lovato, M.E.; Martín, C.A.; Cassano, A.E. Degradation of dichloroacetic acid in homogeneous aqueous media employing ozone and UVC radiation. *Photochem. Photobiol. Sci.* **2011**, *10*, 367–380. [CrossRef]
108. Wang, K.; Guo, J.; Yang, M.; Junji, H.; Deng, R. Decomposition of two haloacetic acids in water using UV radiation, ozone and advanced oxidation processes. *J. Hazard. Mater.* **2009**, *162*, 1243–1248. [CrossRef] [PubMed]
109. Zazouli, M.A.; Kalankesh, L.R. Removal of precursors and disinfection by-products (DBPs) by membrane filtration from water; a review. *J. Environ. Heal. Sci. Eng.* **2017**, *15*, 25. [CrossRef] [PubMed]
110. Zhang, Y.; Lu, Z.; Zhang, Z.; Shi, B.; Hu, C.; Lyu, L.; Zuo, P.; Metz, J.; Wang, H. Heterogeneous Fenton-like reaction followed by GAC filtration improved removal efficiency of NOM and DBPs without adjusting pH. *Sep. Purif. Technol.* **2021**, *260*, 118234. [CrossRef]
111. Zhong, J.; Zhao, Y.; Ding, L.; Ji, H.; Ma, W.; Chen, C.; Zhao, J. Opposite photocatalytic oxidation behaviors of BiOCl and TiO_2: Direct hole transfer vs. indirect OH oxidation. *Appl. Catal. B Environ.* **2019**, *241*, 514–520. [CrossRef]
112. Moussavi, G.; Rezaei, M. Exploring the advanced oxidation/reduction processes in the VUV photoreactor for dechlorination and mineralization of trichloroacetic acid: Parametric experiments, degradation pathway and bioassessment. *Chem. Eng. J.* **2017**, *328*, 331–342. [CrossRef]

113. Park, J.-A.; Nam, H.-L.; Choi, J.-W.; Ha, J.; Lee, S.-H. Oxidation of geosmin and 2-methylisoborneol by the photo-Fenton process: Kinetics, degradation intermediates, and the removal of microcystin-LR and trihalomethane from Nak-Dong River water, South Korea. *Chem. Eng. J.* **2017**, *313*, 345–354. [CrossRef]
114. Aslani, H.; Nasseri, S.; Nabizadeh, R.; Mesdaghinia, A.; Alimohammadi, M.; Nazmara, S. Haloacetic acids degradation by an efficient Ferrate/UV process: Byproduct analysis, kinetic study, and application of response surface methodology for modeling and optimization. *J. Environ. Manage.* **2017**, *203*, 218–228. [CrossRef]
115. Li, Y.; Kemper, J.M.; Datuin, G.; Akey, A.; Mitch, W.A.; Luthy, R.G. Reductive dehalogenation of disinfection byproducts by an activated carbon-based electrode system. *Water Res.* **2016**, *98*, 354–362. [CrossRef]
116. Zhao, B.; Wang, X.; Shang, H.; Li, X.; Li, W.; Li, J.; Xia, W.; Zhou, L.; Zhao, C. Degradation of trichloroacetic acid with an efficient Fenton assisted TiO_2 photocatalytic hybrid process: Reaction kinetics, byproducts and mechanism. *Chem. Eng. J.* **2016**, *289*, 319–329. [CrossRef]
117. Aslani, H.; Nabizadeh, R.; Nasseri, S.; Mesdaghinia, A.; Alimohammadi, M.; Mahvi, A.H.; Rastkari, N.; Nazmara, S. Application of response surface methodology for modeling and optimization of trichloroacetic acid and turbidity removal using potassium ferrate(VI). *Desalin. Water Treat.* **2016**, *57*, 25317–25328. [CrossRef]
118. Zhao, B.; Li, X.; Li, W.; Yang, L.; Li, J.; Xia, W.; Zhou, L.; Wang, F.; Zhao, C. Degradation of trichloroacetic acid by an efficient Fenton/UV/TiO2 hybrid process and investigation of synergetic effect. *Chem. Eng. J.* **2015**, *273*, 527–533. [CrossRef]
119. Alavi, N.; Tahvildarij, K. Removal of trihalomethanes in tehran drinking water by an advanced oxidation process. *Nat. Environ. Pollut. Technol.* **2015**, *14*, 211–216. [CrossRef]
120. Park, B.; Cho, E.; Son, Y.; Khim, J. Distribution of electrical energy consumption for the efficient degradation control of THMs mixture in sonophotolytic process. *Ultrason. Sonochem.* **2014**, *21*, 1982–1987. [CrossRef] [PubMed]
121. Tang, S.; Wang, X.M.; Yang, H.W.; Xie, Y.F. Haloacetic acid removal by sequential zero-valent iron reduction and biologically active carbon degradation. *Chemosphere* **2013**, *90*, 1563–1567. [CrossRef]
122. Radjenović, J.; Farré, M.J.; Mu, Y.; Gernjak, W.; Keller, J. Reductive electrochemical remediation of emerging and regulated disinfection byproducts. *Water Res.* **2012**, *46*, 1705–1714. [CrossRef]
123. Esclapez, M.D.; Tudela, I.; Díez-García, M.I.; Sáez, V.; Rehorek, A.; Bonete, P.; González-García, J. Towards the complete dechlorination of chloroacetic acids in water by sonoelectrochemical methods: Effect of the anodic material on the degradation of trichloroacetic acid and its by-products. *Chem. Eng. J.* **2012**, *197*, 231–241. [CrossRef]
124. Wang, X.; Ning, P.; Liu, H.; Ma, J. Dechlorination of chloroacetic acids by Pd/Fe nanoparticles: Effect of drying method on metallic activity and the parameter optimization. *Appl. Catal. B Environ.* **2010**, *94*, 55–63. [CrossRef]
125. Czili, H.; Horváth, A. Photodegradation of chloroacetic acids over bare and silver-deposited TiO2: Identification of species attacking model compounds, a mechanistic approach. *Appl. Catal. B Environ.* **2009**, *89*, 342–348. [CrossRef]
126. Li, Y.P.; Cao, H.B.; Zhang, Y. Reductive dehalogenation of haloacetic acids by hemoglobin-loaded carbon nanotube electrode. *Water Res.* **2007**, *41*, 197–205. [CrossRef] [PubMed]
127. Shemer, H.; Narkis, N. Trihalomethanes aqueous solutions sono-oxidation. *Water Res.* **2005**, *39*, 2704–2710. [CrossRef] [PubMed]
128. Shemer, H.; Narkis, N. Sonochemical removal of trihalomethanes from aqueous solutions. *Ultrason. Sonochem.* **2005**, *12*, 495–499. [CrossRef] [PubMed]
129. Shemer, H.; Narkis, N. Effects of aqueous solutions composition and acoustic intensity on THM compounds sonolysis. *Environ. Eng. Sci.* **2005**, *22*, 138–144. [CrossRef]
130. Shemer, H.; Narkis, N. Mechanisms and inorganic byproducts of trihalomethane compounds sonodegradation. *Environ. Sci. Technol.* **2004**, *38*, 4856–4859. [CrossRef]
131. Lifongo, L.L.; Bowden, D.J.; Brimblecombe, P. Photodegradation of haloacetic acids in water. *Chemosphere* **2004**, *55*, 467–476. [CrossRef]
132. Zhang, L.; Arnold, W.A.; Hozalski, R.M. Kinetics of haloacetic acid reactions with Fe(O). *Environ. Sci. Technol.* **2004**, *38*, 6881–6889. [CrossRef]
133. Wu, C.; Wei, D.; Fan, J.; Wang, L. Photosonochemical degradation of trichloroacetic acid in aqueous solution. *Chemosphere* **2001**, *44*, 1293–1297. [CrossRef]
134. Hozalski, R.M.; Zhang, L.; Arnold, W.A. Reduction of haloacetic acids by FeO: Implications for treatment and fate. *Environ. Sci. Technol.* **2001**, *35*, 2258–2263. [CrossRef] [PubMed]
135. Tang, W.Z.; Tassos, S. Oxidation kinetics and mechanisms of trihalomethanes by Fenton's reagent. *Water Res.* **1997**, *31*, 1117–1125. [CrossRef]
136. Spangenberg, D.; Mbller, U. Photooxidation and Thermal Decomposition of Trichloroacetic Acid. *Chemosphere* **1996**, *33*, 43–49. [CrossRef]
137. Kasprzyk-Hordern, B.; Ziółek, M.; Nawrocki, J. Catalytic ozonation and methods of enhancing molecular ozone reactions in water treatment. *Appl. Catal. B Environ.* **2003**, *46*, 639–669. [CrossRef]
138. Hill, G.R. The Kinetics of the Oxidation of Cobaltous Ion by Ozone. *J. Am. Chem. Soc.* **1949**, *71*, 2434–2435. [CrossRef]
139. Nawrocki, J.; Kasprzyk-Hordern, B. The efficiency and mechanisms of catalytic ozonation. *Appl. Catal. B Environ.* **2010**, *99*, 27–42. [CrossRef]
140. Yu, D.; Wu, M.; Hu, Q.; Wang, L.; Lv, C.; Zhang, L. Iron-based metal-organic frameworks as novel platforms for catalytic ozonation of organic pollutant: Efficiency and mechanism. *J. Hazard. Mater.* **2019**, *367*, 456–464. [CrossRef]

141. Yu, D.; Wang, L.; Yang, T.; Yang, G.; Wang, D.; Ni, H.; Wu, M. Tuning Lewis acidity of iron-based metal-organic frameworks for enhanced catalytic ozonation. *Chem. Eng. J.* **2021**, *404*, 127075. [CrossRef]
142. Nawrocki, J. Catalytic ozonation in water: Controversies and questions. Discussion paper. *Appl. Catal. B Environ.* **2013**, *142–143*, 465–471. [CrossRef]
143. Wang, J.; Chen, H. Catalytic ozonation for water and wastewater treatment: Recent advances and perspective. *Sci. Total Environ.* **2020**, *704*, 135249. [CrossRef] [PubMed]
144. Rivera-Utrilla, J.; López-Ramón, M.V.; Sánchez-Polo, M.; Álvarez, M.Á.; Velo-Gala, I. Characteristics and behavior of different catalysts used for water decontamination in photooxidation and ozonation processes. *Catalysts* **2020**, *10*, 1485. [CrossRef]
145. Yu, G.; Wang, Y.; Cao, H.; Zhao, H.; Xie, Y. Reactive Oxygen Species and Catalytic Active Sites in Heterogeneous Catalytic Ozonation for Water Purification. *Environ. Sci. Technol.* **2020**, *54*, 5931–5946. [CrossRef]
146. Fogler, H.S. *Elements of Chemical Reaction Engineering*, 3rd ed.; Goodwin, B.M., Ed.; Prentice-Hall: Englewood-Cliffs, NJ, USA, 1999.
147. Liu, Y.; Shen, J.; Chen, Z.; Yang, L.; Liu, Y.; Han, Y. Effects of amorphous-zinc-silicate-catalyzed ozonation on the degradation of p-chloronitrobenzene in drinking water. *Appl. Catal. A Gen.* **2011**, *403*, 112–118. [CrossRef]
148. Andreozzi, R.; Insola, A.; Caprio, V.; Marotta, R.; Tufano, V. The use of manganese dioxide as a heterogeneous catalyst for oxalic acid ozonation in aqueous solution. *Appl. Catal. A Gen.* **1996**, *138*, 75–81. [CrossRef]
149. Martins, R.C.; Quinta-Ferreira, R.M. Catalytic ozonation of phenolic acids over a Mn-Ce-O catalyst. *Appl. Catal. B Environ.* **2009**, *90*, 268–277. [CrossRef]
150. Ikhlaq, A.; Brown, D.R.; Kasprzyk-Hordern, B. Catalytic ozonation for the removal of organic contaminants in water on ZSM-5 zeolites. *Appl. Catal. B Environ.* **2014**, *154–155*, 110–122. [CrossRef]
151. Zhao, L.; Ma, J.; Sun, Z.; Zhai, X. Mechanism of influence of initial pH on the degradation of nitrobenzene in aqueous solution by ceramic honeycomb catalytic ozonation. *Environ. Sci. Technol.* **2008**, *42*, 4002–4007. [CrossRef] [PubMed]
152. Liu, Z.Q.; Ma, J.; Cui, Y.H.; Zhao, L.; Zhang, B.P. Factors affecting the catalytic activity of multi-walled carbon nanotube for ozonation of oxalic acid. *Sep. Purif. Technol.* **2011**, *78*, 147–153. [CrossRef]
153. Restivo, J.; Órfão, J.J.M.; Pereira, M.F.R.; Vanhaecke, E.; Rönning, M.; Iouranova, T.; Kiwi-Minsker, L.; Armenise, S.; Garcia-Bordejé, E. Catalytic ozonation of oxalic acid using carbon nanofibres on macrostructured supports. *Water Sci. Technol.* **2012**, *65*, 1854–1862. [CrossRef] [PubMed]
154. Orge, C.A.; Órfão, J.J.M.; Pereira, M.F.R. Composites of manganese oxide with carbon materials as catalysts for the ozonation of oxalic acid. *J. Hazard. Mater.* **2012**, *213–214*, 133–139. [CrossRef] [PubMed]
155. Orge, C.A.; Órfão, J.J.M.; Pereira, M.F.R. Carbon xerogels and ceria-carbon xerogel materials as catalysts in the ozonation of organic pollutants. *Appl. Catal. B Environ.* **2012**, *126*, 22–28. [CrossRef]
156. Wang, Y.; Duan, X.; Xie, Y.; Sun, H.; Wang, S. Nanocarbon-Based Catalytic Ozonation for Aqueous Oxidation: Engineering Defects for Active Sites and Tunable Reaction Pathways. *ACS Catal.* **2020**, *10*, 13383–13414. [CrossRef]
157. Qu, X.; Zheng, J.; Zhang, Y. Catalytic ozonation of phenolic wastewater with activated carbon fiber in a fluid bed reactor. *J. Colloid Interface Sci.* **2007**, *309*, 429–434. [CrossRef] [PubMed]
158. Sánchez-Polo, M.; Leyva-Ramos, R.; Rivera-Utrilla, J. Kinetics of 1,3,6-naphthalenetrisulphonic acid ozonation in presence of activated carbon. *Carbon N. Y.* **2005**, *43*, 962–969. [CrossRef]
159. Rivera-Utrilla, J.; Sánchez-Polo, M.; Gómez-Serrano, V.; Álvarez, P.M.; Alvim-Ferraz, M.C.M.; Dias, J.M. Activated carbon modifications to enhance its water treatment applications. An overview. *J. Hazard. Mater.* **2011**, *187*, 1–23. [CrossRef] [PubMed]
160. Salla, J.S.; Padoin, N.; Amorim, S.M.; Li Puma, G.; Moreira, R.F.P.M. Humic acids adsorption and decomposition on Mn2O3 and α-Al2O3 nanoparticles in aqueous suspensions in the presence of ozone. *J. Environ. Chem. Eng.* **2020**, *8*, 102780. [CrossRef]
161. Alver, A.; Kılıç, A. Catalytic ozonation by iron coated pumice for the degradation of natural organic matters. *Catalysts* **2018**, *8*, 219. [CrossRef]
162. Zhang, H.; Wang, J. Catalytic Ozonation of Humic Acids by Ce–Ti Composite Catalysts. *Kinet. Catal.* **2017**, *58*, 734–740. [CrossRef]
163. Gümüş, D.; Akbal, F. A comparative study of ozonation, iron coated zeolite catalyzed ozonation and granular activated carbon catalyzed ozonation of humic acid. *Chemosphere* **2017**, *174*, 218–231. [CrossRef]
164. Turkay, O.; Inan, H.; Dimoglo, A. Experimental and theoretical study on catalytic ozonation of humic acid by ZnO catalyst. *Sep. Sci. Technol.* **2017**, *52*, 778–786. [CrossRef]
165. Wang, Q.; Yang, Z.; Chai, B.; Cheng, S.; Lu, X.; Bai, X. Heterogeneous catalytic ozonation of natural organic matter with goethite, cerium oxide and magnesium oxide. *RSC Adv.* **2016**, *6*, 14730–14740. [CrossRef]
166. Wang, J.; Wang, G.; Yang, C.; Yang, S.; Huang, Q. Catalytic ozonation of organic compounds in water over the catalyst of RuO_2/ZrO_2-CeO_2. *Front. Environ. Sci. Eng.* **2015**, *9*, 615–624. [CrossRef]
167. Turkay, O.; Inan, H.; Dimoglo, A. Experimental study of humic acid degradation and theoretical modelling of catalytic ozonation. *Environ. Sci. Pollut. Res.* **2015**, *22*, 202–210. [CrossRef]
168. Turkay, O.; Inan, H.; Dimoglo, A. Experimental and theoretical investigations of CuO-catalyzed ozonation of humic acid. *Sep. Purif. Technol.* **2014**, *134*, 110–116. [CrossRef]
169. Wu, Y.; Wu, C.; Wang, Y.; Hu, C. Inhibition of Nano-Metal Oxides on Bromate Formation during Ozonation Process. *Ozone Sci. Eng.* **2014**, *36*, 549–559. [CrossRef]
170. Chen, K.C.; Wang, Y.H. The effects of Fe-Mn oxide and TiO2/α-Al2O3 on the formation of disinfection by-products in catalytic ozonation. *Chem. Eng. J.* **2014**, *253*, 84–92. [CrossRef]

171. Wang, Y.H.; Chen, K.C. Removal of disinfection by-products from contaminated water using a synthetic goethite catalyst via catalytic ozonation and a biofiltration system. *Int. J. Environ. Res. Public Health* **2014**, *11*, 9325–9344. [CrossRef]
172. Han, Q.; Wang, H.; Dong, W.; Liu, T.; Yin, Y. Suppression of bromate formation in ozonation process by using ferrate(VI): Batch study. *Chem. Eng. J.* **2014**, *236*, 110–120. [CrossRef]
173. Wang, Y.H.; Chen, K.C.; Chen, C.R. Combined catalytic ozonation and membrane system for trihalomethane control. *Catal. Today* **2013**, *216*, 261–267. [CrossRef]
174. Han, Q.; Wang, H.; Dong, W.; Liu, T.; Yin, Y. Formation and inhibition of bromate during ferrate(VI)—Ozone oxidation process. *Sep. Purif. Technol.* **2013**, *118*, 653–658. [CrossRef]
175. Molnar, J.; Agbaba, J.; Dalmacija, B.; Klašnja, M.; Watson, M.; Kragulj, M. Effects of Ozonation and Catalytic Ozonation on the Removal of Natural Organic Matter from Groundwater. *J. Environ. Eng.* **2012**, *138*, 804–808. [CrossRef]
176. Mortazavi, S.B.; Asgari, G.; Hashemian, S.J.; Moussavi, G. Degradation of humic acids through heterogeneous catalytic ozonation with bone charcoal. *React. Kinet. Mech. Catal.* **2010**, *100*, 471–485. [CrossRef]
177. Chen, K.C.; Wang, Y.H.; Chang, Y.H. Using catalytic ozonation and biofiltration to decrease the formation of disinfection by-products. *Desalination* **2009**, *249*, 929–935. [CrossRef]
178. Wang, J.; Zhou, Y.; Zhu, W.; He, X. Catalytic ozonation of dimethyl phthalate and chlorination disinfection by-product precursors over Ru/AC. *J. Hazard. Mater.* **2009**, *166*, 502–507. [CrossRef] [PubMed]
179. Zhang, T.; Lu, J.; Ma, J.; Qiang, Z. Fluorescence spectroscopic characterization of DOM fractions isolated from a filtered river water after ozonation and catalytic ozonation. *Chemosphere* **2008**, *71*, 911–921. [CrossRef] [PubMed]
180. Alsheyab, M.A.; Muñoz, A.H. Comparative study of ozone and MnO2/O3 effects on the elimination of TOC and COD of raw water at the Valmayor station. *Desalination* **2007**, *207*, 179–183. [CrossRef]
181. Karnik, B.S.; Davies, S.H.; Baumann, M.J.; Masten, S.J. Fabrication of catalytic membranes for the treatment of drinking water using combined ozonation and ultrafiltration. *Environ. Sci. Technol.* **2005**, *39*, 7656–7661. [CrossRef]
182. Karnik, B.S.; Baumann, M.J.; Masten, S.J.; Davies, S.H. AFM and SEM characterization of iron oxide coated ceramic membranes. *J. Mater. Sci.* **2006**, *41*, 6861–6870. [CrossRef]
183. Shioyama, M.; Kawanishi, T.; Yokoyama, S.; Nuno, M.; Yamamoto, T. Development of advanced ceramic membrane filtration system combined with ozonation and powdered activated carbon treatment. *Water Sci. Technol. Water Supply* **2001**, *1*, 91–96. [CrossRef]
184. Gracia, R.; Cortés, S.; Sarasa, J.; Ormad, P.; Ovelleiro, J.L. Catalytic ozonation with supported titanium dioxide. The stability of catalyst in water. *Ozone Sci. Eng.* **2000**, *22*, 185–193. [CrossRef]
185. Gracia, R.; Cortes, S.; Sarasa, J.; Ormad, P.; Ovelleiro, J.L. Heterogeneous catalytic ozonation with supported titanium dioxide in model and natural waters. *Ozone Sci. Eng.* **2000**, *22*, 461–471. [CrossRef]
186. Volk, C.; Roche, P.; Joret, J.C.; Paillard, H. Comparison of the effect of ozone, ozone-hydrogen peroxide system and catalytic ozone on the biodegradable organic matter of a fulvic acid solution. *Water Res.* **1997**, *31*, 650–656. [CrossRef]
187. Allemane, H.; Delouane, B.; Legube, B. Comparative Efficiency of Three Systems (Os, O_3/H2O_2 and O_3/TiO_2) for the Oxidation of Natural Organic Matter in Water. *Ozone Sci. Eng.* **1993**, *15*, 419–432. [CrossRef]
188. Bai, Z.Y.; Wang, J.L.; Yang, Q. Catalytic ozonation of dimethyl phthalate by Ce-substituted goethite. *Int. J. Environ. Sci. Technol.* **2017**, *14*, 2379–2388. [CrossRef]
189. Beltrán, F.J.; Pocostales, J.P.; Alvarez, P.M.; Oropesa, A. Diclofenac removal from water with ozone and activated carbon. *J. Haz. Mater.* **2009**, *163*, 768–776. [CrossRef]
190. Beltrán, F.J.; Pocostales, J.P.; Alvarez, P.M.; Jaramillo, J. Mechanism and kinetic considerations of TOC removal from the powdered activated carbon ozonation of diclofenac aqueous solutions. *J. Hazard. Mater.* **2009**, *169*, 532–538. [CrossRef]
191. Tomiyasu, H.; Fukutomi, H.; Gordon, G. Kinetics and Mechanism of Ozone Decomposition in Basic Aqueous Solution. *Inorg. Chem.* **1985**, *24*, 2962–2966. [CrossRef]
192. Staehelin, J.; Buehler, R.E.; Hoigne, J. Ozone decomposition in water studied by pulse radiolysis. 2. Hydroxyl and hydrogen tetroxide (HO_4) as chain intermediates. *J. Phys. Chem.* **1984**, *88*, 5999–6004. [CrossRef]
193. Elovitz, M.S.; von Gunten, U. Hydroxyl Radical/Ozone Ratios During Ozonation Processes. I. The Rct Concept. *Ozone Sci. Eng.* **1999**, *21*, 239–260. [CrossRef]
194. Yuan, R.; Zhou, B.; Zhang, X.; Guan, H. Photocatalytic degradation of humic acids using substrate-supported Fe3+-doped TiO2 nanotubes under UV/O_3 for water purification. *Environ. Sci. Pollut. Res.* **2015**, *22*, 17955–17964. [CrossRef] [PubMed]
195. Agustina, T.E.; Ang, H.M.; Vareek, V.K. A review of synergistic effect of photocatalysis and ozonation on wastewater treatment. *J. Photochem. Photobiol. C Photochem. Rev.* **2005**, *6*, 264–273. [CrossRef]
196. Mehrjouei, M.; Müller, S.; Möller, D. A review on photocatalytic ozonation used for the treatment of water and wastewater. *Chem. Eng. J.* **2015**, *263*, 209–219. [CrossRef]
197. Mecha, A.C.; Chollom, M.N. Photocatalytic ozonation of wastewater: A review. *Environ. Chem. Lett.* **2020**, *18*, 1491–1507. [CrossRef]
198. Beltrán, F.J.; Rey, A. Solar or UVA-Visible Photocatalytic Ozonation of Water Contaminants. *Molecules* **2017**, *22*, 177. [CrossRef]
199. Yu, D.; Li, L.; Wu, M.; Crittenden, J.C. Enhanced photocatalytic ozonation of organic pollutants using an iron-based metal-organic framework. *Appl. Catal. B Environ.* **2019**, *251*, 66–75. [CrossRef]

200. Cao, S.; Low, J.; Yu, J.; Jaroniec, M. Polymeric Photocatalysts Based on Graphitic Carbon Nitride. *Adv. Mater.* **2015**, *27*, 2150–2176. [CrossRef]
201. Xiao, J.; Xie, Y.; Rabeah, J.; Brückner, A.; Cao, H. Visible-Light Photocatalytic Ozonation Using Graphitic C3N4 Catalysts: A Hydroxyl Radical Manufacturer for Wastewater Treatment. *Acc. Chem. Res.* **2020**, *53*, 1024–1033. [CrossRef]
202. Kerc, A.; Bekbolet, M.; Saatci, A.M. Sequential Oxidation of Humic Acids by Ozonation and Photocatalysis. *Ozone Sci. Eng.* **2003**, *25*, 497–504. [CrossRef]
203. Bekbolet, M.; Uyguner, C.S.; Selcuk, H.; Rizzo, L.; Nikolaou, A.D.; Meriç, S.; Belgiorno, V. Application of oxidative removal of NOM to drinking water and formation of disinfection by-products. *Desalination* **2005**, *176*, 155–166. [CrossRef]
204. Uyguner, C.S.; Suphandag, S.A.; Kerc, A.; Bekbolet, M. Evaluation of adsorption and coagulation characteristics of humic acids preceded by alternative advanced oxidation techniques. *Desalination* **2007**, *210*, 183–193. [CrossRef]
205. Li, G.; Li, K.; Liu, A.; Yang, P.; Du, Y.; Zhu, M. 3D Flower-like β-MnO2/Reduced Graphene Oxide Nanocomposites for Catalytic Ozonation of Dichloroacetic Acid. *Sci. Rep.* **2017**, *7*, 1–7. [CrossRef]
206. Shin, D.; Jang, M.; Cui, M.; Na, S.; Khim, J. Enhanced removal of dichloroacetonitrile from drinking water by the combination of solar-photocatalysis and ozonation. *Chemosphere* **2013**, *93*, 2901–2908. [CrossRef]
207. Mehrjouei, M.; Müller, S.; Möller, D. Synergistic effect of the combination of immobilized TiO_2, UVA and ozone on the decomposition of dichloroacetic acid. *J. Environ. Sci. Health Part A Toxic/Hazardous Subst. Environ. Eng.* **2012**, *47*, 1073–1081. [CrossRef]
208. Gu, L.; Yu, X.; Xu, J.; Lv, L.; Wang, Q. Removal of dichloroacetic acid from drinking water by using adsorptive ozonation. *Ecotoxicology* **2011**, *20*, 1160–1166. [CrossRef] [PubMed]
209. Zhai, X.; Chen, Z.; Zhao, S.; Wang, H.; Yang, L. Enhanced ozonation of dichloroacetic acid in aqueous solution using nanometer ZnO powders. *J. Environ. Sci.* **2010**, *22*, 1527–1533. [CrossRef]
210. Yang, J.; Dong, Z.; Jiang, C.; Wang, C.; Liu, H. An overview of bromate formation in chemical oxidation processes: Occurrence, mechanism, influencing factors, risk assessment, and control strategies. *Chemosphere* **2019**, *237*, 124521. [CrossRef]
211. Fischbacher, A.; Löppenberg, K.; von Sonntag, C.; Schmidt, T.C. A New Reaction Pathway for Bromite to Bromate in the Ozonation of Bromide. *Environ. Sci. Technol.* **2015**, *49*, 11714–11720. [CrossRef]
212. Parrino, F.; Camera-Roda, G.; Loddo, V.; Palmisano, G.; Augugliaro, V. Combination of ozonation and photocatalysis for purification of aqueous effluents containing formic acid as probe pollutant and bromide ion. *Water Res.* **2014**, *50*, 189–199. [CrossRef]

catalysts

MDPI

Review

Review on the Visible Light Photocatalysis for the Decomposition of Ciprofloxacin, Norfloxacin, Tetracyclines, and Sulfonamides Antibiotics in Wastewater

Samar Shurbaji [1], Pham Thi Huong [2,3] and Talal Mohammed Altahtamouni [4,*]

1 Biomedical Research Center, Qatar University, Doha 2713, Qatar; ss1104227@qu.edu.qa
2 Institute of Research and Development, Duy Tan University, DaNang 550000, Vietnam; phamthihuong4@duytan.edu.vn or phambary@gmail.com
3 Faculty of Environment and Chemical Engineering, Duy Tan University, Danang 550000, Vietnam
4 Materials Science and Technology Program, College of Arts and Sciences, Qatar University, Doha 2713, Qatar
* Correspondence: taltahtamouni@qu.edu.qa; Tel.: +974-44036809

Abstract: Antibiotics are chemical compounds that are used to kill or prevent bacterial growth. They are used in different fields, such as the medical field, agriculture, and veterinary. Antibiotics end up in wastewater, which causes the threat of developing antibacterial resistance; therefore, antibiotics must be eliminated from wastewater. Different conventional elimination methods are limited due to their high cost and effort, or incomplete elimination. Semiconductor-assisted photocatalysis arises as an effective elimination method for different organic wastes including antibiotics. A variety of semiconducting materials were tested to eliminate antibiotics from wastewater; nevertheless, research is still ongoing due to some limitations. This review summarizes the recent studies regarding semiconducting material modifications for antibiotic degradation using visible light irradiation.

Keywords: antibiotics; photocatalysis; doping; heterojunction; surface plasmon resonance; wastewater

1. Introduction

Benjamin Franklin said once "When the well is dry, we know the worth of water" to encourage us to value what we have before water resources become limited. Although water covers more than two-thirds of the earth's surface, not all of the available water is accessible for use [1]. Water quality is crucial when it comes to human utilization, to sustain good health and a disease-free community. Water has been consumed in many industries, which are all linked to human health, such as the agricultural, electronic, food, and pharmaceutical industries [2]. The increased population growth, along with the plentiful human activities, has led to a shortage of freshwater resources, as water consumption has increased by six-fold in the 20th century [3,4]. As a result, there is extensive research into finding methods to reuse and purify wastewater from different contaminants [5]. There are a variety of water pollutants that come from different resources, such as sewage wastewater, industrial waste, oil pollution, radioactive waste, and pharmaceutical wastes [3,4,6]. Nowadays, pharmaceutical wastes, mainly antibiotics, are considered a major concern. Antibiotics are chemical compounds that are intended to kill or slow the growth of bacteria [7,8]. They are used in a variety of fields, including the medical field, agriculture, and veterinary [9,10]. The overconsumption of antibiotics and their massive presence in wastewater can cause a serious situation where the bacteria induce certain mutations and become antibiotic-resistant [11].

Conventional elimination methods, such as filtration, biodegradation, or reverse osmosis, are limited, due either to their high cost or the incomplete elimination of pollutants. To overcome these limitations, advanced oxidative process (AOP) has been used extensively to degrade different organic pollutants [12]. AOP involves homogenous photocatalysis and heterogeneous photocatalysis. Homogenous photocatalysis has been widely studied, so there is more motive to study heterogeneous photocatalysis, which involves semiconductor-assisted photocatalysis [13]. Semiconductor-assisted photocatalysis involves the photosensitization of semiconducting material to generate free radicals that degrade pollutants in the presence of light [14]. TiO_2 is one of the most semiconducting materials that has been used for photocatalysis for the degradation of different pollutants including antibiotics from wastewater. TiO_2 is safe, efficient, and cost-effective; however, its light-absorption capability in the visible light range is limited, which increases costs and limits its use on an industrial scale. To overcome such limitations, there is intensive research into finding methods to modify or use alternative semiconducting materials, which can degrade antibiotics using solar energy [15].

Although there are a few reviews on semiconductor-assisted photocatalysis for antibiotic or pharmaceuticals removal with visible light irradiation [16,17], a comprehensive review of recent reports is still needed, as the research is still ongoing, which will help to update the readers about this research field.

This review summarizes the basic principle of photocatalysis, different antibiotics decomposition approaches, and the essential characteristics of semiconducting materials for photocatalytic decomposition of antibiotics in wastewater, as well as recent reports on different antibiotic degradation, including the most studied and consumed antibiotics (ciprofloxacin, norfloxacin, tetracyclines, and sulfonamides). Different semiconductor modification approaches are discussed, along with the mechanism of degradation for the most common antibiotics.

2. Antibiotics Routes of Entry into Wastewater and Its Consequences

Human consumption of antibiotics has increased significantly [18], which might not seem alarming, as it is being used for treating diseases [19]. However, the overconsumption of antibiotics is not human-friendly or environmentally safe [20]. Antibiotics' disadvantages are not only limited to their consumption, but also their disposal. In most cases, antibiotics end up in wastewater, which cannot easily be eliminated by different degradation methods [21].

Due to the growing population demands, antibiotics are being used in different fields other than the medical field. In particular, antibiotics are extensively used in livestock and animal veterinary [22]. In some countries, antibiotics are not only used for animal treatment but also to promote animal growth and increase production [23]. However, there will not be considerable antibiotic amounts in animal meat, as all the antibiotics will be completely released from the animal's blood if the treatment is restricted for a few days [24]. Nevertheless, antibiotics will be released as animal waste due to its incomplete digestion, which either used as fertilizers in agriculture or dumped into wastewater [25]. Few studies were conducted to assess the uptake of antibiotics from fertilizers by plants. For example, a study was done concerning assessing the amount of antibiotics in different vegetables after being fertilized with animal waste. They found that the concentration of different antibiotics was less than 10 $\mu g \backslash Kg$, which is below the detection limit. Nevertheless, this result is alarming if antibiotics are used extensively [26]. Further study revealed that chlortetracycline antibiotic was uptaken by onions, cabbage, and corn. However, those vegetables did not take up the tyrosine antibiotic, probably due to its large molecular size [27]. The outcomes of these studies suggest that different plants can uptake different antibiotics at different rates depending on the plant type and the molecule size of the antibiotic.

The fundamental problem of having antibiotics in wastewater is the development of bacterial resistance to antibiotics. Those antibiotic-resistant bacterial strains are very difficult to treat, which causes a health and environmental threat. The bacterial resistance mostly occurs in hospital wastewater, due to antibiotics' extensive usage at high concentrations [28,29]. As bacterial resistance becomes a major concern, antibiotic degradation becomes necessary. Figure 1 summarizes antibiotics' possible routes of entry to wastewater.

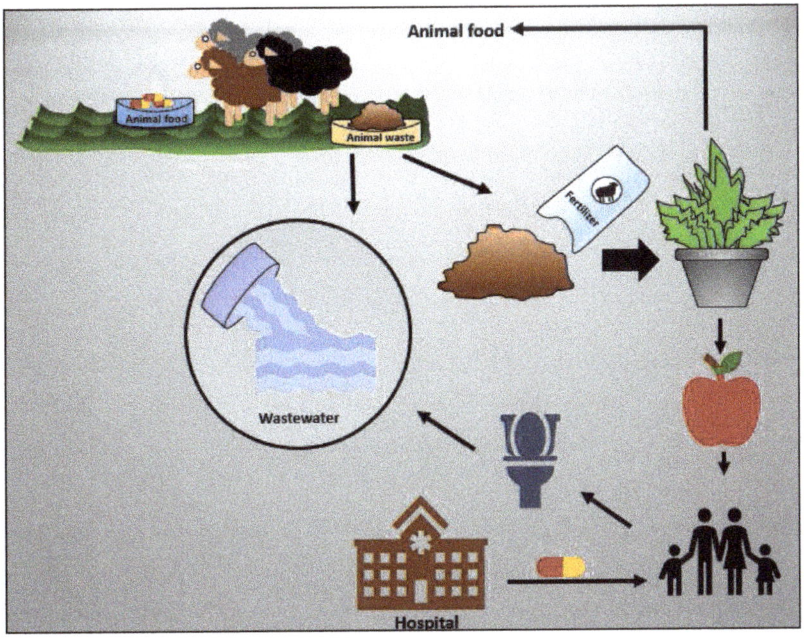

Figure 1. Possible routes of entry of antibiotics into wastewater. Adapted from [30] with modifications.

3. Current and Conventional Elimination Methods of Antibiotics

Biodegradation is one of the methods that were used to eliminate antibiotics from wastewater. The principle of biodegradation depends upon the use of microorganisms to biologically degrade the antibiotics to safer compounds [31]. Some microorganisms showed promising behavior in the degradation of certain antibiotics; conversely, in other cases, antibiotics could not be removed completely. Generally, this approach is not considered to be effective, not only in terms of antibiotics degradation but also due to the associated risks of the direct use of microorganisms [32]. Another method is reverse osmosis, in which water is compressed across a semipermeable membrane (e.g., cellulose acetate membrane), allowing pure water to pass and eliminating other impurities [33]. This method is efficient but not cost-effective [34]. Furthermore, different filtration methods were conventionally used, such as ultra-filtration, which purifies water by using membranes with a pore size from 10 to 1000 Å. This technique is an efficient but unreliable approach to eliminate some organic wastes [35]. The advanced oxidation process is a further approach to eliminate antibiotics from wastewater. This approach depends upon the generation of free radicals, such as hydroxyl radical or singlet oxygen. Free radicals are very reactive; they react with organic compounds, leading to oxidation reaction and degradation of that compound. Free radicals can be produced using a variety of methods; some are photochemical, and others are non-photochemical, such as ozonation [19,35]. The photochemical process involves semiconductor-assisted photocatalysis which is the main topic of this review.

4. Semiconductor Assisted Photocatalysis

Photocatalysis is a term derived from a Greek word, where photo means light and catalysis signifies a process where the rate of a chemical reaction is accelerated by a substance without being altered. Therefore, photocatalysis depends upon the simultaneous use of light and catalysts to speed up a chemical reaction. The concept of photocatalysis was first introduced by Fujishima and Honda in 1972, where water was split using a TiO2 electrode under UV light. The main idea of applying photocatalysis to purify wastewater depends upon the generation of free radicals, which can react with organic components in wastewater, resulting in the degradation of that component [36].

The general mechanism of semiconductor-assisted photocatalysis is represented in Figure 2, when light energy (photons) hit the semiconductor surface with energy equal to or higher than the bandgap energy. Electrons from the valance band will get excited to the conduction band, leaving holes at the valance band. The holes at the valance band can react with water molecules, generating hydroxyl radicals, which have a strong oxidizing capability that is used to degrade organic matters. However, electrons at the conduction band can interact with oxygen-generating superoxide anion. Therefore, alternative oxidation and reduction reactions can occur by the formed electrons and holes [37].

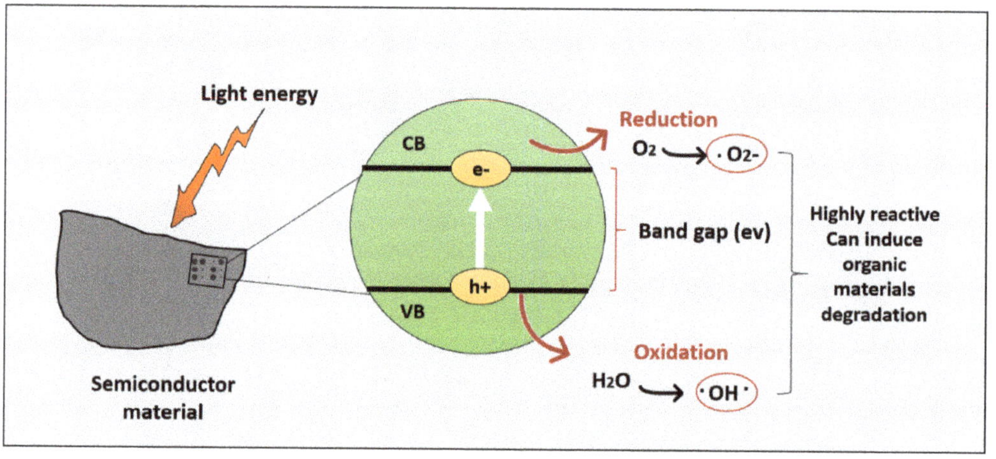

Figure 2. General photocatalytic mechanism. Adapted from [38] with modifications.

Although semiconductor photocatalysis has many advantages compared to other wastewater removal techniques, many aspects need improvement. Particularly, the problem of fast recombination of the electron–hole pair, which significantly affects the photocatalytic performance [39]. Furthermore, the utilization of natural energy resources is limited, since some of the photocatalysts have a maximum absorption wavelength in the UV range [15]. Additionally, the morphology of some nanomaterials might not provide an efficient surface–light interaction; this can be improved by increasing the surface area/volume ratio. It has also been shown that the shape of the nanomaterial can influence the electron mobility, as in TiO_2 nanoparticles, which showed lower electron mobility compared to highly ordered TiO_2 nanotubes [39]. Besides this, the overall cost and safety of the material must be considered [36]; although TiO_2 suffers from many of the mentioned limitations, it is still used due to its affordable cost [15]. TiO_2 is also more chemically stable compared to other semiconducting materials with a lower energy bandgap, such as g-C_3N_4, which is also vulnerable to degradation by hydroxyl radicals [40].

Generally, increasing the overall photocatalytic efficiency would enhance the degradation of antibiotics. There are variety of studies testing different material characteristics to overcome such limitations. For example, surface engineering and morphology optimization can greatly affect photocatalytic performance. The photocatalytic activity would be enhanced by maximizing the active sites, mainly by increasing the surface area and increasing the proportion of crystal facets. [41,42]. Inducing defects such as anion or cation vacancies can affect the photocatalytic activity as well [43]; for example, oxygen vacancy was shown to enhance light absorption and charge separation in BiOBr. [44]. Doping semiconducting materials is another approach used to narrow the bandgap; doping can reduce the recombination rate as well as shift the absorption wavelength to a visible range [17]. Different dopants can be used, such as Ag, Mg, Cu, Rb, S, N and P. Additionally, semiconductor photocatalysts can be either doped with a single dopant or co-doped with two dopants [45]. Heterojunction charge transfer is another method where the photocatalytic performance is improved by lowering the recombination rate of electron–hole pairs. Semiconductor heterojunction depends upon combining the semiconductor material with other material, which can be metallic, carbon group, or a semiconductor. When coupling a semiconductor with another semiconductor, the photocatalytic efficiency would be enhanced, as photoexcited electrons will transfer from the semiconductor with more negative positions in the conduction band to the other semiconductor, whereas holes will be transferred from the more positive positions in the valance band to the other semiconductor with lower positions, which leads to overall enhanced charge separation, and, thus, better photocatalytic activity. [46]. A further approach to improve photocatalytic activity is the use of the plasmonic effect by involving metallic nanoparticles. Metallic nanomaterials such as Au and Ag have a property known as surface plasmon resonance (SPR). Upon irradiation, if the incident energy is enough to prevent the positive nuclei-restoring force, the valance band of the metallic particle will be generated. When metallic nanoparticles are combined with semiconductor material, the light absorption will be enhanced, as electrons will migrate to the conjugated semiconducting material; thus, overall photocatalytic efficiency will be improved. [17]. Non-Nobel metals such as bismuth, aluminum, and copper were also reported to have a plasmonic effect. They can be used as an alternative to nobel metals due to their convenience and low cost [47].

There are further approaches used to enhance the photocatalytic activity; adding a sensitizer is one of them. Organic metal complexes, such as Ru(bpy) 2 + 3 (bpy = 2′,2 bipyridine), can be used as a sensitizer. However, sensitization is highly dependent on the solution pH, as it involves the adsorption of the molecule on the surface. Furthermore, a sensitizer might be degraded if an electron donor is not present. Sensitization has not been commonly used recently; rather, research focuses on other approaches to enhance photocatalytic activity [41,48]. Hydrogenation is another approach to enhancing the photocatalytic performance. Hydrogenation, as the name implies, involves adding hydrogen molecules to the semiconductor's surface by a chemical reaction using catalysts or by applying H_2 plasma treatment [30]. The surface of the semiconducting material can be either fully or partially hydrogenated, i.e., different degrees of hydrogenation can be achieved. A study conducted by Yan, Han et.al revealed that slightly hydrogenated TiO_2 nanoparticles enhanced the photocatalytic degradation of organic dyes by enhancing the light absorption at the UV range. However, highly hydrogenated TiO_2 nanoparticles showed a lower photocatalytic performance compared to the pristine form [49]. Another study showed an enhancement of g-C_3N_4 performance after hydrogenation; this enhancement was mainly attributed to the redistribution of charge density in valance and conduction bands, which leads to efficient charge separation [50].

A novel approach to enhancing photocatalytic performance is coupling the semiconducting material with phosphor. The photocatalytic process involves the use of a photoreactor that ensures an appropriate interaction between photons, the chemical reactant, and the photocatalyst. The photoreactor must be properly designed, so that all the photons emitted from the light source are collected. Unlike classical chemical reactors, different reaction parameters, like pressure, temperature, and fluid dynamics, are less important in the photoreactor, and controlling these parameters will not give the best photocatalytic results. However, the optical path length is an important parameter in the photoreactor's design. This makes sure that the photon flux is uniform inside the reactor. This parameter is critical as, when the light travels through the photoreactor, the photons are absorbed by the photocatalyst. This leads to scattering and decaying of the photon intensity, causing a non-uniform reaction rate in the photoreactor. This problem can be overcome by coupling the photocatalyst with light-emitting phosphorescent particles such as phosphors. Phosphor material is a substance that is based on a lanthanide-doped material that emits light and exhibits luminescence characteristics upon exposure to electromagnetic radiation. The introduction of phosphor to the system can enhance the photoreactivity by transforming the external light 365 nm radiation into 440 nm emission, as well as by allowing the photoexcitation of the fraction of photocatalyst in the core reactor volume and not screened by the photocatalyst itself. There are three types of phosphor materials: down-conversion phosphors, up-conversion phosphors, and long-afterglow phosphors. In the down-conversion phosphor, higher energy photons are absorbed and emitted as lower energy photons [51,52]. Different transition metals, such as Eu^{2+}, Er^{3+}, and Pr^{3+}, showed light emission at different wavelengths. This phenomenon will help to tune the emission spectra to the required specifications. It has been reported that when phosphor is coupled with a dye, the emitted photons from phosphor can be easily absorbed by the dye, leading to excitation and the generation of electron–hole pairs. A commonly studied material is N-doped TiO_2 coupled with phosphor, which showed enhancement in photon transfer in the reactor [53]. The mechanism by which phosopher enhances the photon transfer is illustrated in Figure 3. The down-conversion phosphor material can absorb UV light and emit photons at the visible range inside the photoreactor, which can be absorbed and utilized by the photocatalytic material. This system was shown to enhance the photocatalytic activity and degradation of different organic compounds such as Methylene blue, phenol, and Terephthalic acid [53].

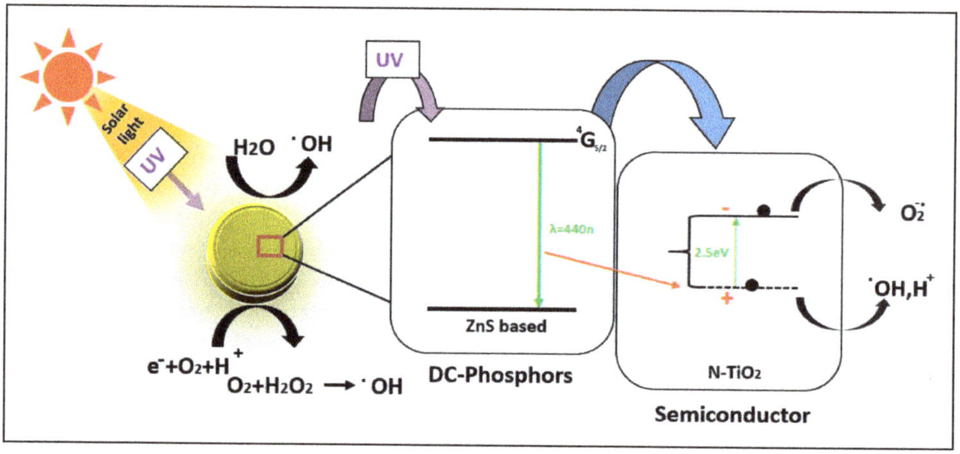

Figure 3. The down-conversion phosphor way of action. Adapted with permission from [53]. Copyright 2015 Elsevier.

An upconversion phosphor is a further approach through which visible light absorption by the photocatalysts is enhanced. In this approach, lower energy photons are transformed into higher energy photons. There are three mechanisms by which the up-conversion phosphor works; these are excited-state absorption, photon avalanche, and energy transfer up-conversion. Sacco, Vaiano and Sannino describe in detail all three approaches in detail [54]. Coupling the photocatalyst with a long persistent phosphor (long afterglow phosphor) is another approach used to enhance photocatalysis. Long afterglow phosphors such as $CaAl_2O_4$: Eu^{2+}, Nd^{3+} can sustain photocatalysis even after the removal of the light source or the photoreactor. Those long afterglow phosphors can store the energy from light and emit it at a slow rate. It has been reported that this approach is sustainable and saves energy, as the photocatalytic process can be done without the presence of light. This approach was used for the degradation of the antibiotic ofloxacin by Alberti et al., when they developed N-P doped TiO_2 for this purpose; the study showed 87% degradation of ofloxacin after 50 min of irradiation. The study showed 95% degradation of the antibiotic after 30 min of total irradiation (i.e., alternating radiation–light cycle followed by dark cycle) [55].

5. Commonly Used Semiconducting Materials for Antibiotic Degradation

5.1. Semiconducting Metal Oxides

Metal oxides have been shown to have certain properties that make them suitable candidates for photocatalysis, for example, light absorption (UV, visible light, or both combined), which initiates charge separation and the photogeneration of electrons and holes. Additionally, they are biocompatible, safe, and stable when exposed to different conditions. Metal oxides include the oxides of titanium, zinc, tin, vanadium, and chromium [56].

TiO_2 is the most studied metal oxide for photocatalysis, due to its superior properties such as good optical and electronic properties, chemical stability and reusability, non-toxicity and low cost. Although TiO_2 is a promising material for commercial use, TiO_2 suffers from limitations, as it has a large bandgap (3.2 eV), thus limiting visible light utilization, which makes its application uneconomic. There are a variety of studies regarding TiO_2 modification to obtain a better photocatalytic performance, which would make its application more feasible [57–61]. ZnO is another semiconducting material that has a better quantum efficiency than TiO_2; furthermore, recent reports suggested that ZnO has higher photocatalytic efficiency compared to TiO_2, especially if used at a neutral pH [62,63]. Nevertheless, the high recombination rate of the photogenerated electron–hole pairs restricts the utilization of ZnO in its pure form. Several studies reported doping ZnO with metals like Ag and Fe [64] or non-metals like N and C. Different studies suggested improvements in photocatalytic properties after doping the ZnO; for instance, a study reported that the ZnO 12 S incorporated nanoparticles for the degradation of organic dyes. This material showed a 100% degradation efficiency that decreased to 92% after five cycles of reuse [65]. Other studies showed enhanced photocatalytic performance of ZnO-based materials [66–68].

Tungsten oxide WO_3 is another metal oxide that has received considerable attention due to its abundance, cost-effectiveness, and non-toxicity [69–71]. Unlike TiO_2, it exhibits 12% absorption of the solar spectrum. Different studies reported a variety of modification approaches. In particular, Huang et al. found that $W_{18}O_{49}$ has higher photocatalytic degradation efficiency compared to WO_3. Different $W_{18}O_{49}$ morphologies exhibit different photocatalytic activities as well. Namely, hollow $W_{18}O_{49}$ spheres exhibit the highest photocatalytic activity compared to other morphologies. Despite the fact that $W_{18}O_{49}$ is more efficient compared to WO_3, it is prone to oxidization to WO_3. To overcome this limitation, a study reported the construction of $W_{18}O_{49}/TiO_2$ hybrid; by this, both stability and efficient photoactivity are achieved [72].

5.2. Bismuth Based Photocatalysts

Bismuth-based semiconductor photocatalysts have been shown to have a good visible light absorption as their valance band electrons form a hybrid of O 2p and Bi 6s, unlike TiO_2 with O 2p orbitals. Those Bi 6s orbitals, when properly dispersed, have been found to increase the mobility of electrons and holes, as well as reduce the bandgap energy. For that, Bi-based semiconductors mostly have a bandgap of less than 3 eV. There are a variety of studies regarding different Bi-based semiconductors for photocatalysis, such as Bi_2O_3, Bi_2CrO_6, Bi_2MoO_6, Bi_2WO_6, $BiVO_4$, BiOCl, and BiOBr [73].

Bi_2O_3 is one of the simplest and most common and major photocatalysts. It has been applied for both water-splitting and water treatment from organic wastes. Bi_2O_3 has a bandgap that ranges from 2.1 to 2.8 eV, which makes its utilization for visible light absorption more efficient. Bi_2O_3 has five different polymorphisms: α, β, γ, δ, and

ω-Bi_2O_3. Metastable phases of Bi_2O_3 can be converted to the α phase at low temperatures or the beta phase at high temperatures [73,74].

A further example is bismuth vanadate ($BiVO_4$), which was reported to have superior physiochemical properties like ferro-elasticity and ionic conductivity [75]. It possesses three phases: monoclinic, fergusonite, and tetragonal. It has a theoretical bandgap of 2.047 eV, which maximizes its visible light utilization. $BiVO_4$ was used in photocatalysis for organic waste treatment as well as water splitting [73,76].

Although bismuth-based photocatalysts have been shown to have good photocatalytic performance as well as efficient utilization of visible light, some parameters need to be addressed, for example, the stability of bismuth-based photocatalysts under different conditions. Their stability is mainly affected by the photocatalyst intrinsic structure and its solubility, as well as the solution's pH. It has been shown that layered bismuth-based photocatalysts with both covalent and van der Waals bonds are more stable than layered bismuth-based photocatalysts with van der Waals bonds only. Nevertheless, bismuth-based photocatalysts usually have low solubility constants, which means that they cannot always be dissolved in the substrate. Furthermore, their photo-corrosion was not sufficiently studied [73,77].

5.3. Graphitic Carbon Nitrides

Graphitic carbon nitride (g-C_3N_4) is a new type of polymeric semiconducting material. It is considered as the most stable allotrope compared to other carbon nitrides at ambient conditions. Unlike TiO_2, g-C_3N_4 can utilize both visible light and UV radiation due to its narrow bandgap (2.7 eV) [78]. However, g-C_3N_4 in its pure form endures many drawbacks, such as the high recombination rate of electron–hole pairs, poor utilization of visible light, and low electrical conductivity [79]. Therefore, material modification is necessary to overcome such limitations. Different studies reported the modification of g-C_3N_4 mainly by doping with transition metals, noble metals, or non-metals [80,81]. Surface modification was reported as well [82]. Based on several reports in the literature, doping with noble metal ions was shown to have the best photocatalytic performance. This was attributed to the higher separation of the photogenerated electrons and holes, due to the ability of electron capture by the noble metallic ions. For example, a study reported 96.8% photocatalytic degradation after 120 h of solar light exposure when using Ag-doped g-C_3N_4. Despite the fact that noble metal doping can enhance the performance of g-C_3N_4 significantly, doping with noble metals is not cost-effective. Consequently, there are other studies concerning the use of noble metal-free composites [83]. Namely, Cao et.al designed g-C_3N_4 coupled with $Ni(dmgH)_2$. This composite showed an efficient ecosystem-friendly photocatalyst with efficient solar-to-hydrogen conversion. Despite this, the photocatalytic degradation of different compounds was not studied [84]. Other studies reported doping g-C_3N_4 with non-noble metals such as nitrogen [85,86]; those studies are limited due to the potential toxicity of the N precursors used, such as hydroxylammonium chloride and hydrazine hydrate. Other environmentally friendly non-noble metallic dopants include carbon; doping g-C_3N_4 with carbon showed enhancement in visible light absorption and charge separation. There are other studies concerning doping with non-noble metals; for more details, a review written by Starukh and Praus summarizes such studies [87].

There is ongoing research regarding material modification to fabricate and design nanomaterials with different properties to obtain the best photocatalytic performance, and thus, better elimination of antibiotics. The next section summarizes the recent studies related to antibiotic degradation by using semiconductor-assisted photocatalysis.

6. Recent Reports on Antibiotics Degradation with Photocatalysis

6.1. Photocatalytic Degradation of Ciprofloxacin (CIP)

Ciprofloxacin (CIP) $C_{17}H_{18}FN_3O_3$ is a second-generation fluoroquinolone antibiotic that is used to kill and prevent the replication of bacteria. It is used in both humans and animals to treat many medical cases, like skin and urinary tract infections [88]. In a study conducted in 2016, it was reported that CIP represents 73% of the total consumption among fluoroquinolone antibiotics. It was further reported by another study that the detected amount of CIP in hospital wastewater can reach up to 150 µg/L, which is considerably high, and has a potential risk if not eliminated [89]. Considering the frequent detection of CIP in the environment and water, researchers have been working to improve its elimination. CIP removal by the photocatalytic reaction can be achieved by attacking different sites in the CIP structure. Figure 4 shows the chemical structure of CIP and Figure 5 shows different sites that can be attacked by generated free radicals to decompose the CIP structure. For more details of CIP degradation pathways, Huo et.al summarize a variety of approaches and pathways for CIP degradation by photocatalysis [90].

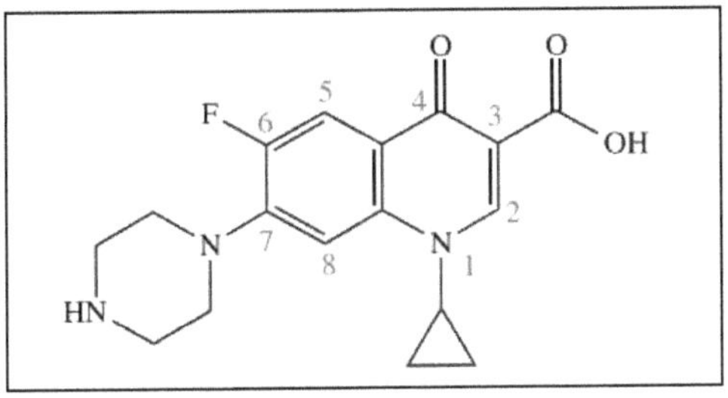

Figure 4. Ciprofloxacin (CIP) chemical structure.

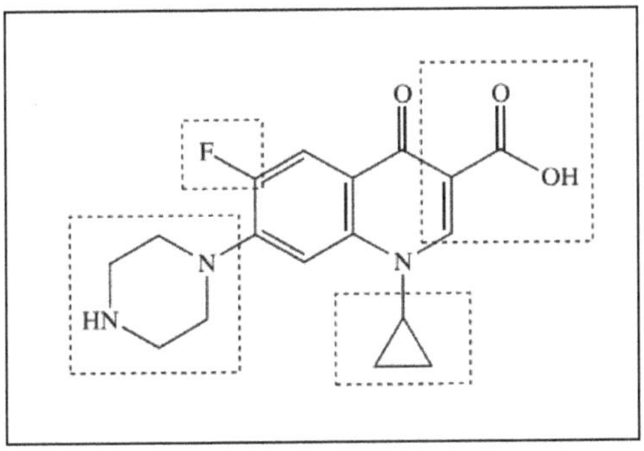

Figure 5. Different attack sites in CIP structure by free radicals generated by photocatalytic reaction.

Different studies reported the use of photocatalysts which have been modified to achieve the highest photocatalytic efficiency. Researchers usually start with doping with a transition metal. Transition metals were shown to enhance photocatalysis when added, for example, in a study conducted by Das, Ghosh, Misra, et al., where Fe-doped ZnO was prepared to degrade CIP under visible light irradiation. Here, the authors prepared Fe-doped ZnO following the precipitation route; after that, they induced the photocatalytic reaction by dispersing three different concentrations (100, 150 and 200 mg/L) of the photocatalyst with 10 mg/L of Ciprofloxacin followed by 3.5 h of visible light exposure. The optimum results were obtained when 150 mg/L of Fe-doped ZnO was used with 66% degradation efficiency, which is five times more than the degradation of the negative control. The improved photocatalytic activity was suggested to be due to the delay in recombination that Fe offers, thus the increment in the produced free radicals [91]. Figure 6 represents the degradation of CIP under different conditions. Although Fe, when added, enhances the degradation, its photocatalytic activity is impaired by its thermal instability and fast recombination rate. To overcome such limitations, metalloids such as Boron are used. In one study, Şimşek synthesized boron-doped TiO_2 using a solvothermal method to degrade CIP under visible light irradiation. TiO_2 was doped with different dopant concentrations (2,4,6,8%) which were then dispersed in 20 mg/L antibiotic solutions. The degradation efficiency was positively correlated with the dopant concentration with the maximum degradation rate of B/TiO_2 8% of around 88.32%, which is four times higher than the undoped TiO_2. The optimal reaction conditions were achieved when 1.1 g/L of 8% B/TiO_2 was used with a solution pH of 7.1 and with the presence of 7.23 mM H_2O_2. The produced material showed a lower bandgap, and thus better photocatalytic activity, as well as stability, when being reused five times [92]. Non-metallic dopants like graphene oxide (GO) were also used to improve the photocatalytic performance, where sphere-like copper tungstate ($CuWO_4$) deposited on 2D GO was produced by the hydrothermal method. Doping with GO decreased the bandgap of $CuWO_4$ from 2.5 to 2.09 eV, which enhanced the degradation of CIP, as more than 97% of CIP was degraded after one hour under visible light irradiation [93]. Co-doping with non-metals was also reported due to the superior photocatalytic enhancement that it offers. For that, Qu, Xu, and others developed N,S-doped carbon quantum dots embedded with ZNO nanoflowers (N, S-CQDs). A hydrothermal method was used to produce this material, which was tested for the degradation of CIP under visible light. It was found that 92.9% of CIP was degraded after being exposed to simulated sunlight for 20 min, whereas, 85.8% of CIP was degraded when exposed to natural sunlight for 50 min. The photocatalytic degradation efficiency of un-doped CQDs material was tested under simulated sunlight irradiation. The degradation rate was 8.6% after only 20 min of exposure. This is 10 times less efficient than the co-doped material [94]. The utilization of transition metal oxides was also reported, $LaNiO_3/TiO_2$ heterojunction was produced by the sol-gel method to degrade organic dyes and the antibiotic CIP under visible light irradiation. The performance of $LaNiO_3/TiO_2$ heterojunction was compared to pure $LaNiO_3$ and TiO_2. A 70% degradation of 10 mg/L CIP after 3 h of exposure to simulated sunlight was reported, which is significantly more than the pure components. The improvement in photocatalytic activity was attributed to the good separation of electron–hole pairs offered by the heterojunction. Metallic nanoparticles doping, such as Ag, was shown to increase the duration of the photogenerated electron–hole pairs. For that, Alvarez and others developed a $AgBr/Ag/Bi_2WO_6$ heterostructure to degrade CIP under visible light irradiation. The authors reported that 57% of CIP was degraded after 5 h of irradiation. The addition of Ag nanoparticles, even at a low weight %, showed a significant change in the material's photocatalytic behavior. The produced heterostructure showed an improvement in photocatalytic activity compared to the unmodified Bi_2WO_6. This is mainly due to the presence of Ag as well as the heterojunction formed by the two semiconducting materials AgBr and Bi_2WO_6 [95]. While the addition of Ag nanoparticles enhanced the photocatalytic activity significantly, it is not cost-effective; other lower-cost materials are suggested for utilization, such as doping with S or B. To overcome the drawbacks, such as material agglomeration, that can happen in some of the modified materials, Zheng et.al. designed graphitized mesoporous carbon TiO_2 nanocomposite to overcome material aggregation. It was reported that the large surface area and the mesoporous structure enhance the absorption of the antibiotic in the case of its presence at low concentrations, which enables a convenient interaction with the photocatalyst.

In addition, the graphitized carbon sheets inhibit TiO_2 aggregation. The produced material was tested to degrade CIP. The results suggest a quick and complete degradation of CIP in 1.5 h, along with inhibition of *Vibrio Fischer* bacteria [96]. These findings are important, as there are limited reports on bacterial experimentation. Concerning the enhancement of photocatalysis by morphology modification, a study reported the production of double-shelled $ZnSnO_3$ hollow cubes (ZSO-C), which was prepared by the co-precipitation method and its effect on the degradation of two antibiotics: CIP and sulfamonomethoxine. The ZSO-C was compared to another two materials: one of them was prepared by hydrothermal method with hollow polyhedral shape (ZSO-H), while the other was prepared with the template method (ZSO-T). It was found that

ZSO-C has the best photocatalytic efficiency and stability compared to the other two materials, not only in the degradation of the two antibiotics, but also in other dyes such as methylene blue and methyl orange, with a degradation rate of 85.9%, which is 1.5 times better than ZSO-H and two times better than ZSO-T. The authors attributed that to the narrower bandgap and larger surface area of ZSO-C compared to the other two materials [97]. A further study was conducted by Li, Fu, and Zhu; they produced 3D tri pyramid TiO_2 by the hydrothermal method to degrade CIP under UV–visible irradiation. They reported that the tri pyramid morphology significantly enhanced the degradation of CIP due to its high specific surface area compared to rod-shaped TiO_2. Correspondingly, the tri-pyramid morphology provided more active sites for catalytic interactions [98]. Green, environmentally friendly methods were also utilized. Particularly, the recent photocatalyst "layered double hydroxides" (LDHs), was modified by making an LDHs Pt hybrid, as reported by Z. Li et al. They found that doping LDHs with Pt leads to significant improvement in photocatalytic activity especially in the visible light range [99]. The mechanism by which CIP was degraded by LDHs/Pt hybrid is presented in Figure 7.

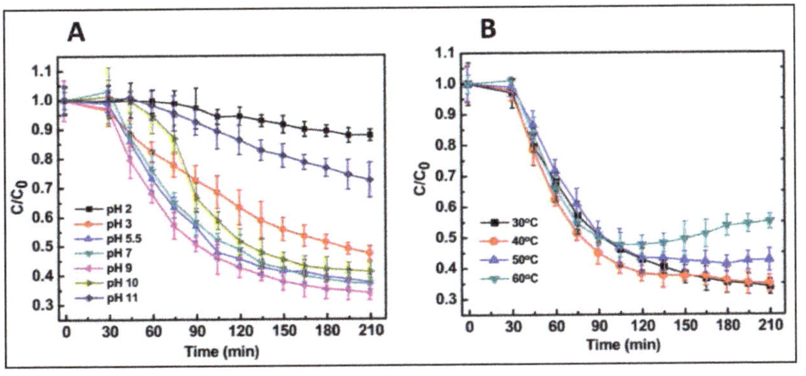

Figure 6. Degradation of CIP over time under different conditions. (**A**) the photocatalytic degradation of 10 mg/L CIP using 150 mg/L Fe-ZnO when irradiated with sunlight. The degradation was tested using different pH levels. The best photocatalytic efficiency is near a pH = 9. (**B**) represents the effect of temperature on CIP photocatalytic degradation at pH = 9. An unexpected trend was obtained as a decreasing in degradation efficiency when increasing the temperature up to 60 °C. C/C0 represents the relative change in CIP with respect to the initial concentration. Adapted from [91].

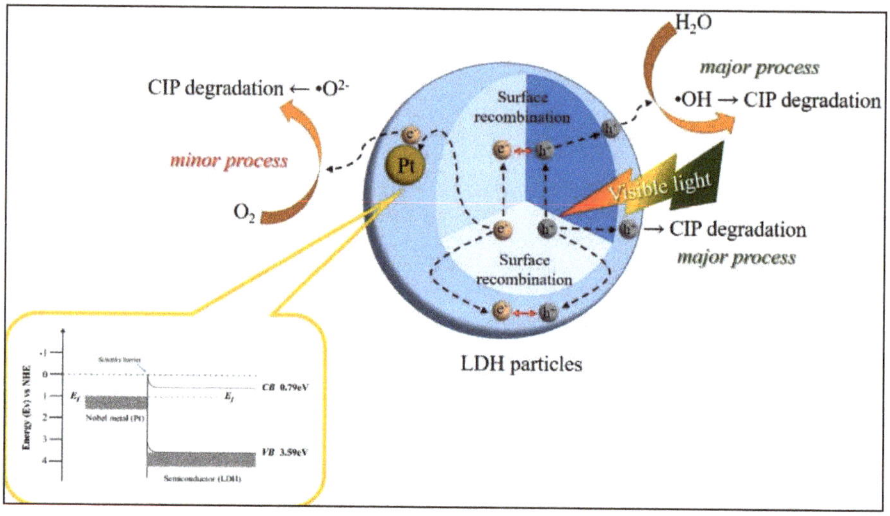

Figure 7. The proposed mechanism for the degradation of CIP using layered double hydroxides (LDHs) Pt hybrid. ESR spin trap experiment was used to determine the dominant free radicals. Scavengers like ascorbic acid, EDTA-2Na and tert-butyl alcohol were used to trap superoxide anion radicals, photogenerated holes, and hydroxyl radicals. Here, metallic Pt acts as springboard that facilitates charge separation and electron movement through Pt/LDH heterojunction interfaces. When Pt is attached to the LDH surface, surface recombination is reduced as the photogenerated electrons transfer to Pt and cannot return due to the presence of Schottky barrier. Simultaneously, the photogenerated holes migrates to LDH external surface and degrades CIP by reacting with H_2O to form. OH. Adapted with permission from [99]. Copyright 2020 Elsevier.

6.2. Photocatalytic Degradation of Norfloxacin

Norfloxacin is a broadly administered antibiotic that belongs to the fluoroquinolone antibiotic class. In the past few years, fluoroquinolone antibiotics have caused a major concern due to their extensive use and environmental toxicity. There are few reports on the degradation of norfloxacin by photocatalysis using different materials. The mechanism of degradation have been proposed as well; there are four proposed mechanisms, all sharing the same attack sites (piperazinyl and quinolone moieties). These mechanisms were proposed based on the recognition of the degraded intermediates, as shown in Figure 8. Where Mn:ZnS Quantum Dots were used to degrade norfloxacin Figure 9 represents different parameters redarding the degredation of norfloxacin using Mn:ZnS Quantum Dots [100].

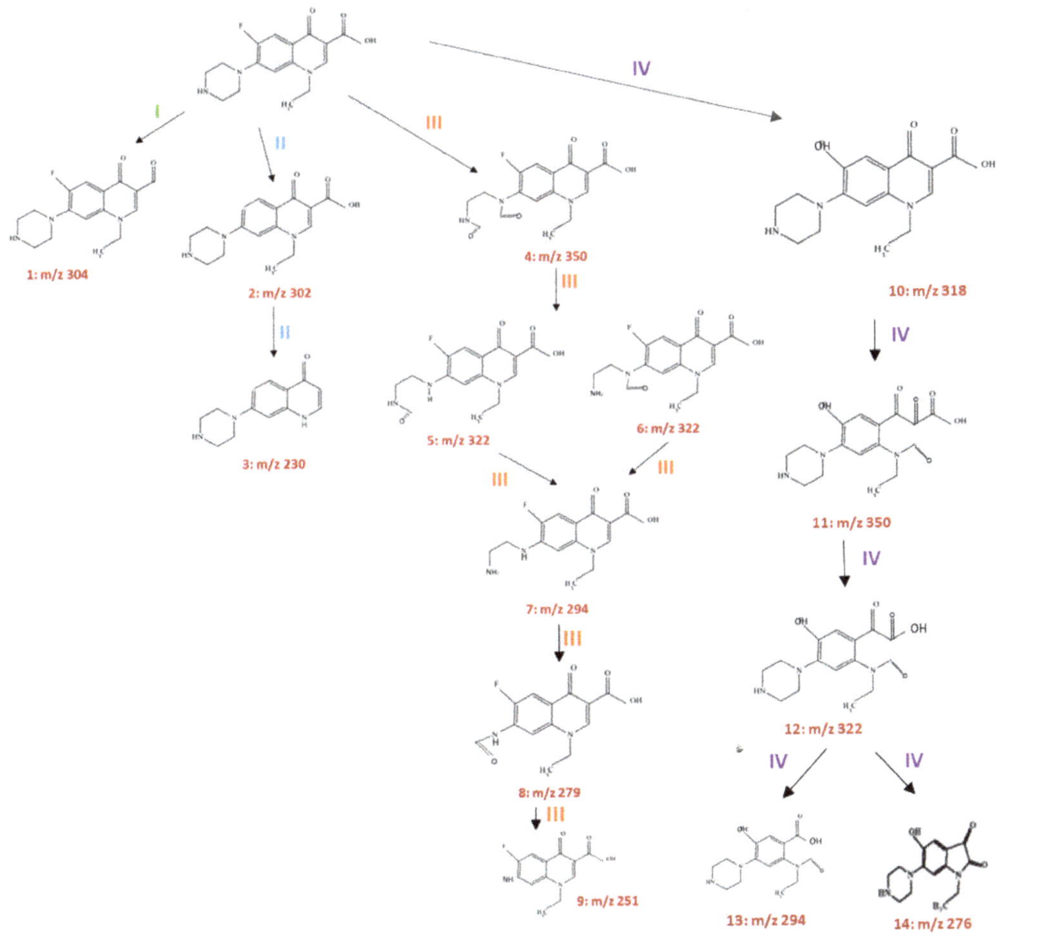

Figure 8. Proposed mechanism for the degradation of norfloxacin; four main pathways are presented. High-resolution liquid chromatography mass spectrometry was used to determine the produced degradation intermediates. There are 4 suggested pathways; in the first pathway, the norfloxacin is degraded by dehydroxylation reaction. In the second pathway, structure 2 is produced by defluorination reaction, which is transmitted to structure 3 by decarboxylation and deethylation reaction. The third pathway is characterized mainly by the removal of piperazinyl group; in this pathway, ring opening occurs, and 6 intermediates are produced. The fourth pathway is characterized by the opening of the quinolone substituent and the benzene moiety. Adapted from [100].

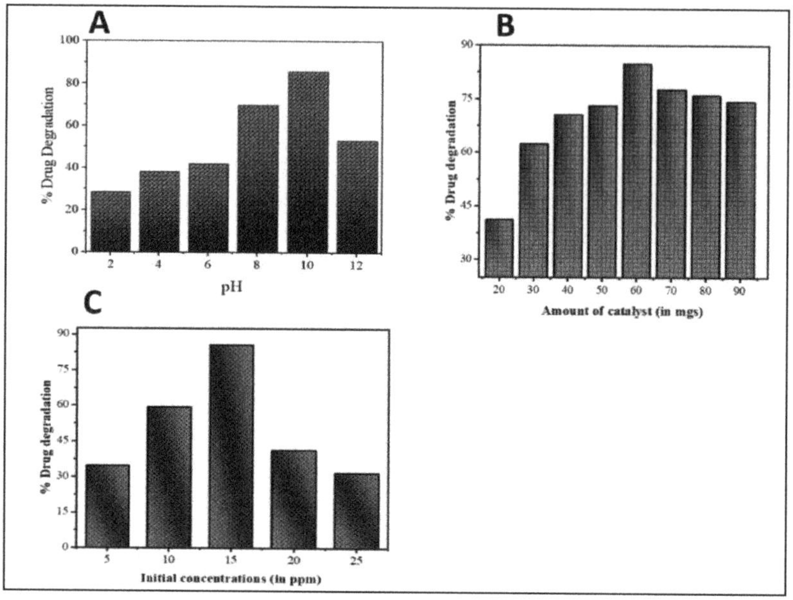

Figure 9. Different parameters were applied to degrade the antibiotic norfloxacin using Mn:ZnS Quantum Dots. (**A**) showing the effect of pH on norfloxacin, with pH = 10 owining the best photocatalytic efficiency(**B**) represents a bar chart with the effect of the number of catalysts used. (**C**) signifies the effect of the initial antibiotic concentration. Adapted from [100].

Nevertheless, it is still necessary to summarize what has been done in the literature to know where we stand in terms of different materials' fabrication and the efficiency of elimination. For this, one study reported the preparation of the In_2O_3/TiO_2 heterostructure by Yu, Chen and others, by the polymeric precursor method to degrade norfloxacin under visible-light irradiation. The antibiotic was completely degraded after 10 min of irradiation. This desirable photocatalytic activity is mainly due to the narrow bandgap that is offered by the heterostructure, which enhanced absorption at the visible range and the large specific surface area that allows for better surface–light interaction [101]. Another study was reported by Jin, Zhou, et al., where they produced spherical-shaped N-doped TiO_2 to degrade norfloxacin under visible light irradiation. Within 30 min, 99.53% of the antibiotic was degraded, which cannot be achieved when using TiO_2 alone due to the reduction in the bandgap to 2.92 eV offered by the dopant. The authors further assessed the toxicity after norfloxacin degradation against *E. coli* bacteria; they found that the norfloxacin toxicity reduced to less than 50% after degradation. Moreover, 25% of mineralization was achieved from the photocatalysis [102]. These findings are important, since the antibiotic toxicity against bacteria after the antibiotic degradation was studied; many studies lack this type of experimentation, disregarding its importance. The application of bismuth-based nanocomposites was reported by many studies. One study was done to investigate the effectiveness of $Bi_2WO_6/rGO/Bi_{25}FeO_{40}$ heterojunction in the degradation of norfloxacin antibiotics under visible light irradiation. Different ratios of $Bi_2WO_6/rGO/Bi_{25}FeO_{40}$ were produced by the hydrothermal method. A total of 8% of $Bi_2WO_6/rGO/Bi_{25}FeO_{40}$ showed the highest ability to degrade norfloxacin. This enhancement in photocatalytic properties is due to the efficient charge transfer between Bi_2WO_6 and $Bi_{25}FeO_{40}$ and the smaller bandgap in this nanocomposite compared to a single component. Additionally, the high surface area and enhanced light absorption of the produced heterojunction contribute to its superior photocatalytic properties [103]. Besides this, $Bi_2MoO_6/rGO/BiOBr$ heterostructure was developed by Zhang et al. to degrade norfloxacin under visible light irradiation. The heterostructure showed a superior degradation rate of the antibiotic, with 78.12% degradation. This is approximately four times more than the degradation of Bi_2MoO_6 and BiOBr alone, and one time better than the degradation of Bi_2MoO_6/BiOBr heterojunction. The enhancement in the photocatalytic property was attributed to the formation of heterojunction as well as the presence of rGo, which speeds up the electron transfer [104]. A further study was conducted to test the efficiency of the $COFe_2O_4$-rGO-BiOBr magnetic nanocomposite for the degradation of norfloxacin antibiotics under visible light irradiation using a mercury lamp. The nanocomposite was prepared by the hydrothermal method. The authors found that $COFe_2O_4$-rGO-BiOBr has higher photocatalytic efficiency compared to single or binary systems. This improvement was attributed to the formation of a heterojunction between $COFe_2O_4$ and BiOBr. rGO was used to increase the electron transfer rate. The overall photocatalysis rate was intensified due to the faster transfer rate and lower recombination rate. This makes $COFe_2O_4$-rGO-BiOBr a candidate material for photocatalysis [105]. There are very few studies regarding the utilization of quantum dots to degrade norfloxacin. One recent study reported the fabrication of novel Mn-doped ZnS quantum dots. The produced material exhibited an excellent degradation property; however, the material was tested only under UV irradiation [100]. A less recent study reported the design of carbon quantum dots loaded with mesoporous g-C_3N_4. This material showed an enhanced photocatalytic performance for fluoroquinolone antibiotics including norfloxacin compared to the bulk g-C_3N_4 or the unloaded. It was revealed by scavenging experiments that superoxide anion radical (O_2^-) and photo-hole (h^+) are the main active species. The degradation mechanism was proposed based on the produced intermediates [106].

6.3. Photocatalytic Degradation of Tetracyclines

Tetracyclines are a class of broad-spectrum antibiotics that was first utilized in 1940. They are used for the management of bacterial infections caused by Gram-positive and negative bacteria. Tetracyclines are usually consumed by a human for disease management, or animals as a growth-promoter. Tetracyclines cannot be completely metabolized by humans or animals, leading to their entry into the environment, causing many adverse effects on the food chain, and thus affecting human and animal health. Removing such antibiotics by photocatalysis was reported, using different materials and mechanisms of degradation [107,108]. Figure 10 represents the chemical structure of Tetracycline.

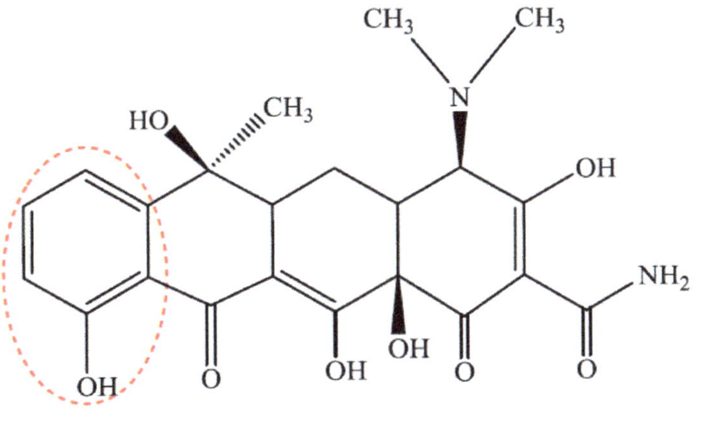

Figure 10. Tetracycline chemical structure.

There are different proposed mechanisms on the photocatalytic degradation of tetracyclines; some of the reported mechanisms did not show a ring-opening in tetracycline structure. It was reported that degradation without ring opening is not efficient as the antibiotic would still retain its function. It has been reported by some studies that tetracyclines can be degraded by UV or simulated sunlight [109]. Figure 11 represents degradation mechanisms by photolysis and photocatalysis.

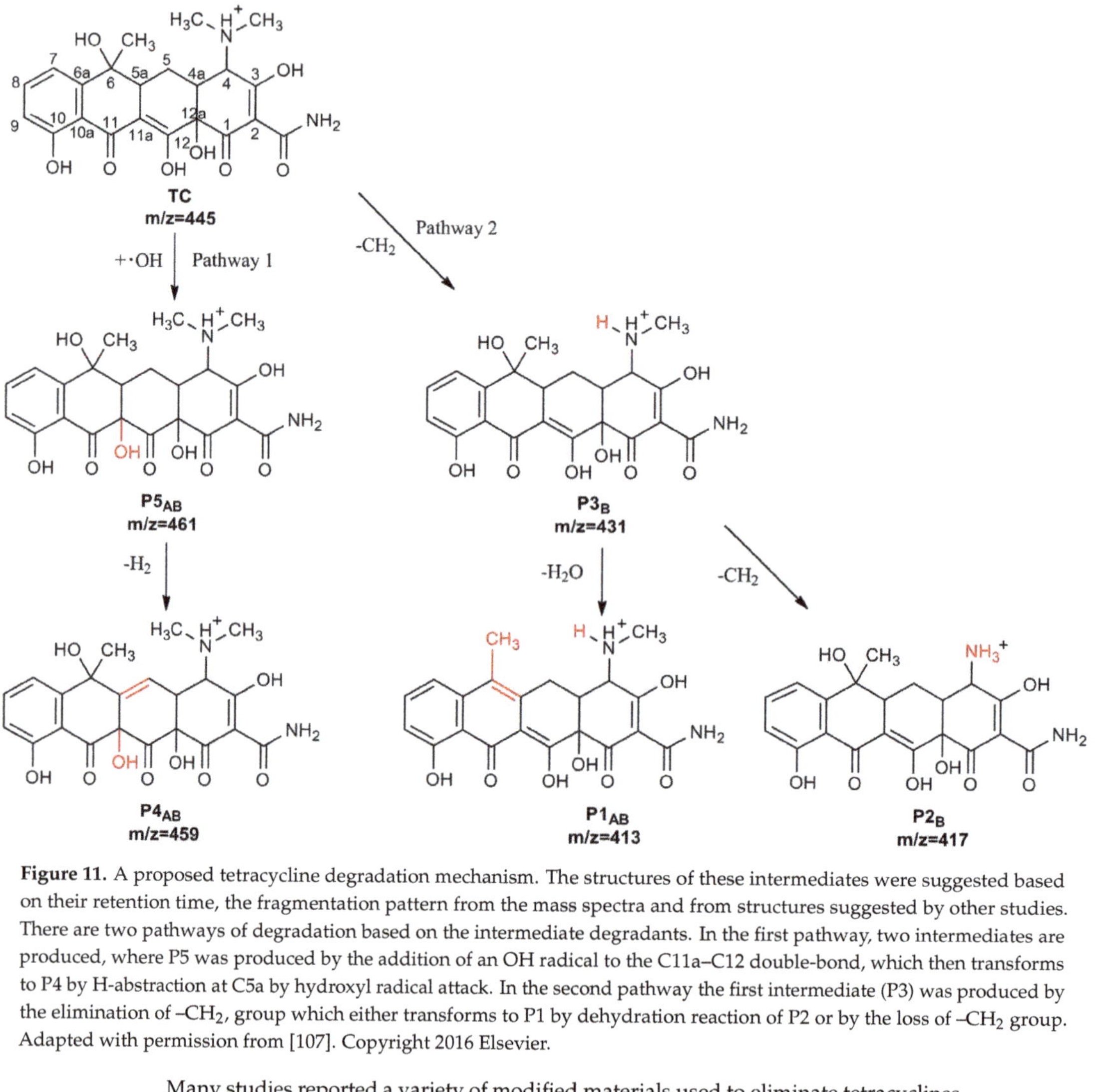

Figure 11. A proposed tetracycline degradation mechanism. The structures of these intermediates were suggested based on their retention time, the fragmentation pattern from the mass spectra and from structures suggested by other studies. There are two pathways of degradation based on the intermediate degradants. In the first pathway, two intermediates are produced, where P5 was produced by the addition of an OH radical to the C11a–C12 double-bond, which then transforms to P4 by H-abstraction at C5a by hydroxyl radical attack. In the second pathway the first intermediate (P3) was produced by the elimination of $-CH_2$, group which either transforms to P1 by dehydration reaction of P2 or by the loss of $-CH_2$ group. Adapted with permission from [107]. Copyright 2016 Elsevier.

Many studies reported a variety of modified materials used to eliminate tetracyclines from wastewater. Various studies reported the modification of graphitic carbon nitrides to achieve the best photocatalytic efficiency. One of these studies reported the fabrication of Ag-doped graphitic carbon nitride (Ag-doped g-C_3N_4). Tri, Kim et al. tested Ag-doped g-C_3N_4 against the degradation of the tetracycline under solar light irradiation. The best photocatalytic degradation was when 3 mmol of Ag was used, with a 96.8% degradation rate after 2 h of exposure to solar irradiation. The same experiment was done, but to degrade antibiotics from hospital water; the material showed 89.6% degradation efficiency after 2 h of solar exposure. The produced material showed stability and reserved its degradation ability after six cycles of reuse. The good performance of the doped g-C_3N_4

is attributed to the efficient charge separation and charge transfer [83]. In another study by Jodeyri et al., they developed an Ag/C_3N_4-Clinoptilolite nanomaterial to eliminate tetracycline antibiotics under simulated solar light irradiation. They found that 90% of the antibiotics were degraded after 3 h of exposure when Ag/C_3N_4-Clinoptilolite nano-photocatalyst was used, which showed superior efficiency when compared to Ag/C_3N_4, C_3N_4-Clinoptilolite, and C_3N_4. This enhancement is attributed to the improvement in the specific surface area that Clinoptilolite offers, as well as the SPR, due to the presence of Ag [110]. Further study was done using urea-derived graphitic carbon nitride (g-C_3N_4) doped with four different metals (K, Na, Mg, and Ca) to find the effect of different metal dopants on morphology, structure, and photocatalytic activity. The produced materials were tested to degrade three different antibiotics (tetracycline, Enrofloxacin ENR, and sulfamethoxazole) under visible light irradiation. All the produced photocatalysts were compared to undoped g-C_3N_4. Undoped g-C_3N_4 showed a degradation ability, with a removal rate constant of 0.014/minutes after 2 h of exposure. This rate was significantly increased to 3.3 when Ca was used as a dopant, and 5.4 when K was used as a dopant. The best photocatalytic performance was for g-C_3N_4 doped with K > Na > Mg > Ca > undoped g-C_3N_4. The improvement in photocatalytic performance was attributed to the narrowing bandgap, which thus increased charge separation efficiency and extended charge lifetimes [111].

The co-doping of g-C_3N_4 was also reported. In one study, g-C_3N_4 was co-doped with Phosphorus and Sulfur by in situ thermal copolymerizations to degrade tetracycline under visible light irradiation. In their experiment, they dispersed 0.5 and 1 g/L of the photocatalyst with 10 mg/L tetracycline, which was then exposed to 100 mW/Cm^2 of visible light for 2 h. It is shown that the Co-Doped g-C_3N_4 is 5.9 times more efficient than the undoped material, with a 70.33% degradation rate. It is suggested that the better performance of the codoped material compared to single-doped or undoped material is due to enhancements in light trapping, surface area and charge separation [112]. An additional study where g-C_3N_4 was co-doped with Na and Cl was reported. The study showed that the codoped material exhibited enhanced visible light absorption compared to the non-doped. Moreover, the degradation efficiency of TC-HCl was improved due to the reduction in the bandgap region as well as the lowered recombination rate [113]. The plasmonic effect of Au nanomaterials was also utilized, as reported by Xue, Ma, Zhou, et al., and the plasmonic photocatalyst Au/Pt/g-C_3N_4 showed a 3.4 higher degradation rate than the pure g-C_3N_4. In this study, the photocatalytic activity of Au/Pt/g-C_3N_4 was tested for the degradation of the antibiotic tetracycline hydrochloride under visible-light irradiation. The enhanced activity was attributed to the plasmonic effect of gold nanoparticles, which also increased the optical absorption range. Additionally, the electron sink effect of platinum contributed to improvements in the photocatalytic activity [114].

Fewer studies on the degradation of tetracycline using bismuth-based materials were reported. For instance, Wang and others utilized the molten salt method to fabricate $Bi_4Ti_3O_{12}$ nanosheets with moderate ratios of top and lateral facets. They found that the best photocatalytic ability was for $Bi_4Ti_3O_{12}$, with {001} top facets to {010}/{100} lateral facets. This material showed an improvement in photocatalysis for the degradation of tetracycline hydrochloride (TC-HCl) under Xenon lamp irradiation. The mechanism by which TC-HCl was degraded is mainly by the generation of h^+, where O^{2-} plays only a marginal role in photocatalytic degradation [115]. Owing to the superior properties of bismuth ferrite -$BiFeO_3$ (BFO), such as the suitable bandgap (2.2 eV), the low cost and the chemical stability, Zhou, Jiang, Chen, et al. modified BFO to enhance the photocatalytic activity. For that, they produced Er-doped BFO nanoparticles by the sole gel method, which was tested for the degradation of tetracycline hydrochloride under visible-light irradiation with a wavelength of >420 nm. Different concentrations of Er-doped in BFO were prepared (1%, 3%, and 5% Er) and compared to undoped BFO. The photocatalytic activity was tested by dispersing 200 mg of doped or undoped BFO on 100 mL of a 30 mg/L tetracycline solution, which was then irradiated to assess the photocatalytic degradation efficiency. A linear increment in the surface area of BFO with increasing Er concentration was observed, with the highest surface area for the 5% Er-doped BFO. However, the highest photocatalytic degradation was for the 3% Er-doped BFO with a 75.8% degradation rate which is approximately 3 times higher than the undoped BFO. The enhanced photocatalytic degradation of the doped BFO was attributed to the enhanced absorption wavelength of the doped material especially at the visible range spectra when assessed by UV–vis DRS spectra. Additionally, the doped material revealed a good charge separation and migration, as well as a low recombination rate, compared to the undoped material [116]. Moreover, Wang et al. fabricated a $BiVO_4/TiO_2/RGO$ nanocomposite for the photocatalytic degradation of four types of tetracyclines antibiotics upon visible light irradiation. They tracked the degradation of antibiotics by detecting the intermediate products that are formed due to antibiotic degradation. They achieved a homogenous distribution of $BiOV_4$ and TiO_2 particles when increasing the GO content to 0.3% and 0.5%. The greatest photocatalytic degradation was achieved when 0.5% of GO was used, with 80% antibiotic degradation after 1 h. The photocatalytic efficiency of the produced composite was the best when compared to two produced materials: $BiVO_4/RGO$ and TiO_2/RGO. The authors claimed that the presence of GO is advantageous, as it separates the photogenerated electron–hole pairs and thus reduces the recombination chances. Hence, the photogenerated electrons will move from the $BiVO_4$ conduction band to TiO_2, which improves the degradation rate. Furthermore, the decoration with GO leads to a drop in the bandgap of $BiVO_4$ and TiO_2, which also prevents the fast recombination of electron–hole pairs [117].

The photocatalytic activity of Bi_2O_3/montmorillonite was enhanced by the addition of Ag nanoparticles, as reported by Tun, Wnag, et al., where they produced Ag Bi_2O_3/montmorillonite nanocomposite by thermal and impregnation method. They tested the produced nanocomposite for the degradation of 20 mg/L tetracycline under visible light irradiation. The results showed significantly enhanced photocatalytic activity when silver is used compared to unmodified Bi_2O_3/montmorillonite. The dosage of the catalyst was directly related to the degradation rate, as the degradation rate increased from 78.1% to 90% when increasing the dosage from 0.5 to 1 g/L. The enhanced photocatalytic activity was attributed to the presence of more reaction sites, enhanced charge separation, and improved absorption at the visible light spectral range. Additionally, the produced nanocomposite showed good stability and reproducibility when reduced four times, without any significant loss in photocatalytic activity [118].

CdS is another material that was utilized due to its suitable band gap that allows it to absorb light in the visible region (2.4 eV). Nevertheless, its modification is needed to enhance material stability, as, in aqueous solution, CdS is associated with photo-corrosion and the release of toxic Cd^{+2} ions. For this, the authors designed a CdS/N-doped Carbon composite using the in situ carbonization method to degrade tetracycline antibiotics under visible light irradiation. The produced material achieved 83% degradation of the antibiotic within one hour, which is considered to be effective and comparable to the best performing photocatalysts. A trapping experiment was done to understand the mechanism of degradation. It was found that the degradation of antibiotic occurs in two steps: the first degradation step involves oxidation by three dominant radicals, h^+, $\cdot OH$ and O_2^-, which prompts ring opening followed by a consequent degradation, where complete decomposition is done, with the production of H_2O and CO_2 as end products. Furthermore, cyclic experiments showed that the material is stable and able to be reused, which makes it a candidate material for future wastewater treatment [119].

An additional study was conducted by Xue, Ma, Zhou et al., where they designed an Ag/ZnO/C nanocomposite. Here, Au and ZnO nanoparticles were deposited on carbonaceous layers by facile calcination and the photo-deposition method. They utilized the produced material to degrade the antibiotic tetracycline hydrochloride under UV and visible-light irradiation. The photocatalytic activity of Ag/ZnO/C was compared to ZnO and ZnO/C, where 100 mg of each photocatalyst was dispersed on 100 mL of

20 mg/L of the antibiotic then exposed to visible light irradiation for 280 min. The best photocatalytic activity (90.6% degradation) was for Ag/ZnO/C, which has enhanced visible light absorption, and due to the presence of Au nanoparticles, which exhibit a plasmonic effect, compared to ZnO alone (15% degradation), due to the large bandgap and low absorption at the visible light spectrum, and ZnO/C (81% degradation rate), which has no SPR effect [120]. Additionally, Semeraro et.al. reported the use of a ZnO/γ-Fe_2O_3 Paramagnetic Nanocomposite Material, where 88.52% degradation of tetracycline was achieved. Figure 12 represents different experimental parameters in which ZnO/γ-Fe_2O_3 was tested [68].

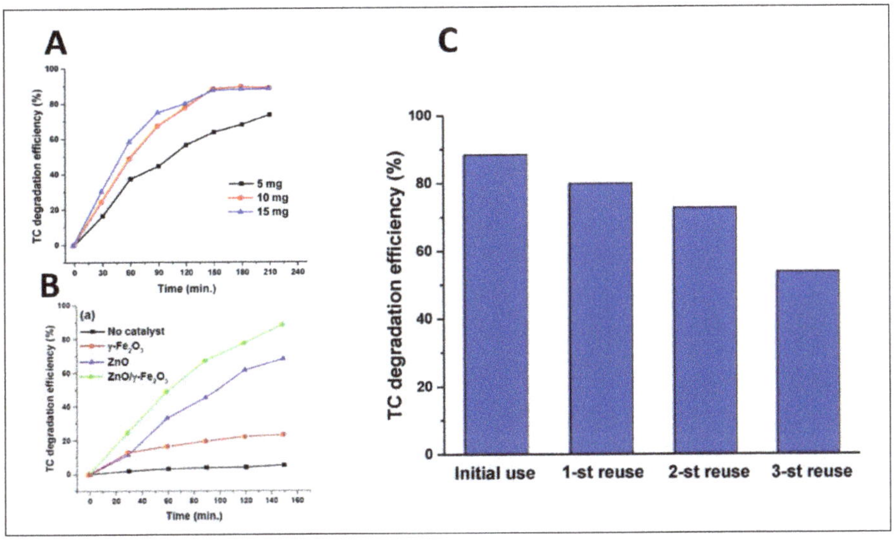

Figure 12. (**A**) shows the effect of different ZnO/γ-Fe_2O_3 concentrations of tetracycline degradation, in which 15 mg showed the highest degradation efficiency (**B**) represents the TC (%) vs. irradiation time (min) graph with different materials with the best tetracycline degradation efficiency of ZnO/γ-Fe_2O_3 (**C**) represents the tetracycline degradation efficiency with tr, using the prepared ZnO/γ-Fe_2O_3 with slight degradation in the material efficiency after 3 cycles of re-use. Adapted from [68].

6.4. Photocatalytic Degradation of Sulfamethoxazole & Other Antibiotics

Sulfamethoxazole is a widely used antibiotic that belongs to the sulfonamides antibiotic group, which is the oldest antibiotic used in human and veterinary. These antibiotics pose thermal and photostability as well as high solubility in water. This enabled these antibiotics to spread in the environment at a high rate [121]. There are some reports regarding the use of TiO_2 to degrade sulfamethoxazole; from older reports, it was already shown that TiO_2 can degrade sulfamethoxazole, forming products that are less toxic and biodegradable compared to sulfamethoxazole [122]. Although toxicity with bacterial experimentation was reported, the utilization of visible light instead of UV radiation was not studied. Novel materials, such as TiO_2 immobilized on expanded perlite (EP) (EP-TiO_2-773), were tested as well by Długosz and colleagues; they reported a noticeable photocatalytic enhancement of EP-TiO_2-773 using UV–vis irradiation [121]. The mechanism of degradation was proposed, as shown in Figure 13 A more recent study utilized iron-doped TiO_2 (Fe-TiO_2) under simulated sunlight irradiation. Different Fe/TiO_2 molar ratios were used (0–2%); however, the best photocatalytic degradation was achieved when 1 g/L of 0.04% of Fe/TiO_2 molar ratio was tested. A 95% degradation of the antibiotic was achieved after 90 min of exposure to visible light, where hydroxyl radicals were the major contributors to the degradation. It was unexpectedly shown that the presence of bicarbonates at concentrations from 0.125 to 2 g/L have a synergistic effect in the degradation of the antibiotic, along with the photocatalyst. Nevertheless, increasing the complexity of the water matrix was inversely proportional to the degradation rate and reduced the efficiency of the photocatalyst [123]. Figure 14 represents the degradation of sulfamethoxazole under different conditions.

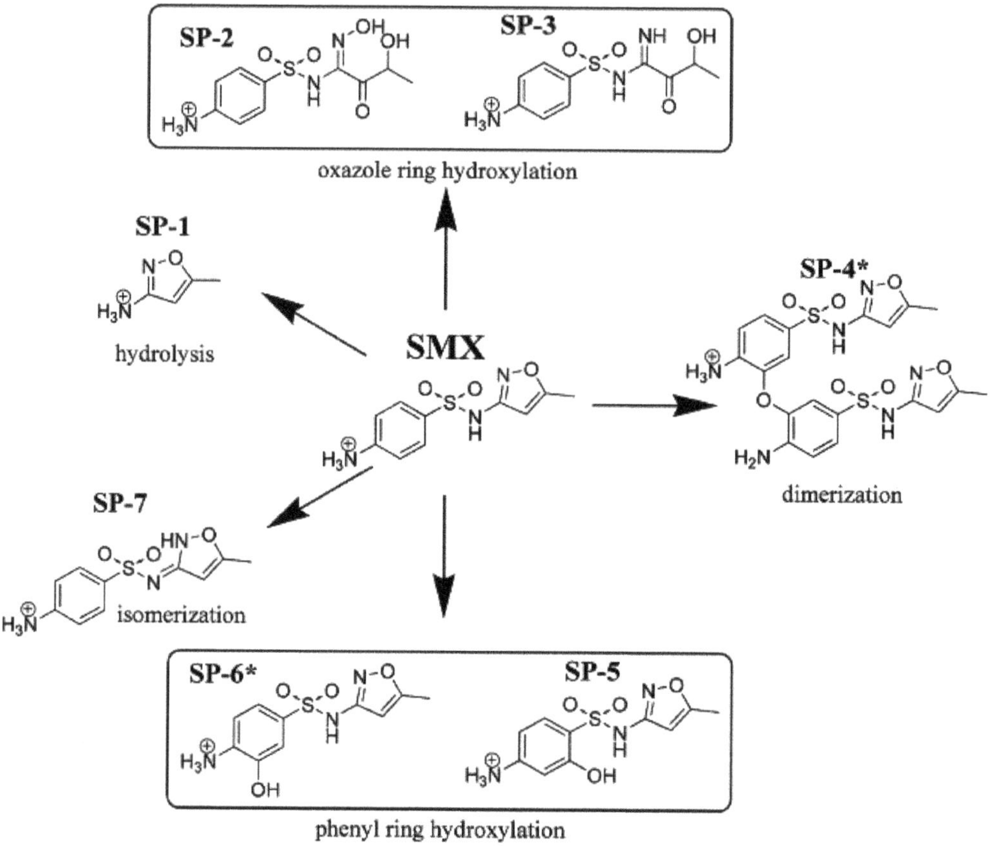

Figure 13. A proposed mechanism for the degradation of sulfamethoxazole antibiotic using (EP-TiO2-773). The degradation was determined by HPLC analysis, which showed the disappearance of sulfamethoxazole peaks and the appearance of new peaks. Two pathways are mainly involved. The first pathway involves the photogeneration of hydroxyl radicals which resulted from the interaction of holes with water. The second pathway involves the electron transfer from the excited organic molecule to TiO_2 conduction band. Adapted with permission from [121]. Copyright 2015 Elsevier.

A further example is bismuth-based photocatalyst, which was used in a study done by Ling, Yue, Yuan et al. Dot-shaped TiO_2 is produced on microrod Bi_2O_4 with different molar ratios of Ti to Bi by the two-step hydrothermal method. These photocatalysts were tested for the degradation of Sulfamethoxazole (SMZ) under simulated sunlight as well as natural sunlight. A photochemical reactor was used to experiment with a xenon lamp to simulate the sunlight. A total of 25 mg of the photocatalysts was dispersed in a 60 mL quartz tube, with from 50 of 10 mg/L of the antibiotic. It was found that the best degradation was achieved with the Ti: Bi molar ratio (2.0), where the photocatalytic activity was enhanced by 64% under simulated sunlight and 112% under visible sunlight compared to pure Bi_2O_4. Moreover, 90% of 10 mg/L of SMZ was degraded under natural sunlight after 2 h of exposure. Scavenging experiments revealed that the main contributors in photodegradation are h^+ and $O_2^{\cdot -}$, and the enhanced photocatalytic behavior is due to the effective heterojunction and charge transfer [124]. Moreover, Alvarez and others designed TiO_2 doped with boron with different weight % by the sol gel method. The produced material was subjected to surface modification by Au nanoparticles which exhibit SPR effect. The produced material was tested for the degradation of sulfamethoxazole antibiotics. The best mineralization was for 0.5 wt% Au/0.25 wt% $BTiO_2$; this nanocomposite exhibited better mineralization and photocatalytic activity compared to the single components. This is mainly due to the boron doping. which reduced the particle size, thus increasing the surface area and allowing for better light interaction. Additionally, the introduction of Au nanoparticles acted to reduce the recombination rate of the photogenerated electron–hole pairs [125]. Further study with toxicity assessment was performed by Naraginti et al. The authors fabricated Ag_3PO_4 integrated with N-doped rGO. The degradation efficiency of this composite was assessed against sulfamethoxazole antibiotic, under visible light irradiation. The photocatalytic performance was dependent on variants like pH, the dopant content (N), and catalyst dosage. The best activity was seen when a pH of 5.8, N content of 5.14%, and 0.2 g/L of catalyst were used. These conditions lead to 93.8% degradation of sulfamethoxazole after one hour. It was further reported that the toxicity against bacteria is reduced after the photodegradation [126].

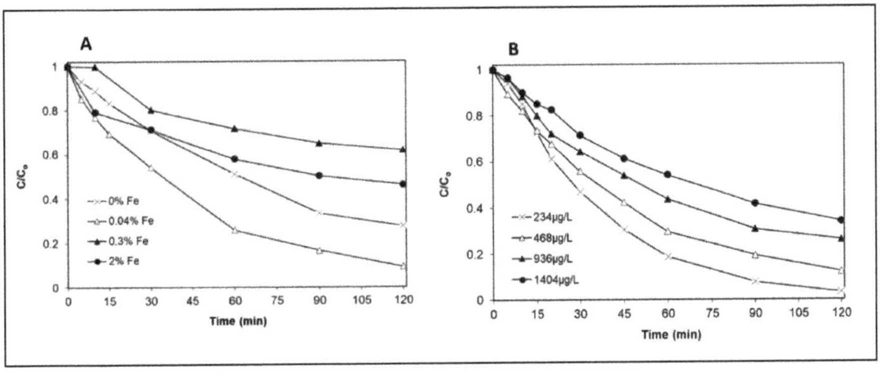

Figure 14. (**A**) Signifies the effect of iron doping level, where 0.5% Fe/TiO_2 showed the most efficient photocatalytic degradation. (**B**) represents the effect of initial SMZ concertation; as noted, there is a decrement in the rate constant as the concentration of SMZ increases. Adapted with permission from [123]. Copyright 2019 Elsevier.

Another antibiotic that belongs to the sulfonamides class is sulfadiazine (SDZ). These drugs have been known to be resistant to degradation by different methods, including biological oxidation. One study was conducted by Dhiman, Dhiman, Kumar, et al., where they used spinal ferrites with the general formula of MFe_2O_4 due to its comparatively narrow bandgap as well as good electrical and magnetic properties. The authors prepared nano-Zn_1−x$Mg_xFe_2O_4$ by the facile sole gel method to find the effect of using Mg-substituted nano-Zinc ferrite in the photocatalytic degradation of pharmaceuticals under visible light irradiation. In their experiment, they dispersed 0.3 g/L of the photocatalyst to 10 mg/L of the antibiotic solution; the reaction was done in a glass photochemical reactor where the samples were exposed either to artificial visible light (400 mW/cm^2) or natural sunlight. They found that 99.1% of SDZ was degraded after 90 min of visible light exposure. Scavenging experiments confirmed that hydroxyl radicals are the major ROS that contribute to the degradation of SDZ. The efficient photocatalytic activity was attributed to the introduction of Mg ions, which reduced the bandgap of the photocatalyst, thus offering higher visible light absorption as well as better charge flow [127].

Another great study of the degradation of eight antibiotics (amoxicillin, ampicillin, doxycycline, oxytetracycline, lincomycin, vancomycin, sulfamethazine, and sulfamethoxazole) was done by Do, Nguyen et al., who added Au nanoparticles to enhance the photocatalytic activity of TiO_2. They prepared Au nanoparticles NP-decorated TiO_2 nanotubes arrays (TNAs) and Au NP-decorated TiO_2 nanowire on nanotube arrays (TNWs/TNAs). the degradation was assessed under different light wavelengths (UV and visible light irradiation). The photocatalytic behavior of the produced materials was compared to TiO_2-decorated TNAs and TNWs/TNAs without the presence of Au. They found that Au-TiO_2-decorated TNWs/TNAs have the highest photocatalytic degradation for the eight antibiotics, with enhanced visible light absorption after 20 min of exposure. The enhanced behavior was attributed to the large surface area of TNWs/TNAs. Additionally, the presence of Au nanoparticles has a plasmonic effect [128]. The degradation of the antibiotic amoxicillin was studied using TiO_2 co-doped with Pt and Bi. Salimi, Behbahani, Sobhi, et al. They prepared multiple materials with different concentrations of the dopants. The maximum photocatalytic degradation efficiency was achieved when 1 Pt-5 Bi-TiO_2 is used, with a degradation efficiency of 87.67%, which is 21 times more efficient than using undoped TiO_2. The authors credited the enhanced efficiency of narrowing the bandgap that the dopants provided (2.7 eV). Additionally, quenching experiments showed that h^+ and superoxide radicals are the key players in photocatalytic antibiotic degradation. The reusability and stability for 1 Pt-5 Bi-TiO_2 were assessed; there were no changes in photocatalytic ability when recycled for more than four cycles. Moreover, photo corrosion was not detected after being recycled, which signifies the stability of the material [129]. A further study was conducted by Eswar, Ramamurthy, and Madras, where they produced vanadium and nitrogen co-doped TiO_2 (V-N TiO_2) by the hydrothermal method for the degradation of the antibiotic chloramphenicol, as well as bacterial degradation under visible light irradiation, using a halide lamp with 220 $\mu W/cm^2$. They used a photochemical reactor to experiment by dispersing 1 g/L of the photocatalyst in 25 ppm aqueous solution of the antibiotic with bacterial suspension (E.Coli). The novelty of this study is in assessing the photocatalytic degradation efficiency of the produced V-N-TiO_2 to decompose both the antibiotic and the bacteria when they are in proximity. They found that the best degradation efficiency is achieved when the concentration of V:N is 2 atom%:1 atom%. The authors reported an enhancement of the photocatalytic activity after co-doping, due to the enhancement in the visible region that co-doping offered when enhancing the charge transfer mechanism [128].

7. Conclusions and Perspectives

Semiconductor-assisted photocatalysis is a candidate approach for wastewater treatment from different pharmaceuticals, mainly antibiotics. The degradation of antibiotics is achieved by the formation of free radicals and reactive oxygen species. Different semiconducting materials were studied for antibiotic degradation under visible light irradiation. Semiconducting material modification was achieved, applying different modification methods such as doping, manipulating the morphology and surface area, heterojunction, and SPR. Although different materials showed enhancement in photocatalytic activity after modification, there are only a few studies regarding the safety of some materials using animal models. Furthermore, most, if not all, of the studies are being done at the laboratory scale. The transition from laboratory beaker to a larger scale is definitely complex, but it will provide a more realistic indication of how the behavior will be at the industrial scale. Moreover, most of the studies reported the stability of the photocatalysts under different conditions. However, the elimination of the photocatalysts from the wastewater after the treatment was not fully studied. Another major problem is material stability and reusability at a larger scale. Filling those research gaps will help to guarantee the overall photocatalytic safety and efficiency, and thus improve application.

Author Contributions: S.S., writing and original draft preparation, review and editing; P.T.H., conceptualization; T.M.A., conceptualization, comprehensive revision, approval for submission, supervision. All authors have read and agreed to the published version of the manuscript.

Funding: The publication of this article was funded by the Qatar National Library.

Data Availability Statement: Data available in a publicly accessibility.

Conflicts of Interest: The authors declare no conflict of interest.

References

1. FAO. *Review of World Water Resources by Country: 2. Concepts and Definitions*; FAO: Rome, Italy, 2003.
2. Ahmed, S.N.; Haider, W. Heterogeneous photocatalysis and its potential applications in water and wastewater treatment: A review. *Nanotechnology* **2018**, *29*, 342001. [CrossRef] [PubMed]
3. Engelman, R.; LeRoy, P. Sustaining water: Population and the future of renewable water supplies. *J. Chem. Inf. Model.* **1993**, *53*, 1689–1699.
4. World Water Council. World Water Vision—Making Water Everybody's Business. *J. Chem. Inf. Model.* **2000**, *53*, 1689–1699.
5. Topare, N.S.; Attar, S.J.; Manfe, M.M. Sewage/Wastewater treatment technologies: A review. *Sci. Rev. Chem. Commun.* **2011**, *1*, 18–24.
6. Tran, N.H.; Reinhard, M.; Khan, E.; Chen, H.; Nguyen, V.T.; Li, Y.; Goh, S.G.; Nguyen, Q.; Saeidi, N.; Gin, K.Y.-H. Emerging contaminants in wastewater, stormwater runoff, and surface water: Application as chemical markers for diffuse sources. *Sci. Total Environ.* **2019**, *676*, 252–267. [CrossRef] [PubMed]
7. Abbassi, B.E.; Saleem, M.A.; Zytner, R.G.; Gharabaghi, B.; Rudra, R. Antibiotics in wastewater: Their degradation and effect on wastewater treatment efficiency. *J. Food Agric. Environ.* **2016**, *14*, 95–99.
8. Kurt, A.; Mert, B.K.; Özengin, N.; Sivrioğlu, Ö.; Yonar, T. Treatment of Antibiotics in Wastewater Using Advanced Oxidation Processes (AOPs). In *Physico-Chemical Wastewater Treatment and Resource Recovery*; IntechOpen: London, UK, 2017. [CrossRef]
9. Chang, Q.; Wang, W.; Regev-Yochay, G.; Lipsitch, M.; Hanage, W.P. Antibiotics in agriculture and the risk to human health: How worried should we be? *Evol. Appl.* **2015**, *8*, 240–247. [CrossRef]
10. Wegener, H.C. Antibiotics in animal feed and their role in resistance development. *Curr. Opin. Microbiol.* **2003**, *6*, 439–445. [CrossRef] [PubMed]
11. Barancheshme, F.; Munir, M. Development of Antibiotic Resistance in Wastewater Treatment Plants. In *Antimicrobial Resistance—A Global Threat*; IntechOpen: London, UK, 2019. [CrossRef]
12. Deng, Y.; Zhao, R. Advanced Oxidation Processes (AOPs) in Wastewater Treatment. *Curr. Pollut. Rep.* **2015**, *1*, 167–176. [CrossRef]
13. Amor, C.; Marchão, L.; Lucas, M.; Peres, J.A. Application of Advanced Oxidation Processes for the Treatment of Recalcitrant Agro-Industrial Wastewater: A Review. *Water* **2019**, *11*, 205. [CrossRef]
14. Mills, A.; Le Hunte, S. An overview of semiconductor photocatalysis. *J. Photochem. Photobiol. A Chem.* **1997**, *108*, 1–35. [CrossRef]
15. Dong, H.; Zeng, G.; Tang, L.; Fan, C.; Zhang, C.; He, X.; He, Y. An overview on limitations of TiO_2-based particles for photocatalytic degradation of organic pollutants and the corresponding countermeasures. *Water Res.* **2015**, *79*, 128–146. [CrossRef] [PubMed]
16. Majumdar, A.; Pal, A. Recent advancements in visible-light-assisted photocatalytic removal of aqueous pharmaceutical pollutants. *Clean Technol. Environ. Policy* **2020**, *22*, 11–42. [CrossRef]
17. Li, D.; Shi, W. Recent developments in visible-light photocatalytic degradation of antibiotics. *Cuihua Xuebao/Chin. J. Catal.* **2016**, *37*, 792–799. [CrossRef]
18. Ecantas, L.; Shah, S.Q.A.; Cavaco, L.M.; Manaia, C.M.; Walsh, F.; Popowska, M.; Garelick, H.; Bürgmann, H.; Esørum, H. A brief multi-disciplinary review on antimicrobial resistance in medicine and its linkage to the global environmental microbiota. *Front. Microbiol.* **2013**, *4*, 96. [CrossRef]
19. Calvete, M.J.; Piccirillo, G.; Vinagreiro, C.S.; Pereira, M.M. Hybrid materials for heterogeneous photocatalytic degradation of antibiotics. *Coord. Chem. Rev.* **2019**, *395*, 63–85. [CrossRef]
20. *Bad Medicine: How the Pharmaceutical Industry Is Contributing to the Global Rise of Antibiotic-Resistant Superbugs*; SumOfUs: New York, NY, USA, 2015. [CrossRef]
21. Serna-Galvis, E.A.; Silva-Agredo, J.; Giraldo, A.L.; Flórez-Acosta, O.A.; Torres-Palma, R.A. Comparative study of the effect of pharmaceutical additives on the elimination of antibiotic activity during the treatment of oxacillin in water by the photo-Fenton, TiO_2-photocatalysis and electrochemical processes. *Sci. Total Environ.* **2016**, *541*, 1431–1438. [CrossRef]
22. Van Boeckel, T.P.; Pires, J.; Silvester, R.; Zhao, C.; Song, J.; Criscuolo, N.G.; Gilbert, M.; Bonhoeffer, S.; Laxminarayan, R. Global trends in antimicrobial resistance in animals in low- and middle-income countries. *Science* **2019**, *365*, eaaw1944. [CrossRef]
23. Adesokan, H.K.; Akanbi, I.O.; Akanbi, I.M.; Obaweda, R.A. Pattern of antimicrobial usage in livestock animals in south-western Nigeria: The need for alternative plans. *Onderstepoort J. Vet. Res.* **2015**, *82*, 1–6. [CrossRef]
24. Bacanlı, M.; Başaran, N. Importance of antibiotic residues in animal food. *Food Chem. Toxicol.* **2019**, *125*, 462–466. [CrossRef]
25. Michael, I.; Rizzo, L.; McArdell, C.S.; Manaia, C.M.; Merlin, C.; Schwartz, T.; Dagot, C.; Fatta-Kassinos, D. Urban wastewater treatment plants as hotspots for the release of antibiotics in the environment: A review. *Water Res.* **2013**, *47*, 957–995. [CrossRef]
26. Kang, D.H.; Gupta, S.; Rosen, C.J.; Fritz, V.; Singh, A.; Chander, Y.; Murray, H.; Rohwer, C. Antibiotic Uptake by Vegetable Crops from Manure-Applied Soils. *J. Agric. Food Chem.* **2013**, *61*, 9992–10001. [CrossRef]
27. Kang, D.H.; Gupta, S.C.; Rosen, C.J.; Fritz, V.; Singh, A.; Chander, Y.; Murray, H. *Antibiotic Uptake by Vegetable Crops from Manure-Applied Soils*; Technical Report submitted to the North Central Region Sustainable Agricultural Research and Extension (SARE) Program; University of Minnesota: St. Paul, MN, USA, 2012; p. 129.

28. Duong, H.A.; Pham, N.H.; Nguyen, H.T.; Hoang, T.T.; Pham, H.V.; Pham, V.C.; Berg, M.; Giger, W.; Alder, A.C. Occurrence, fate and antibiotic resistance of fluoroquinolone antibacterials in hospital wastewaters in Hanoi, Vietnam. *Chemosphere* **2008**, *72*, 968–973. [CrossRef]
29. Sinthuchai, D.; Boontanon, S.K.; Boontanon, N.; Polprasert, C. Evaluation of removal efficiency of human antibiotics in wastewater treatment plants in Bangkok, Thailand. *Water Sci. Technol.* **2016**, *73*, 182–191. [CrossRef] [PubMed]
30. Hoyett, Z. Pharmaceuticals and Personal Care Products: Risks, Challenges, and Solutions. In *Risk Assessment*; IntechOpen: London, UK, 2018. [CrossRef]
31. Ingerslev, F.; Halling-Sørensen, B. Biodegradability of Metronidazole, Olaquindox, and Tylosin and Formation of Tylosin Degradation Products in Aerobic Soil–Manure Slurries. *Ecotoxicol. Environ. Saf.* **2001**, *48*, 311–320. [CrossRef] [PubMed]
32. Reis, A.C.; Kolvenbach, B.A.; Nunes, O.C.; Corvini, P.F.-X. Biodegradation of antibiotics: The new resistance determinants—Part I. *New Biotechnol.* **2020**, *54*, 34–51. [CrossRef] [PubMed]
33. Directorate, W. *Guidelines on Water Purification by Reverse Osmosis (RO)*; FDA: Silver Spring, MD, USA, 2015.
34. Kosutic, K.; Dolar, D.; Asperger, D.; Kunst, B. Removal of antibiotics from a model wastewater by RO/NF membranes. *Sep. Purif. Technol.* **2007**, *53*, 244–249. [CrossRef]
35. Derakhshan, Z.; Mokhtari, M.; Babaei, F.; Ahmadi, R.M.; Ehrampoush, M.H.; Faramarzian, M. Removal Methods of Antibiotic compounds from Aqueous Environments—A Review. *J. Environ. Health Sustain. Dev.* **2016**, *1*, 51–74.
36. Rajendran, S.; Gracia, F.; Stephen, A. Basic Principles, Mechanism, and Challenges of Photocatalysis. In *Nanocomposites for Visible Light-Induced Photocatalysis*; Springer: Cham, Switzerland, 2017.
37. Pawar, M.; Sendoğdular, S.T.; Gouma, P. A Brief Overview of TiO_2 Photocatalyst for Organic Dye Remediation: Case Study of Reaction Mechanisms Involved in Ce-TiO_2 Photocatalysts System. *J. Nanomater.* **2018**, *2018*, 1–13. [CrossRef]
38. Feng, T.; Feng, G.S.; Yan, L.; Pan, J.H. One-Dimensional Nanostructured TiO_2 for Photocatalytic Degradation of Organic Pollutants in Wastewater. *Int. J. Photoenergy* **2014**, *2014*, 1–14. [CrossRef]
39. Peighambardoust, N.S.; Asl, S.K.; Maghsoudi, M. The effect of doping concentration of TiO2 nanotubes on energy levels and its direct correlation with photocatalytic activity. *Thin Solid Films* **2019**, *690*, 137558. [CrossRef]
40. Xiao, J.; Han, Q.; Xie, Y.; Yang, J.; Su, Q.; Chen, Y.; Cao, H. Is C_3N_4 Chemically Stable toward Reactive Oxygen Species in Sunlight-Driven Water Treatment? *Environ. Sci. Technol.* **2017**, *51*, 13380–13387. [CrossRef]
41. Kumar, S.G.; Rao, K.K. Comparison of modification strategies towards enhanced charge carrier separation and photocatalytic degradation activity of metal oxide semiconductors (TiO_2, WO_3 and ZnO). *Appl. Surf. Sci.* **2017**, *391*, 124–148. [CrossRef]
42. Ge, M.; Cao, C.; Huang, J.; Li, S.; Chen, Z.; Zhang, K.-Q.; Al-Deyab, S.S.; Lai, Y. A review of one-dimensional TiO2nanostructured materials for environmental and energy applications. *J. Mater. Chem. A* **2016**, *4*, 6772–6801. [CrossRef]
43. Huang, J.; Yu, H.; Wang, H.; Yu, H.; Cao, Y.; Zou, H.; Liu, Z. Revealing active-site structure of porous nitrogen-defected carbon nitride for highly effective photocatalytic hydrogen evolution. *Chem. Eng. J.* **2019**, *373*, 687–699. [CrossRef]
44. Wang, H.; Yong, D.; Chen, S.; Jiang, S.; Zhang, X.; Shao, W.; Zhang, Q.; Yan, W.; Pan, B.; Xie, Y. Oxygen-Vacancy-Mediated Exciton Dissociation in BiOBr for Boosting Charge-Carrier-Involved Molecular Oxygen Activation. *J. Am. Chem. Soc.* **2018**, *140*, 1760–1766. [CrossRef] [PubMed]
45. Zhang, W.; Li, Y.; Fan, X.; Zhang, F.; Zhang, G.; Zhu, Y.-A.; Peng, W.; Wang, S.; Duan, X. Synergy of nitrogen doping and structural defects on hierarchically porous carbons toward catalytic oxidation via a non-radical pathway. *Carbon* **2019**, *155*, 268–278. [CrossRef]
46. Wang, H.; Zhang, L.; Chen, Z.; Hu, J.; Li, S.; Wang, Z.; Liu, J.; Wang, X. Semiconductor heterojunction photocatalysts: Design, construction, and photocatalytic performances. *Chem. Soc. Rev.* **2014**, *43*, 5234–5244. [CrossRef]
47. Djurišić, A.B.; He, Y.; Ng, A.M.C. Visible-light photocatalysts: Prospects and challenges. *APL Mater.* **2020**, *8*, 030903. [CrossRef]
48. Wang, S.; Guan, B.Y.; Lou, X.W. (David) Rationally designed hierarchical N-doped carbon@$NiCo_2O_4$ double-shelled nanoboxes for enhanced visible light CO_2 reduction. *Energy Environ. Sci.* **2018**, *11*, 306–310. [CrossRef]
49. Yan, Y.; Han, M.; Konkin, A.; Koppe, T.; Wang, D.; Andreu, T.; Chen, G.; Vetter, U.; Morante, J.R.; Schaaf, P. Slightly hydrogenated TiO_2 with enhanced photocatalytic performance. *J. Mater. Chem. A* **2014**, *2*, 12708–12716. [CrossRef]
50. Gong, Y.; Li, H.; Jiao, C.; Xu, Q.; Xu, X.; Zhang, X.; Liu, Y.; Dai, Z.; Liu, X.Y.; Chen, W.; et al. Effective hydrogenation of g-C3N4 for enhanced photocatalytic performance revealed by molecular structure dynamics. *Appl. Catal. B Environ.* **2019**, *250*, 63–70. [CrossRef]
51. Vaiano, V.; Sacco, O.; Pisano, D.; Sannino, D.; Ciambelli, P. From the design to the development of a continuous fixed bed photoreactor for photocatalytic degradation of organic pollutants in wastewater. *Chem. Eng. Sci.* **2015**, *137*, 152–160. [CrossRef]
52. Van Gerven, T.; Mul, G.; Moulijn, J.; Stankiewicz, A. A review of intensification of photocatalytic processes. *Chem. Eng. Process. Process. Intensif.* **2007**, *46*, 781–789. [CrossRef]
53. Sacco, O.; Vaiano, V.; Hana, C.; Sannino, D.; Dionysiou, D.D. Photocatalytic removal of atrazine using N-doped TiO_2 supported on phosphors. *Appl. Catal. B Environ.* **2015**, *164*, 462–474. [CrossRef]
54. Sacco, O.; Vaiano, V.; Sannino, D. Phosphors-Based Photocatalysts for Wastewater Treatment. In *Nanophotocatalysis and Environmental Applications*; Springer International Publishing: New York, NY, USA, 2019; pp. 119–138. [CrossRef]
55. Alberti, S.; Locardi, F.; Sturini, F.; Speltini, A.; Maraschi, F.; Costa, G.A.; Ferretti, M.; Caratto, V. Photocatalysis in Darkness: Optimization of Sol-Gel Synthesis of NP-TiO_2 Supported on a Persistent Luminescence Material and its Application for the Removal of Ofloxacin from Water. *J. Nanomed. Nanotechnol.* **2018**, *9*, 1–6. [CrossRef]

56. Khan, M.M.; Adil, S.F.; Al-Mayouf, A.; Adil, S.F. Metal Oxides as Photocatalysts. *J. Saudi Chem. Soc.* **2015**, *19*, 462–464. [CrossRef]
57. Kowsari, E. Carbon-Based Nanocomposites for Visible Light-Induced Photocatalysis. In *Nanocomposites for Visible Light-Induced Photocatalysis*; Springer: Cham, Switzerland, 2017.
58. Ahmad, R.; Ahmad, Z.; Khan, A.U.; Mastoi, N.R.; Aslam, M.; Kim, J. Photocatalytic systems as an advanced environmental remediation: Recent developments, limitations and new avenues for applications. *J. Environ. Chem. Eng.* **2016**, *4*, 4143–4164. [CrossRef]
59. Malakootian, M.; Nasiri, A.; Gharaghani, M.A. Photocatalytic degradation of ciprofloxacin antibiotic by TiO_2 nanoparticles immobilized on a glass plate. *Chem. Eng. Commun.* **2020**, *207*, 56–72. [CrossRef]
60. Bobirică, C.; Bobirică, L.; Râpă, M.; Matei, E.; Predescu, A.M.; Orbeci, C. Photocatalytic Degradation of Ampicillin Using PLA/TiO2 Hybrid Nanofibers Coated on Different Types of Fiberglass. *Water* **2020**, *12*, 176. [CrossRef]
61. Duong, T.-N.-B.; Le, M.-V. High efficiency degradation of tetracycline antibiotic with TiO_2 - SiO_2 photocatalyst under low power of simulated solar light irradiation. *AIP Conf. Proc.* **2019**, *2085*. [CrossRef]
62. Qiu, R.; Zhang, D.; Mo, Y.; Song, L.; Brewer, E.; Huang, X.; Xiong, Y. Photocatalytic activity of polymer-modified ZnO under visible light irradiation. *J. Hazard. Mater.* **2008**, *156*, 80–85. [CrossRef] [PubMed]
63. Chen, X.; He, Y.; Zhang, Q.; Li, L.; Hu, D.; Yin, T. Fabrication of sandwich-structured ZnO/reduced graphite oxide composite and its photocatalytic properties. *J. Mater. Sci.* **2010**, *45*, 953–960. [CrossRef]
64. Chen, T.; Zheng, Y.; Lin, J.-M.; Chen, G. Study on the Photocatalytic Degradation of Methyl Orange in Water Using Ag/ZnO as Catalyst by Liquid Chromatography Electrospray Ionization Ion-Trap Mass Spectrometry. *J. Am. Soc. Mass Spectrom.* **2008**, *19*, 997–1003. [CrossRef] [PubMed]
65. Mirzaeifard, Z.; Shariatinia, Z.; Jourshabani, M.; Darvishi, S.M.R. ZnO Photocatalyst Revisited: Effective Photocatalytic Degradation of Emerging Contaminants Using S-Doped ZnO Nanoparticles under Visible Light Radiation. *Ind. Eng. Chem. Res.* **2020**, *59*, 15894–15911. [CrossRef]
66. He, J.; Zhang, Y.; Guo, Y.; Rhodes, G.; Yeom, J.; Li, H.; Zhang, W. Photocatalytic degradation of cephalexin by ZnO nanowires under simulated sunlight: Kinetics, influencing factors, and mechanisms. *Environ. Int.* **2019**, *132*, 105105. [CrossRef]
67. Chavoshan, S.; Khodadadi, M.; Nasseh, N. Photocatalytic degradation of penicillin G from simulated wastewater using the UV/ZnO process: Isotherm and kinetic study. *J. Environ. Health Sci. Eng.* **2020**, *18*, 107–117. [CrossRef]
68. Semeraro, P.; Bettini, S.; Sawalha, S.; Pal, S.; Licciulli, A.; Marzo, F.; Lovergine, N.; Valli, L.; Giancane, G. Photocatalytic Degradation of Tetracycline by ZnO/γ-Fe_2O_3 Paramagnetic Nanocomposite Material. *Nanomaterials* **2020**, *10*, 1458. [CrossRef]
69. Tahir, M.B.; Nabi, G.; Rafique, M.; Khalid, N.R. Nanostructured-based WO_3 photocatalysts: Recent development, activity enhancement, perspectives and applications for wastewater treatment. *Int. J. Environ. Sci. Technol.* **2017**, *14*, 2519–2542. [CrossRef]
70. Nguyen, T.T.; Nam, S.-N.; Son, J.; Oh, J. Tungsten Trioxide (WO_3)-assisted Photocatalytic Degradation of Amoxicillin by Simulated Solar Irradiation. *Sci. Rep.* **2019**, *9*, 1–18. [CrossRef]
71. Gholamiabc, P.; Khataeeade, A.; Bhatnagarb, A. Photocatalytic degradation of antibiotic and hydrogen production using diatom-templated 3D WO_{3-x}@mesoporous carbon nanohybrid under visible light irradiation. *J. Clean. Prod.* **2020**, *275*, 124157. [CrossRef]
72. Huang, Z.-F.; Song, J.; Pan, L.; Lv, F.; Wang, Q.; Zou, J.-J.; Zhang, X.; Wang, L. Mesoporous $W_{18}O_{49}$ hollow spheres as highly active photocatalysts. *Chem. Commun.* **2014**, *50*, 10959–10962. [CrossRef] [PubMed]
73. Zhang, Z.; Zhang, Z. Bismuth-based photocatalytic semiconductors: Introduction, challenges and possible approaches. *J. Mol. Catal. A Chem.* **2016**, *423*, 533–549. [CrossRef]
74. Oudghiri-Hassani, H.; Rakass, S.; Al Wadaani, F.; Al-Ghamdi, K.J.; Omer, A.; Messali, M.; Abboudi, M. Synthesis, characterization and photocatalytic activity of α-Bi_2O_3 nanoparticles. *J. Taibah Univ. Sci.* **2015**, *9*, 508–512. [CrossRef]
75. Bierlein, J.D.; Sleight, A.W. Ferroelasticity in $BiVO_4$. *Solid State Commun.* **1975**, *16*, 69–70. [CrossRef]
76. Gotić, M.; Musić, S.; Ivanda, M.; Šoufek, M.; Popović, S. Synthesis and characterisation of bismuth(III) vanadate. *J. Mol. Struct.* **2005**, *744*, 535–540. [CrossRef]
77. Ye, L.; Su, Y.; Jin, X.; Xie, H.; Zhang, C. Recent advances in BiO_X (X = Cl, Br and I) photocatalysts: Synthesis, modification, facet effects and mechanisms. *Environ. Sci. Nano* **2014**, *1*, 90–112. [CrossRef]
78. Zhang, C.; Li, Y.; Shuai, D.; Shen, Y.; Xiong, W.; Wang, L. Graphitic carbon nitride (g-C_3N_4)-based photocatalysts for water disinfection and microbial control: A review. *Chemosphere* **2019**, *214*, 462–479. [CrossRef]
79. Cao, S.; Low, J.; Cheng, B.; Jaroniec, M. Polymeric Photocatalysts Based on Graphitic Carbon Nitride. *Adv. Mater.* **2015**, *27*, 2150–2176. [CrossRef]
80. Chen, W.; Jiang, D.; Zhu, M.; Shi, T.; Li, H.; Wang, K. An effective strategy for fabricating highly dispersed nanoparticles on O-C_3N_4 with enhanced electrocatalytic activity and stability. *J. Alloys Compd.* **2018**, *741*, 1203–1211. [CrossRef]
81. Zhang, M.; Dai, Y.; Zhang, S.; Chen, W. Highly efficient photocatalytic activity of boron-doped TiO_2 for gas phase degradation of benzene. *Rare Met.* **2011**, *30*, 243–248. [CrossRef]
82. Gao, B.; Wang, J.; Dou, M.; Huang, X.; Yu, X. Novel nitrogen-rich g-C_3N_4 with adjustable energy band by introducing triazole ring for cefotaxime removal. *Sep. Purif. Technol.* **2020**, *241*, 116576. [CrossRef]
83. Nguyen-Le, M.-T.; Jitae, K.; Giang, B.L.; Al Tahtamouni, T.; Huong, P.T.; Lee, C.; Nguyen, M.V.; Trung, D.Q. Ag-doped graphitic carbon nitride photocatalyst with remarkably enhanced photocatalytic activity towards antibiotic in hospital wastewater under solar light. *J. Ind. Eng. Chem.* **2019**, *80*, 597–605. [CrossRef]

84. Cao, S.-W.; Yuan, Y.; Barber, J.; Loo, J.S.C.; Xue, C. Noble-metal-free g-C_3N_4/Ni(dmgH)$_2$ composite for efficient photocatalytic hydrogen evolution under visible light irradiation. *Appl. Surf. Sci.* **2014**, *319*, 344–349. [CrossRef]
85. Miller, D.R.; Swenson, D.C.; Gillan, E.G. Synthesis and Structure of 2,5,8-Triazido-s-Heptazine: An Energetic and Luminescent Precursor to Nitrogen-Rich Carbon Nitrides. *J. Am. Chem. Soc.* **2004**, *126*, 5372–5373. [CrossRef]
86. Fang, J.; Fan, H.; Li, M.; Long, C. Nitrogen self-doped graphitic carbon nitride as efficient visible light photocatalyst for hydrogen evolution. *J. Mater. Chem. A* **2015**, *3*, 13819–13826. [CrossRef]
87. Starukh, H.; Praus, P. Doping of Graphitic Carbon Nitride with Non-Metal Elements and Its Applications in Photocatalysis. *Catalysts* **2020**, *10*, 1119. [CrossRef]
88. Imam, S.S.; Adnan, R.; Kaus, N.H.M. Photocatalytic degradation of ciprofloxacin in aqueous media: A short review. *Toxicol. Environ. Chem.* **2018**, *100*, 518–539. [CrossRef]
89. Ahmadzadeh, S.; Asadipour, A.; Pournamdari, M.; Behnam, B.; Rahimi, H.R.; Dolatabadi, M. Removal of ciprofloxacin from hospital wastewater using electrocoagulation technique by aluminum electrode: Optimization and modelling through response surface methodology. *Process. Saf. Environ. Prot.* **2017**, *109*, 538–547. [CrossRef]
90. Huo, P.; Lu, Z.; Liu, X.; Wu, D.; Liu, X.; Pan, J.; Gao, X.; Guo, W.; Li, H.; Yan, Y. Preparation photocatalyst of selected photodegradation antibiotics by molecular imprinting technology onto TiO_2/fly-ash cenospheres. *Chem. Eng. J.* **2012**, *189–190*, 75–83. [CrossRef]
91. Das, S.; Ghosh, S.; Misra, A.J.; Tamhankar, A.J.; Mishra, A.; Lundborg, C.S.; Tripathy, S.K. Sunlight Assisted Photocatalytic Degradation of Ciprofloxacin in Water Using Fe Doped ZnO Nanoparticles for Potential Public Health Applications. *Int. J. Environ. Res. Public Health* **2018**, *15*, 2440. [CrossRef] [PubMed]
92. Şimşek, E.B. Doping of boron in TiO_2 catalyst: Enhanced photocatalytic degradation of antibiotic under visible light irradiation. *J. Boron* **2017**, *2*, 18–27.
93. Thiruppathi, M.; Kumar, J.V.; Vahini, M.; Ramalingan, C.; Nagarajan, E. A study on divergent functional properties of sphere-like $CuWO_4$ anchored on 2D graphene oxide sheets towards the photocatalysis of ciprofloxacin and electrocatalysis of methanol. *J. Mater. Sci. Mater. Electron.* **2019**, *30*, 10172–10182. [CrossRef]
94. Qu, Y.; Xu, X.; Huang, R.; Qi, W.; Su, R.; He, Z. Enhanced photocatalytic degradation of antibiotics in water over functionalized N,S-doped carbon quantum dots embedded ZnO nanoflowers under sunlight irradiation. *Chem. Eng. J.* **2020**, *382*, 123016. [CrossRef]
95. Durán-Álvarez, J.C.; Méndez-Galván, M.; Lartundo-Rojas, L.; Rodríguez-Varela, M.; Ramírez-Ortega, D.; Guerrero-Araque, D.; Zanella, R. Synthesis and Characterization of the All Solid Z-Scheme Bi_2WO_6/Ag/AgBr for the Photocatalytic Degradation of Ciprofloxacin in Water. *Top. Catal.* **2019**, *62*, 1011–1025. [CrossRef]
96. Xu, S.; Xu, S.; Wang, Y.; Sun, X.; Gao, Y.; Gao, B. Enhanced degradation of ciprofloxacin by graphitized mesoporous carbon (GMC)-TiO_2 nanocomposite: Strong synergy of adsorption-photocatalysis and antibiotics degradation mechanism. *J. Colloid Interface Sci.* **2018**, *527*, 202–213. [CrossRef]
97. Dong, S.; Cui, L.; Zhang, W.; Xia, L.; Zhou, S.; Russell, C.K.; Fan, M.; Feng, J.; Sun, J. Double-shelled $ZnSnO_3$ hollow cubes for efficient photocatalytic degradation of antibiotic wastewater. *Chem. Eng. J.* **2020**, *384*, 123279. [CrossRef]
98. Li, Y.; Fu, Y.; Zhu, M. Green synthesis of 3D tripyramid TiO_2 architectures with assistance of aloe extracts for highly efficient photocatalytic degradation of antibiotic ciprofloxacin. *Appl. Catal. B Environ.* **2020**, *260*, 118149. [CrossRef]
99. Li, Z.; Chen, M.; Hu, H.; Zhang, Q.; Tao, D. Mechanochemical synthesis of novel Pt modified ZnAl-LDH for effective ciprofloxacin photodegradation. *J. Solid State Chem.* **2020**, *290*, 121594. [CrossRef]
100. Patel, J.; Singh, A.K.; Carabineiro, S. Assessing the Photocatalytic Degradation of Fluoroquinolone Norfloxacin by Mn:ZnS Quantum Dots: Kinetic Study, Degradation Pathway and Influencing Factors. *Nanomaterials* **2020**, *10*, 964. [CrossRef]
101. Yu, H.; Chen, F.; Ye, L.; Zhou, H.; Zhao, T. Enhanced photocatalytic degradation of norfloxacin under visible light by immobilized and modified In_2O_3/TiO_2 photocatalyst facilely synthesized by a novel polymeric precursor method. *J. Mater. Sci.* **2019**, *54*, 10191–10203. [CrossRef]
102. Jin, X.; Zhou, X.; Sun, P.; Lin, S.; Cao, W.; Li, Z.; Liu, W. Photocatalytic degradation of norfloxacin using N-doped TiO_2: Optimization, mechanism, identification of intermediates and toxicity evaluation. *Chemosphere* **2019**, *237*, 124433. [CrossRef]
103. Zhang, R.; Zhao, C.; Zhang, T.; Han, Q.; Li, Y.; Liu, Y.; Zeng, K. Ternary Z-Scheme Heterojunction of Bi_2WO_6 with Reduced Graphene Oxide (rGO) and $Bi_{25}FeO_{40}$ for Enhanced Visible-Light Photocatalysis. *J. Inorg. Organomet. Polym. Mater.* **2019**, *30*, 2152–2162. [CrossRef]
104. Zhang, R.; Han, Q.; Li, Y.; Zhang, T.; Liu, Y.; Zeng, K.; Zhao, C. Fabrication and characterization of high efficient Z-scheme photocatalyst Bi_2MoO_6/reduced graphene oxide/BiOBr for the degradation of organic dye and antibiotic under visible-light irradiation. *J. Mater. Sci.* **2019**, *54*, 14157–14170. [CrossRef]
105. Zhang, R.; Li, Y.; Han, Q.; Zhang, T.; Liu, Y.; Zeng, K.; Zhao, C. Investigation the High Photocatalytic Activity of Magnetically Separable Graphene Oxide Modified BiOBr Nanocomposites for Degradation of Organic Pollutants and Antibiotic. *J. Inorg. Organomet. Polym. Mater.* **2020**, *30*, 1703–1715. [CrossRef]
106. Liu, G.; Wang, F.; Feng, Y.; Xie, Z.; Zhang, Q.; Jin, X.; Liu, H.; Liu, Y.; Lv, W.; Liu, G. Facile synthesis of carbon quantum dots loaded with mesoporous g-C_3N_4 for synergistic absorption and visible light photodegradation of fluoroquinolone antibiotics. *Dalton Trans.* **2018**, *47*, 1284–1293. [CrossRef]

107. Li, S.; Hu, J. Photolytic and photocatalytic degradation of tetracycline: Effect of humic acid on degradation kinetics and mechanisms. *J. Hazard. Mater.* **2016**, *318*, 134–144. [CrossRef]
108. Wu, S.; Hu, H.; Lin, Y.; Zhang, J.; Hu, Y.H. Visible light photocatalytic degradation of tetracycline over TiO_2. *Chem. Eng. J.* **2020**, *382*, 122842. [CrossRef]
109. Lv, C.; Lan, X.; Wang, L.; Dai, X.; Zhang, M.; Cui, J.; Yuan, S.; Wang, S.; Shi, J. Rapidly and highly efficient degradation of tetracycline hydrochloride in wastewater by 3D IO-TiO_2-CdS nanocomposite under visible light. *Environ. Technol.* **2019**, 1–11. [CrossRef] [PubMed]
110. Jodeyri, M.; Haghighi, M.; Shabani, M. Enhanced-photoreduction deposition of Ag over sono-dispersed C_3N_4-Clinoptilolite used as nanophotocatalyst for efficient photocatalytic degradation of tetracycline antibiotic under simulated solar-light. *J. Mater. Sci. Mater. Electron.* **2019**, *30*, 13877–13894. [CrossRef]
111. Yan, W.; Yan, L.; Jing, C. Impact of doped metals on urea-derived g-C_3N_4 for photocatalytic degradation of antibiotics: Structure, photoactivity and degradation mechanisms. *Appl. Catal. B Environ.* **2019**, *244*, 475–485. [CrossRef]
112. Jiang, L.; Yuan, X.; Zeng, G.; Chen, X.; Wu, Z.; Liang, J.; Zhang, J.; Wang, H.; Wang, H. Phosphorus- and Sulfur-Codoped g-C_3N_4: Facile Preparation, Mechanism Insight, and Application as Efficient Photocatalyst for Tetracycline and Methyl Orange Degradation under Visible Light Irradiation. *ACS Sustain. Chem. Eng.* **2017**, *5*, 5831–5841. [CrossRef]
113. Wang, K.-L.; Li, Y.; Sun, T.; Mao, F.; Wu, J.-K.; Xue, B. Fabrication of Na, Cl co-doped graphitic carbon nitride with enhanced photocatalytic activity for degradation of dyes and antibiotics. *J. Mater. Sci. Mater. Electron.* **2019**, *30*, 4446–4454. [CrossRef]
114. Xue, J.; Ma, S.; Zhou, Y.; Zhang, Z.; He, M. Facile Photochemical Synthesis of Au/Pt/g-C_3N_4 with Plasmon-Enhanced Photocatalytic Activity for Antibiotic Degradation. *ACS Appl. Mater. Interfaces* **2015**, *7*, 9630–9637. [CrossRef]
115. Wang, J.; Liu, W.; Zhong, D.; Ma, Y.; Ma, Q.; Wang, Z.; Pan, J. Fabrication of bismuth titanate nanosheets with tunable crystal facets for photocatalytic degradation of antibiotic. *J. Mater. Sci.* **2019**, *54*, 13740–13752. [CrossRef]
116. Zhou, J.; Jiang, L.; Chen, D.; Liang, J.; Qin, L.; Bai, L.; Sun, X.; Huang, Y. Facile synthesis of Er-doped $BiFeO_3$ nanoparticles for enhanced visible light photocatalytic degradation of tetracycline hydrochloride. *J. Sol-Gel Sci. Technol.* **2019**, *90*, 535–546. [CrossRef]
117. Wang, W.; Han, Q.; Zhu, Z.; Zhang, L.; Zhong, S.; Liu, B. Enhanced photocatalytic degradation performance of organic contaminants by heterojunction photocatalyst $BiVO_4/TiO_2$/RGO and its compatibility on four different tetracycline antibiotics. *Adv. Powder Technol.* **2019**, *30*, 1882–1896. [CrossRef]
118. Tun, P.P.; Wang, J.; Khaing, T.T.; Wu, X.; Zhang, G. Fabrication of functionalized plasmonic Ag loaded Bi_2O_3/montmorillonite nanocomposites for efficient photocatalytic removal of antibiotics and organic dyes. *J. Alloy. Compd.* **2020**, *818*, 152836. [CrossRef]
119. Cao, H.-L.; Cai, F.-Y.; Yu, K.; Zhang, Y.-Q.; Lü, J.; Cao, R. Photocatalytic Degradation of Tetracycline Antibiotics over CdS/Nitrogen-Doped–Carbon Composites Derived from in Situ Carbonization of Metal–Organic Frameworks. *ACS Sustain. Chem. Eng.* **2019**, *7*, 10847–10854. [CrossRef]
120. Xue, J.; Ma, S.; Zhou, Y.; Zhang, Z.; Jiang, P. Synthesis of Ag/ZnO/C plasmonic photocatalyst with enhanced adsorption capacity and photocatalytic activity to antibiotics. *RSC Adv.* **2015**, *5*, 18832–18840. [CrossRef]
121. Długosz, M.; Żmudzki, P.; Kwiecień, A.; Szczubiałka, K.; Krzek, J.; Nowakowska, M. Photocatalytic degradation of sulfamethoxazole in aqueous solution using a floating TiO_2-expanded perlite photocatalyst. *J. Hazard. Mater.* **2015**, *298*, 146–153. [CrossRef] [PubMed]
122. Baran, W.; Sochacka, J.; Wardas, W. Toxicity and biodegradability of sulfonamides and products of their photocatalytic degradation in aqueous solutions. *Chemosphere* **2006**, *65*, 1295–1299. [CrossRef] [PubMed]
123. Tsiampalis, A.; Frontistis, Z.; Binas, V.; Kiriakidis, G.; Mantzavinos, D. Degradation of Sulfamethoxazole Using Iron-Doped Titania and Simulated Solar Radiation. *Catalysts* **2019**, *9*, 612. [CrossRef]
124. Ling, C.; Yue, C.; Yuan, R.; Qiu, J.; Liu, F.; Zhu, J.-J. Enhanced removal of sulfamethoxazole by a novel composite of TiO_2 nanocrystals in situ wrapped-Bi_2O_4 microrods under simulated solar irradiation. *Chem. Eng. J.* **2020**, *384*, 123278. [CrossRef]
125. Durán-Álvarez, J.C.; Santiago, A.L.; Ramírez-Ortega, D.; Acevedo-Peña, P.; Castillón, F.; Ramírez-Zamora, R.M.; Zanella, R. Surface modification of B–TiO_2 by deposition of Au nanoparticles to increase its photocatalytic activity under simulated sunlight irradiation. *J. Sol-Gel Sci. Technol.* **2018**, *88*, 474–487. [CrossRef]
126. Naraginti, S.; Yu, Y.-Y.; Fang, Z.; Yong, Y.-C. Novel tetrahedral Ag_3PO_4@N-rGO for photocatalytic detoxification of sulfamethoxazole: Process optimization, transformation pathways and biotoxicity assessment. *Chem. Eng. J.* **2019**, *375*, 122035. [CrossRef]
127. Dhiman, P.; Dhiman, N.; Kumar, A.; Sharma, G.; Naushad, M.; Ghfar, A.A. Solar active nano-$Zn_{1-x}Mg_xFe_2O_4$ as a magnetically separable sustainable photocatalyst for degradation of sulfadiazine antibiotic. *J. Mol. Liq.* **2019**, *294*, 111574. [CrossRef]
128. Do, T.C.M.V.; Nguyen, D.Q.; Nguyen, K.T.; Le, P.H. TiO_2 and Au-TiO_2 Nanomaterials for Rapid Photocatalytic Degradation of Antibiotic Residues in Aquaculture Wastewater. *Materials* **2019**, *12*, 2434. [CrossRef]
129. Salimi, M.; Behbahanib, M.; Sobhi, H.R.; Gholami, M.; Jafari, A.J.; Kalantari, R.R.; Farzadkia, M.; Esrafili, A. A new nano-photocatalyst based on Pt and Bi co-doped TiO_2 for efficient visible-light photo degradation of amoxicillin. *New J. Chem.* **2019**, *43*, 1562–1568. [CrossRef]

MDPI AG
Grosspeteranlage 5
4052 Basel
Switzerland
Tel.: +41 61 683 77 34

Catalysts Editorial Office
E-mail: catalysts@mdpi.com
www.mdpi.com/journal/catalysts

www.ingramcontent.com/pod-product-compliance
Lightning Source LLC
Chambersburg PA
CBHW061613210926
51850CB00009B/2308